FOUNDATIONS OF VIBROACOUSTICS

FOUNDATIONS OF VIBROACOUSTICS

COLIN H. HANSEN

CRC Press
Taylor & Francis Group
Boca Raton London New York

CRC Press is an imprint of the
Taylor & Francis Group, an **informa** business

CRC Press
Taylor & Francis Group
6000 Broken Sound Parkway NW, Suite 300
Boca Raton, FL 33487-2742

Printed on acid-free paper
Version Date: 20180117

International Standard Book Number-13: 978-1-138-09381-2 (Hardback)

Visit the Taylor & Francis Web site at
http://www.taylorandfrancis.com

and the CRC Press Web site at
http://www.crcpress.com

Dedication

This book is dedicated to Susan, to Kristy, to Laura and to Vladimir.

Contents

Preface

Acoustics and vibration are inextricably linked as acoustic fields can excite structures into vibration and vibrating structures can radiate sound. This structural/acoustic interaction has come to be known as vibroacoustics and there is considerable complex literature available on many phenomena that fall into this class of problem. It is hoped that this book provides the foundations necessary for readers to be able to fully appreciate the extensive and sometimes convoluted vibroacoustic literature and that some instructors in advanced undergraduate and Masters courses in Physics and Mechanical Engineering find it useful as a textbook.

I have attempted to cover all fundamental aspects of vibroacoustics, including both low- and high-frequency ranges, analysis of simple structures from first principles and numerical and experimental analysis of more complex structures. Acoustic and structural wave equations and corresponding solutions are discussed in detail for plane and spherical acoustic waves and for simple structures such as beams, plates and shells. Measurement of sound and structural intensity and power transmission are also discussed at length. Tools such as statistical energy analysis and modal analysis are also discussed in detail and the use of numerical techniques such as finite element analysis is mentioned with references provided for further guidance.

Chapter 1 outlines basic acoustics concepts, the derivation of the wave equation for both stationary and moving media and the propagation of sound in porous media. Chapter 2 is focused on the fundamentals of structural mechanics and the vibration of continuous systems with a discussion of wave equations and solutions for beams, plates and cylinders, which can be used to determine resonance frequencies and the response to external forces and moments. Chapter 3 is concerned with the interaction of vibrating structures with surrounding media, sound radiation and sound propagation, and the use of Green's functions to characterise these phenomena. Chapter 4 is a discussion of the characterisation and quantification of sound energy propagating in air and in structures as a result of applied structural forces or acoustic sources, using structural input impedances, sound intensity and structural intensity. Means for measuring these quantities and associated errors are discussed in detail. Chapter 5 describes the theoretical basis of modal analysis and its experimental implementation for the purpose of characterising low-frequency structural vibration and low-frequency acoustic fields in enclosures, in terms of normal modes. This characterisation can be used to analyse the radiation of sound from vibrating structures or the sound field generated by a sound source in an enclosure. Chapter 6 outlines the principles of statistical energy analysis (SEA) and how it may be applied to the analysis of the vibration response of structures to applied forces or moments as well as the analysis of sound radiation from structures in the high-frequency region, where a modal description of the structural vibration is impractical. The analysis of sound fields in enclosed spaces at high frequencies is also discussed using SEA. Chapter 7 is concerned with frequency analysis, which is an important tool for identifying and quantifying noise sources. Appendix A provides some basic mathematics which is essential for understanding the analyses discussed in the rest of the book, while Appendix B provides properties of materials, useful for undertaking the calculations discussed in the book.

Where possible, references are given within the text to books, reports and technical papers which may provide the reader with a more detailed treatment of the subject matter than is possible here. The reference list at the end of the book is intended as a first source for further reading and is by no means claimed to be comprehensive; thus, omission of a reference is not intended as a reflection of its value.

I sincerely hope that my goal of making the subject of vibroacoustics accessible to advanced students and practising acousticians has been achieved.

Colin H. Hansen
June, 2017

Acknowledgements

I would like to express my deep appreciation to my wife Susan for her patience and support which was freely given during the many days, nights and weekends that were needed to complete this book.

1

Basic Concepts and Acoustic Fundamentals

1.1   Introduction

This book is intended to address the fundamental concepts that underpin the theoretical analyses of problems involving sound radiation from vibrating structures and the propagation of the generated sound through air. Sound generated as a result of aerodynamic sources (such as propellers and aircraft wings moving through the air) and sound propagation in the ocean are not discussed here, as these topics are extensive and are discussed in detail in many excellent books.

The aim of conducting a vibroacoustic analysis of any system is generally to determine the system response due to excitation by known forces or to find the relative magnitudes of subsystem responses caused by arbitrary exciting forces. Examples include the vibration response of aircraft and rockets to estimated excitation forces and the acoustic response of enclosures excited by an attached structure, including a partition separating the two enclosures. There are two fundamentally different approaches used in these analyses. The first is known as a 'wave' approach in which it is assumed that the subsystem being considered is semi-infinite in nature so that the amplitudes of any waves reflected back to the source of excitation (energy source) are negligible compared to the amplitudes of non-reflected waves. The second approach is known as a 'modal' approach in which the subsystem is small enough that reflected waves have a significant amplitude. In this case the system response is described in terms of its normal modes of vibration. The vibration can be either structural, where a solid structure vibrates, or an enclosed space where the air in the enclosure vibrates to produce sound.

The starting point for fundamental analyses in acoustics is usually the wave equation. The derivation and application of this equation is discussed at length in this chapter, following the discussion of some basic concepts such as acoustic field variables, mean square quantities, decibels, addition of incoherent sounds, subtraction of incoherent sounds, sound pressure magnitudes, speed of sound, dispersion and basic frequency analysis. Understanding these fundamental concepts is an important prerequisite to following the subsequent discussion in this and following chapters.

The derivation of the wave equation is discussed here because it is important to understand how it is derived, and the assumptions that are made in the linearisation of it, so that it is not used inappropriately. The wave equation is then derived for a medium with a mean flow, as this is relevant to a large number of practical problems, such as sound propagation in a windy atmosphere and sound propagation in ducts in which there is a mean air flow in one direction. Following the derivation of the wave equation from first principles, solutions to the equation are then found for plane and spherical waves and for sound radiation from point, line and plane sources. This is followed by a discussion of sound propagation in porous media such as rockwool and fibreglass.

The principles that underpin structural mechanics are discussed in Chapter 2. An understanding of these principles is important, as they underpin the discussion of structural intensity and sound propagation in structures (see Chapter 4). The fundamental principles discussed in this chapter are also used as a basis to derive the wave equation for various wave types in beams, plates and cylinders. Solutions to the wave equations can then be used to find resonance frequencies and mode shapes. Analyses outlined in Chapter 4 can then be used to determine vibration amplitudes in simple structures caused by a known excitation force. The analysis of more complex structures requires the use of finite element analysis (FEA) or boundary element analysis (BEA), which is discussed in detail by Howard and Cazzolato (2015). However, these analyses are only applicable in the frequency range where the resonant modes are well-separated in frequency. At higher frequencies, where the modes are difficult to separate in frequency, statistical techniques must be used and these are outlined in Chapter 6.

1.2   Notation

Notation consistency is important in any book and in this book, the use of bar and hat symbols, and bold font is defined in detail below to save having to repeat definitions each time they are used.

The bar symbol above a variable is used to denote an averaged quantity. It usually represents the average of a large number of values. It can also represent the distance from a reference line to the centroid of a body.

The hat symbol above a variable is used to denote the amplitude of a quantity, which does not vary with time. The quantity may be real or it may be complex. Thus, a displacement, y, that varies with time, t, at radian frequency, ω, may be written as:

$$y = \hat{y}\,e^{j\omega t} \tag{1.1}$$

Variables in bold font represent vector quantities, which may be real or complex. A vector quantity is one that is defined by two or more scalar elements. In a 3D coordinate system these elements would be components of the vector in a direction along each of the coordinate axes and the resulting vector would be defined by an amplitude and a direction with respect to each axis. For example the location, x is used to denote a coordinate location, (x, y) in two dimensions (2D) or (x, y, z) in three dimensions (3D). A bold symbol for force or displacement indicates that the quantity is composed of two or more scalar elements, each of which represents contributions (usually from different directions corresponding to particular coordinate system axes) to the total. Thus, the resultant total force amplitude and the direction of its action would be defined by the vector of elements. This is most easily illustrated using a vector of two elements. If a two-element force vector consists of N_x Newtons in the x-axis direction and N_y Newtons in the y-axis direction, then the total force would be $\sqrt{N_x^2 + N_y^2}$ Newtons and the direction of its action would be at an angle, $\theta = \tan^{-1}(N_y/N_x)$, from the x-axis. Of course, a vector can have more than 3 elements, such as a number of forces acting at different points and in different directions on a body or the acoustic pressure at different locations in an enclosure being represented as a vector. A matrix is made up of columns of vectors and these are represented as upper case bold variables. Although vectors are traditionally represented as lower case bold variables, some are represented as upper case in this book, as upper case reflects the traditional symbol used to describe the quantity. Examples are $\boldsymbol{F}$ for a force vector and $\boldsymbol{M}$ for a moment vector. For 1D problems, such as sound propagation in a duct or longitudinal wave propagation in a beam, bold font is not used to denote quantities such as location, displacement or force, as only a single element (representing a single direction) is involved. This reasoning also applies in all situations where the direction associated with a quantity is defined (such as the vibration amplitude of a plate in the direction normal to the plate).

1.3 Basic Concepts

Acoustic disturbances travel through fluid media in the form of longitudinal waves and are generally regarded as small amplitude perturbations to an ambient state. For a fluid such as air or water, the ambient state is characterised by the values of the physical variables (pressure, P_s, velocity, U, and density, ρ) which exist in the absence of the disturbance. The ambient state defines the medium through which the sound propagates. A homogeneous medium is one in which all ambient quantities are independent of position. A quiescent medium is one in which they are independent of time and in which $U = 0$. The idealisation of homogeneity and quiescence will be assumed in the following derivation of the wave equation, as this generally provides a satisfactory quantitative description of acoustic phenomena. As acoustic wave propagation is often associated with fluid flow, the wave equation for the condition $U \neq 0$ will also be discussed.

1.3.1 Acoustic Field Variables

For a fluid, expressions for the pressure, P_{tot}, velocity, U_{tot}, temperature, T_{tot}, and density, ρ_{tot}, may be written in terms of the steady-state (mean values), shown as P_s, U, T and ρ, and the variable (perturbation) values, p, u, τ and σ, as follows:

$$
\begin{aligned}
\text{Pressure}: && P_{\text{tot}} &= P_s + p(\boldsymbol{r}, t) && \text{(Pa)} \\
\text{Velocity}: && \boldsymbol{U}_{\text{tot}} &= \boldsymbol{U} + \boldsymbol{u}(\boldsymbol{r}, t) && \text{(m/s)} \\
\text{Temperature}: && T_{\text{tot}} &= T + \tau(\boldsymbol{r}, t) && (^\circ\text{C}) \\
\text{Density}: && \rho_{\text{tot}} &= \rho + \sigma(\boldsymbol{r}, t) && (\text{kg/m}^3)
\end{aligned}
$$

where $\boldsymbol{r}$ is the position vector, t is time and the variables in bold font are vector quantities.

Pressure, temperature and density are familiar scalar quantities that do not require discussion. However, an explanation is required for the particle velocity $\boldsymbol{u}(\boldsymbol{r}, t)$ and the vector equation above that involves it. The notion of particle velocity is based on the assumption of a continuous rather than a molecular medium. The term 'particle' refers to a small part of the assumed continuous medium and not to the molecules of the medium. Thus, even though the actual motion associated with the passage of an acoustic disturbance through the conducting medium, such as air at high frequencies, may be of the order of the molecular motion, the particle velocity describes a macroscopic average motion superimposed upon the inherent Brownian motion of the medium. In the case of a convected medium moving with a mean velocity, $\boldsymbol{U}$, which itself may be a function of the position vector, $\boldsymbol{r}$, and time, t, the perturbing particle velocity, $\boldsymbol{u}(\boldsymbol{r}, t)$, associated with the passage of an acoustic disturbance may be thought of as adding to the mean velocity to give the total velocity.

Any variable could be chosen for the description of a sound field, but it is easiest to measure pressure in a fluid and strain, or more generally acceleration, in a solid. Consequently, these are the variables usually considered. These choices have the additional advantage of providing a scalar description of the sound field from which all other variables may be derived. For example, the particle velocity is important for the determination of sound intensity, but it is a vector quantity and requires three measurements as opposed to one for pressure. Nevertheless, instrumentation (Microflown) is available that allows the instantaneous measurement of particle velocity along all three Cartesian coordinate axes at the same time. In solids, it is generally easiest to measure acceleration, especially in thin panels, although strain might be preferred as the measured variable in some special cases. If non-contact measurement is necessary, then instrumentation known as laser vibrometers are available that can measure vibration velocity along all three Cartesian coordinate axes at the same time and also allow scanning of the surface being measured, so that a complete picture of the surface vibration response can be obtained for any frequency of interest.

It is of interest to consider the nature of an acoustic disturbance and the speed with which it propagates. To begin, it should be understood that small perturbations of the acoustic field may always be described as the sum of cyclic disturbances of appropriate frequencies, amplitudes and relative phases. In a fluid, a sound field will be manifested by variations in local pressure of generally very small amplitude with associated variations in density, displacement, particle velocity and temperature. Thus in a fluid, a small compression, perhaps followed by a compensating rarefaction, may propagate away from a source as a sound wave. The associated particle velocity lies parallel to the direction of propagation of the disturbance, the local particle displacement being first in the direction of propagation, then reversing to return the particle to its initial position after passage of the disturbance. This is a description of a compressional or longitudinal wave.

The viscosity of the fluids of interest in this text is sufficiently small for shear forces to play a very small part in the propagation of acoustic disturbances. A solid surface, vibrating in its plane without any normal component of motion, will produce shear waves in the adjacent fluid in which the local particle displacement is parallel to the exciting surface, but normal to the direction of propagation of the disturbance. However, such motion is always confined to a very narrow region near to the vibrating surface and does not result in energy transport away from the near field region. Alternatively, a compressional wave propagating in a fluid, parallel to an adjacent, bounding, solid surface will give rise to a similar type of disturbance at the fixed boundary, but again the shear wave will be confined to a very thin viscous boundary layer in the fluid. Temperature variations associated with the passage of an acoustic disturbance through a fluid next to a solid boundary, which is characterised by a very much greater thermal capacity, will likewise give rise to a thermal wave propagating into the boundary; but again, as with the shear wave, the thermal wave will be confined to a very thin thermal boundary layer of the same order of size as the viscous boundary layer. Such viscous and thermal effects, generally referred to as the acoustic boundary layer, are usually negligible for energy transport, and are generally neglected, except in the analysis of sound propagation in tubes and porous media, where they provide the energy dissipation mechanisms.

It has been mentioned that sound propagates in liquids and gases predominantly as longitudinal compressional waves; shear and thermal waves play no significant part. In solids, however, the situation is much more complicated, as shear stresses are readily supported. Not only are longitudinal waves possible, but so are transverse shear and torsional waves. In addition, the types of waves that propagate in solids depend strongly on boundary conditions. In thin plates, for example, bending waves, which are really a mixture of longitudinal and shear waves, predominate, with important consequences for acoustics and noise control. Bending waves are of importance in the consideration of sound radiation from extended surfaces, and the transmission of sound from one space to another through an intervening partition.

1.3.2   Mean Square Quantities

In Section 1.3.1 the variables of acoustics were listed and discussed. For the case of fluids, they were shown to be small perturbations in steady-state quantities, such as pressure, density, velocity and temperature. Alternatively, in solids they are small perturbations in displacement, stress and strain variables. In all cases, acoustic fields are concerned with time-varying quantities with mean values of zero; thus, the variables of acoustics are commonly determined by measurement as mean square or as root-mean-square (RMS) quantities. In some cases, however, we are concerned with the product of two time-varying quantities. For example, sound intensity will be discussed in Section 4.3, where it will be shown that the quantity of interest is the product of the two time-varying quantities, acoustic pressure and acoustic particle velocity averaged over time. The time average of the product of two time-dependent variables, $F(t)$ and $G(t)$, will be referred to in the following text and will be indicated by the notation $\langle F(t)G(t)\rangle$. Sometimes the time dependence indicated by (t) will be suppressed to simplify the notation. The time average of the product of $F(t)$ and $G(t)$, averaged over time, T_A, is defined as:

$$\langle F(t)G(t)\rangle = \langle FG\rangle = \lim_{T_A \to \infty} \frac{1}{T_A} \int_0^{T_A} F(t)G(t)\,\mathrm{d}t \tag{1.2}$$

When $F(t) = G(t)$, the mean square of the variable is obtained. Thus the mean square sound pressure $\langle p^2(\boldsymbol{r},t)\rangle$ at position, $\boldsymbol{r}$, is:

$$\langle p^2(\boldsymbol{r},t)\rangle = \lim_{T_A \to \infty} \frac{1}{T_A} \int_0^{T_A} p(\boldsymbol{r},t)p(\boldsymbol{r},t)\,\mathrm{d}t \tag{1.3}$$

The root-mean-square (RMS) sound pressure at location $\boldsymbol{r}$, which will be shown later is used to evaluate the sound pressure level, is calculated as

$$p_{\mathrm{RMS}} = \sqrt{\langle p^2(\boldsymbol{r},t)\rangle} \tag{1.4}$$

The angled brackets, $\langle\ \rangle$, were used in the previous example to indicate the time average of the function within the brackets. They are sometimes used to indicate other types of averages of the function within the brackets; for example, the space average of the function. Where there may be a possibility of confusion, the averaging variable is added as a subscript; for example, the mean-square sound pressure averaged over space and time may also be written as $\langle p^2(\boldsymbol{r},t)\rangle_{S,t}$.

Sometimes the amplitude, G_A, of a single frequency quantity is of interest. In this case, the following useful relation between the amplitude and the RMS value of a sinusoidally varying single frequency quantity is:

$$G_A = \sqrt{2\langle G^2(t)\rangle} \tag{1.5}$$

1.3.3 Decibels

Pressure is an engineering unit, which is measured relatively easily; however, the ear responds approximately logarithmically to energy input, which is proportional to the square of the sound pressure. The minimum sound pressure that the ear may detect is less than $20\,\mu\mathrm{Pa}$, while the greatest sound pressure before pain is experienced is 60 Pa. A linear scale based on the square of the sound pressure would require 10^{13} unit divisions to cover the range of human experience; however, the human brain is not organised to encompass such an enormous range in a linear way. The remarkable dynamic range of the ear suggests that some kind of compressed scale should be used. A scale suitable for expressing the square of the sound pressure in units best matched to subjective response is logarithmic rather than linear. The logarithmic scale provides a convenient way of comparing the sound pressure of one sound with another. To avoid a scale that is too compressed, a factor of 10 is introduced, giving rise to the decibel. The level of sound pressure, p, is then said to be L_p decibels (dB) greater than or less than a reference sound pressure, p_{ref}, according to the following equation:

$$L_p = 10\log_{10}\frac{\langle p^2\rangle}{p_{\mathrm{ref}}^2} = 10\log_{10}\langle p^2\rangle - 10\log_{10}p_{\mathrm{ref}}^2 \quad (\mathrm{dB}) \tag{1.6}$$

For the purpose of absolute level determination, the sound pressure is expressed in terms of a datum pressure corresponding approximately to the lowest sound pressure which the young normal ear can detect. The result is called the sound pressure level, L_p (or SPL), which has the units of decibels (dB) and should be written as 'dB re 20 µPa' when referring to measurements conducted in air. When it can be assumed that a discussion concerns sound pressure level measurements in air, the reference value 're 20 µPa' is dropped and the sound pressure level is simply written with units of 'dB'. This is the quantity that is measured with a sound level meter.

The sound pressure is a measured root-mean-square (RMS) value and the reference pressure $p_{ref} = 2 \times 10^{-5}$ N/m^2 or 20 µPa. When this value for the reference pressure is substituted into Equation (1.6), the following convenient alternative form is obtained:

$$L_p = 10 \log_{10} \langle p^2 \rangle + 94 \quad \text{(dB re 20 µPa)} \tag{1.7}$$

In Equation (1.7), the acoustic pressure, p, is measured in pascals.

1.3.4   Addition of Incoherent Sounds (Logarithmic Addition)

When two separate signals containing bands of noise are added and the phases are random, the resulting sound pressure can be obtained using:

$$\langle p_t^2 \rangle = \langle p_1^2 \rangle + \langle p_2^2 \rangle \tag{1.8}$$

which, by use of Equation (1.6), may be written in a general form for the addition of N incoherent sounds as:

$$L_{pt} = 10 \log_{10} \left(10^{L_1/10} + 10^{L_2/10} + \ldots + 10^{L_N/10} \right) \tag{1.9}$$

Incoherent sounds add together on a linear energy (pressure squared) basis. The simple procedure embodied in Equation (1.9) may easily be performed on a standard calculator. The procedure accounts for the addition of sounds on a linear energy basis and their representation on a logarithmic basis. Note that the division by 10, rather than 20 in the exponent, is because the process involves the addition of squared pressures.

1.3.5   Subtraction of Sound Pressure Levels

Sometimes it is necessary to subtract one noise from another; for example, when background noise must be subtracted from the total noise to obtain the sound produced by a machine alone. The method used is similar to that described in the addition of levels and is illustrated here with an example.

Example 1.1

The sound pressure level measured at a particular location in a factory with a noisy machine operating nearby is 92.0 dB. When the machine is turned off, the sound pressure level measured at the same location is 88.0 dB. What is the sound pressure level due to the machine alone?

Solution 1.1

$$L_{pm} = 10 \log_{10} \left(10^{92/10} - 10^{88/10} \right) = 89.8 \text{ dB}$$

For noise-testing purposes, this procedure should be used only when the total sound pressure level exceeds the background noise by 3 dB or more. If the difference is less than 3 dB a valid sound test probably cannot be made. Note that here subtraction is between squared pressures.

1.3.6 Magnitudes

The minimum acoustic pressure audible to the young human ear judged to be in good health, and unsullied by too much exposure to excessively loud music, is approximately 20 $\times 10^{-6}$ Pa, or 2 $\times 10^{-10}$ atmospheres (since one atmosphere equals 101.3 $\times 10^3$ Pa). The minimum audible level occurs between 3000 and 4000 Hz and is a physical limit; lower sound pressure levels would be swamped by thermal noise due to molecular motion in air. For the normal human ear, pain is experienced at sound pressures of the order of 60 Pa or 6 $\times 10^{-4}$ atmospheres. Evidently, acoustic pressures ordinarily are quite small fluctuations about the mean.

1.3.7 Speed of Sound

Sound is conducted to the ear through the surrounding medium, which in general will be air and sometimes water, but sound may be conducted by any fluid or solid. In fluids, which readily support compression, sound is transmitted as longitudinal waves and the associated particle motion in the transmitting medium is parallel to the direction of wave propagation. However, as fluids support shear very weakly, waves dependent on shear are weakly transmitted, but usually they may be neglected. Consequently, longitudinal waves are often called sound waves. For example, the speed of sound waves travelling in plasma has provided information about the interior of the sun. In solids, which can support both compression and shear, energy may be transmitted by all types of waves, but only longitudinal wave propagation is referred to as 'sound'.

The concept of an 'unbounded medium' will be introduced as a convenient and often used idealisation. In practice, the term 'unbounded medium' has the meaning that longitudinal wave propagation may be considered sufficiently remote from the influence of any boundaries that such influence may be neglected. The concept of an unbounded medium is generally referred to as 'free field' and this alternative expression will also be used, where appropriate, in this text. The propagation speed, c, of sound waves, called the phase speed, in any conducting medium (solid or fluid) is dependent on the stiffness, D, and the density, ρ, of the medium. The phase speed, c, takes the following simple form:

$$c = \sqrt{D/\rho} \quad (\text{m/s}) \tag{1.10}$$

For longitudinal wave propagation in solids, the stiffness, D, depends on the ratio of the dimensions of the solid to the wavelength of a propagating longitudinal wave. Let the solid be characterised by three orthogonal dimensions $h_i, i = 1, 2, 3$, which determine its overall size. Let h be the greatest of the three dimensions of the solid, where E denotes Young's modulus and f denotes the frequency of a longitudinal wave propagating in the solid. Then the criterion proposed for determining D is that the ratio of the dimension, h, to the half wavelength of the propagating longitudinal wave in the solid is greater than or equal to one. For example, wave propagation may take place along dimension h when the half wavelength of the propagating wave is less than or just equal to the dimension, h. This observation suggests that the following inequality must be satisfied for wave propagation to take place.

$$2hf \geq \sqrt{D/\rho} \tag{1.11}$$

For the case that only one dimension, h, satisfies the inequality and two dimensions do not then the solid must be treated as a wire or thin rod along dimension, h, on which waves may travel. In this case, the stiffness coefficient, D, is that of a rod, D_r, and takes the following form:

$$D_r = E \quad (\text{Pa}) \tag{1.12}$$

The latter result constitutes the definition of Young's modulus of elasticity, E.

For the case that two dimensions satisfy the inequality and one dimension does not, the solid must be treated as a plate over which waves may travel. In this case, where ν is Poisson's ratio (ν is approximately 0.3 for steel), the stiffness, $D = D_p$, takes the following form:

$$D_p = E/(1 - \nu^2) \quad \text{(Pa)} \tag{1.13}$$

For a material for which Poisson's ratio is equal to 0.3, $D = 1.099E$.

If all three dimensions, h_i, satisfy the criterion, then wave travel may take place in all directions in the solid. In this case, the stiffness coefficient, $D = D_s$, takes the following form:

$$D_s = \frac{E(1 - \nu)}{(1 + \nu)(1 - 2\nu)} \quad \text{(Pa)} \tag{1.14}$$

For fluids, the stiffness, D_F, is the bulk modulus or the reciprocal of the more familiar compressibility, given by:

$$D_F = -V(\partial V/\partial P_s)^{-1} = \rho(\partial P_s/\partial \rho) \quad \text{(Pa)} \tag{1.15}$$

where V is a unit volume and $\partial V/\partial P_s$ is the incremental change in volume associated with an incremental change in static pressure, P_s.

The effect of boundaries on the longitudinal wave speed in fluids will now be considered. For fluids (gases and liquids) in pipes at frequencies below the first higher order mode cut-on frequency, where only plane waves propagate, the close proximity of the wall of the pipe to the fluid within may have a very strong effect in decreasing the medium stiffness. The cut-on frequency, f_{co}, for a rectangular-section pipe may be calculated using:

$$f_{co} = \frac{c}{2L_y} \tag{1.16}$$

where L_y is the largest duct cross-sectional dimension. For circular-section pipes:

$$f_{co} = 0.586 \frac{c}{d} \tag{1.17}$$

where d is the pipe diameter.

The stiffness of a fluid in a pipe, tube or more generally, a conduit will be written as D_C. The difference between D_F and D_C represents the effect of the pipe wall on the stiffness of the contained fluid. This effect will depend on the ratio of the mean pipe radius, R, to wall thickness, t, the ratio of the density, ρ_w, of the pipe wall to the density, ρ, of the fluid within it, Poisson's ratio, ν, of the pipe wall material, as well as the ratio of the fluid stiffness, D_F, to the Young's modulus, E, of the pipe wall. The expression for the stiffness, D_C, of a fluid in a conduit is (Pavic, 2006):

$$D_C = \frac{D_F}{1 + \dfrac{D_F}{E}\left(\dfrac{2R}{t} + \dfrac{\rho_w}{\rho}\nu^2\right)} \tag{1.18}$$

The compliance of a pipe wall will tend to increase the effective compressibility of the fluid and thus decrease the speed of longitudinal wave propagation in pipes. Generally, the effect will be small for gases, but for water in plastic pipes, the effect may be large. In liquids, the effect may range from negligible in heavy-walled, small-diameter pipes to large in large-diameter conduits.

For fluids (gases and liquids), thermal conductivity and viscosity are two other mechanisms, besides chemical processes, by which fluids may interact with boundaries. Generally, thermal conductivity and viscosity in fluids are very small, and such acoustical effects as may arise from them are only of importance very close to boundaries and in consideration of damping mechanisms. Where a boundary is at the interface between fluids or between fluids and a solid, the effects may be large, but as such effects are often confined to a very thin layer at the boundary, they are commonly neglected.

Variations in pressure are associated with variations in temperature as well as density; thus, heat conduction during the passage of an acoustic wave is important. In gases, for acoustic waves at frequencies ranging from infrasonic up to well into the ultrasonic frequency range, the associated gradients are so small that pressure fluctuations may be considered to be essentially adiabatic; that is, no sensible heat transfer takes place between adjacent gas particles and, to a very good approximation, the process is reversible. However, at very high frequencies, and in porous media at low frequencies, the compression process tends to be isothermal. In the latter cases, heat transfer tends to be complete and in phase with the compression.

For gases, use of Equation (1.10), the equation for adiabatic compression (which gives $D = \gamma P_s$), and the equation of state for gases gives the following for the speed of sound:

$$c = \sqrt{\gamma P_s / \rho} = \sqrt{\gamma R T / M} \quad \text{(m/s)} \tag{1.19}$$

where P_s is the static pressure ($=$ atmospheric pressure, P_0 in an unconfined space), γ is the ratio of specific heats (1.40 for air), T is the temperature in Kelvin (K), R is the universal gas constant which has the value 8.314 J mol^{-1} K^{-1} and M is the molecular weight, which for air is 0.029 kg/mol. Equations (1.10) and (1.19) are derived in many standard texts: for example Morse (1948); Pierce (1981); and Kinsler et al. (1982).

For gases, the speed of sound depends on the temperature of the gas through which the acoustic wave propagates. For sound propagating in air at audio frequencies, the process is adiabatic. In this case, for temperature, T, in degrees Celsius (and not greatly different from 20°C), the speed of sound may be calculated to three significant figures using the following approximation:

$$c = 331 + 0.6T \quad \text{(m/s)} \tag{1.20}$$

For calculations in this text, unless otherwise stated, a temperature of 20°C for air will be assumed, resulting in a speed of sound of 343 m/s and an air density of 1.206 kg/m^3 at sea level, thus giving $\rho c = 414$.

1.3.8 Dispersion

The speed of sound wave propagation, as given by Equation (1.10), is quite general and permits the possibility that the stiffness, D, may either be constant or a function of frequency. For the cases considered thus far, it has been assumed that the stiffness, D, is independent of the frequency of the sound wave, with the consequence that all associated wave components of whatever frequency will travel at the same speed and thus the wave will propagate without dispersion, meaning wave travel takes place without changing the wave shape.

On the other hand, there are many cases where the stiffness, D, is a function of frequency and in such cases, the associated wave speed will also be a function of frequency. A familiar example is that of an ocean wave, the speed of which is dependent on the ocean depth. As a wave advances into shallow water, its higher frequency components travel faster than the lower frequency components, as the speed of each component is proportional to the depth of water relative to its wavelength. The greater the depth of water relative to the component wavelength, the greater the component speed. In deep water, the relative difference in the ratio of water depth to wavelength between low- and high-frequency components is small. However, as the water becomes shallow near the shore, this difference becomes larger and eventually causes the wave to break. A dramatic example is that of an ocean swell produced by an earthquake deep beneath the ocean far out to sea, which becomes the excitement of a tsunami on the beach.

Bending waves that occur in panels, which are a combination of longitudinal and shear waves, play an important role in sound transmission through and from panels. Bending wave speed is dependent on the frequency of the disturbance and thus bending waves are dispersive. A dispersive wave means that it will propagate at a *phase velocity* that depends on its wavelength. Particle motion associated with bending waves is normal to the direction of propagation, in contrast to longitudinal waves, for which it is in the same direction.

In liquids and gases, dispersive propagation is observed above the audio frequency range at ultrasonic frequencies, where relaxation effects are encountered. Such effects make longitudinal wave propagation frequency dependent and consequently dispersive. Although not strictly dispersive, the speed of propagation of longitudinal waves associated with higher order modes in ducts is an example of a case where the effective wave speed along the duct axis is frequency dependent. However, this is because the number of reflections of the wave from the duct wall per unit axial length is frequency dependent, rather than the speed of propagation of the wave through the medium in the duct.

When an acoustic disturbance is produced, some work must be done on the conducting medium to produce the disturbance. Furthermore, as the disturbance propagates, energy stored in the field is convected with the advancing disturbance. When the wave propagation is non-dispersive, the energy of the disturbance propagates with the speed of sound; that is, with the phase speed of the longitudinal compressional waves. On the other hand, when propagation is dispersive, the frequency components of the disturbance all propagate with different phase speeds; the energy of the disturbance, however, propagates with a speed that is referred to as the group speed. Thus in the case of dispersive propagation, one might imagine a disturbance that changes its shape as it advances, while at the same time maintaining a group identity, and travelling at a group speed different from that of any of its frequency components. The group speed is quantified later in Equation (1.107). For non-dispersive wave propagation, the group speed, c_g, is equal to the phase speed, c. For bending waves in beams, plates or cylinders, the group speed is twice the phase speed (Fahy and Gardonio, 2007).

1.3.9   Basic Frequency Analysis

A propagating sound wave can be described either as an undefined disturbance as in Equation (1.91), or as a single frequency disturbance as given, for example, by Equation (1.95). Here it is shown how an undefined disturbance may be described conveniently as composed of narrow frequency bands, each characterised by a range of frequencies. There are various such possible representations and all are referred to broadly as spectra. It is customary to refer to spectral density level when the measurement band is one hertz wide, to 1/3-octave or octave band level when the measurement band is 1/3-octave or one octave wide, respectively, and to spectrum level for measurement bands of other frequency widths. In air, sound is transmitted in the form of a longitudinal wave. To illustrate longitudinal wave generation, as well as to provide a model for the discussion of sound spectra, the example of a vibrating piston at the end of a very long tube filled with air is used, as illustrated in Figure 1.1.

Let the piston in Figure 1.1(a) move to the right. Since the air has inertia, only the air immediately next to the face of the piston moves at first, resulting in an increase in the pressure in the element of air next to the piston. This element will then expand forward, displacing the next layer of air and compressing the next elemental volume. A pressure pulse is formed which travels along the tube with the speed of sound, c. Let the piston stop and subsequently move to the left. This results in the formation of a rarefaction next to the surface of the piston, which follows the previously formed compression down the tube. If the piston again moves to the right, the process is repeated with the net result being a 'wave' of positive and negative pressure transmitted along the tube. If the piston moves with simple harmonic motion, a sine wave is produced; that is, at any instant, the pressure distribution along the tube will have the form of a sine wave, or at any fixed point in the tube, the pressure disturbance, displayed as a function of time, will have a sine wave appearance. Such a disturbance is characterised by a

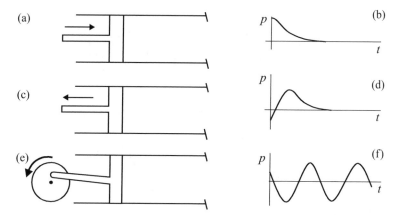

FIGURE 1.1 Sound generation illustrated. (a) The piston moves right, compressing air as in (b). (c) The piston stops and reverses direction, moving left and decompressing air in front of the piston, as in (d). (e) The piston moves cyclically back and forth, producing alternating compressions and rarefactions, as in (f). In all cases, disturbances move to the right with the speed of sound.

single frequency. The sound pressure variations at a single location as a function of time, together with the corresponding spectra, are illustrated in Figures 1.2(a) and (b), respectively. Although the sound pressure at any location varies sinusoidally with time, the particle motion is parallel to the direction of propagation of the wave, resulting in a longitudinal wave, as illustrated in Figure 1.3(a). Such a wave consists of compressions and rarefactions where the distance between particles is smaller in the compression part and larger in the rarefaction part.

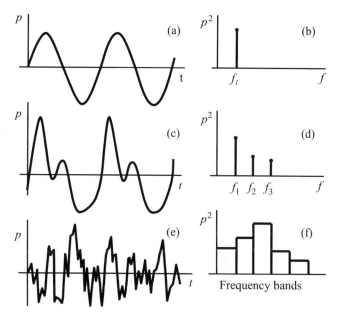

FIGURE 1.2 Spectral analysis illustrated. (a) Disturbance, p, varies sinusoidally with time, t, at a single frequency, f_1, as in (b). (c) Disturbance, p, varies cyclically with time, t, as a combination of three sinusoidal disturbances of fixed relative amplitudes and phases; the associated spectrum has three single-frequency components, f_1, f_2 and f_3, as in (d). (e) Disturbance, p, varies erratically with time, t, with a frequency band spectrum as in (f).

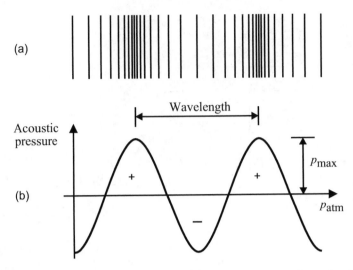

FIGURE 1.3 Representation of a sound wave: (a) compressions and rarefactions of particles, whose motion produces a sound wave, as a function of spatial location at a fixed instance in time; (b) graphical representation of the sound pressure variation.

If the piston moves irregularly but cyclically, for example, so that it produces the waveform shown in Figure 1.2(c), the resulting sound field will consist of a combination of sinusoids of several frequencies. The spectral (or frequency) distribution of the energy in this particular sound wave is represented by the frequency spectrum of Figure 1.2(d). As the motion is cyclic, the spectrum consists of a set of discrete frequencies. Although some sound sources have single-frequency components, most sound sources produce a very disordered and random waveform of pressure versus time, as illustrated in Figure 1.2(e). Such a wave has no periodic component, but by Fourier analysis, it may be shown that the resulting waveform may be represented as a collection of waves of many different frequencies. For a random type of wave, the sound pressure squared in a band of frequencies is plotted as shown, for example, in the frequency spectrum of Figure 1.2(f). Two special kinds of spectra are commonly referred to as white random noise and pink random noise. White random noise contains equal energy per Hertz and thus has a constant spectral density level. Pink random noise contains equal energy per measurement band and thus has an octave or 1/3-octave band level that is constant with frequency.

Frequency analysis is a process by which a time-varying signal is transformed into its frequency components. When tonal components are identified by frequency analysis, it may be advantageous to treat these somewhat differently than broadband noise. Frequency analysis serves the important function of determining the effects of noise or vibration control and it may aid in the identification of sources of noise or vibration.

To facilitate comparison of measurements between instruments, frequency analysis bands have been standardised. The International Standards Organisation has agreed on 'preferred' frequency bands for sound measurement and by agreement, the octave band is the widest band usually considered for frequency analysis. The upper-frequency limit of each octave band is approximately twice its lower-frequency limit and each band is identified by its geometric mean called the band centre frequency. When more detailed information about a noise is required, standardised 1/3-octave band analysis may be used. The preferred frequency bands for octave and 1/3-octave band analysis are summarised in Table 1.1. Reference to the table shows that all information is associated with a band number, BN, listed in column one on the left. In turn, the band number is related to the centre band frequencies, f, of either the octaves or the 1/3-octaves listed in columns two and three. The respective band limits are listed in columns four and five as the lower- and upper-frequency limits, f_ℓ and f_u. These observations may be summarised as:

TABLE 1.1 Preferred octave and 1/3-octave frequency bands

Band number	Octave band centre frequency	1/3-octave band centre frequency	Band limits Lower	Upper
−1		0.8	0.7	0.9
0	1	1	0.9	1.1
1		1.25	1.1	1.4
2		1.6	1.4	1.8
3	2	2	1.8	2.2
4		2.5	2.2	2.8
5		3.15	2.8	3.5
6	4	4	3.5	4.4
7		5	4.4	5.6
8		6.3	5.6	7
9	8	8	7	9
10		10	9	11
11		12.5	11	14
12	16	16	14	18
13		20	18	22
14		25	22	28
15	31.5	31.5	28	35
16		40	35	44
17		5	44	57
18	63	63	57	71
19		80	71	88
20		100	88	113
21	125	125	113	141
22		160	141	176
23		200	176	225
24	250	250	225	283
25		315	283	353
26		400	353	440
27	500	500	440	565
28		630	565	707
29		800	707	880
30	1 000	1 000	880	1 130
31		1 250	1 130	1 414
32		1 600	1 414	1 760
33	2 000	2 000	1 760	2 250
34		2 500	2 250	2 825
35		3 150	2 825	3 530
36	4 000	4 000	3 530	4 400
37		5 000	4 400	5 650
38		6 300	5 650	7 070
39	8 000	8 000	7 070	8 800
40		10 000	8 800	11 300
41		12 500	11 300	14 140
42	16 000	16 000	14 140	17 600
43		20 000	17 600	22 500

$$BN = 10 \log_{10} f \quad \text{and} \quad f = \sqrt{f_\ell f_u} \tag{1.21}$$

A clever manipulation has been used in the construction of Table 1.1. By small adjustments in the calculated values recorded in the table, it has been possible to arrange the 1/3-octave centre frequencies so that ten times their logarithms are the band numbers of column one on the left of the table. Consequently, as may be observed, the 1/3-octave centre frequencies repeat every decade in the table.

In Table 1.1, the frequency band limits have been defined so that they are functions of the analysis band number, BN, and the ratios of the upper to lower frequencies, and are:

$$f_u/f_\ell = 2^{1/N}; \quad N = 1, 3 \tag{1.22}$$

where $N = 1$ for octave bands and $N = 3$ for 1/3-octave bands.

The information provided thus far allows calculation of the bandwidth, Δf, of every band, using the following equation:

$$\Delta f = f_C \frac{2^{1/N} - 1}{2^{1/2N}} \tag{1.23}$$

$$= 0.2316 f_C \quad \text{for 1/3-octave bands}$$

$$= 0.7071 f_C \quad \text{for octave bands}$$

It will be found that the above equations give calculated numbers that are always close to those given in the table. When logarithmic scales are used in plots, the centre frequencies of the 1/3-octave bands between 12.5 Hz and 80 Hz inclusive will lie, respectively, at 0.1, 0.2, 0.3, 0.4, 0.5, 0.6, 0.7, 0.8 and 0.9 of the distance on the scale between 10 and 100. The latter two numbers, in turn, will lie at 0 and 1.0, respectively, on the same scale.

1.4   Acoustic Wave Equation

The derivation of the acoustic wave equation is based on three fundamental fluid dynamical equations: the continuity (or conservation of mass) equation, Euler's equation (or the equation of motion) and the equation of state. Each of these equations are discussed separately in Sections 1.4.1, 1.4.2 and 1.4.3.

1.4.1   Conservation of Mass

Consider an arbitrary volume, V, as shown in Figure 1.4.

The total mass contained in this volume is $\iiint_V \rho_{\text{tot}} \, dV$. The law of conservation of mass states that the rate of mass leaving the volume, V, must equal the rate of change of mass in the volume. That is:

$$\iint_S \rho_{\text{tot}} \boldsymbol{U}_{\text{tot}} \cdot \boldsymbol{n} \, dS = -\frac{d}{dt} \iiint_V \rho_{\text{tot}} \, dV \tag{1.24}$$

where ρ_{tot} is the total (mean plus time varying component) of the density of the fluid contained in the enclosed space of volume, V, at time, t, $\boldsymbol{U}_{\text{tot}}$ is the total velocity of fluid at time, t, and $\boldsymbol{n}$ is the unit vector normal to the surface, S, at location, dS.

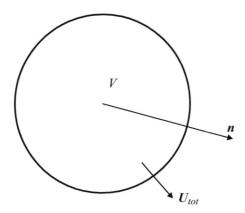

FIGURE 1.4 Arbitrary volume for illustrating conservation of mass.

At this stage, it is convenient to transform the area integral on the left-hand side of Equation (1.24) to a volume integral by use of Gauss' integral theorem, which is written as:

$$\iint_S \boldsymbol{\psi} \cdot \boldsymbol{n} \ \mathrm{d}S = \iiint_V \nabla \cdot \boldsymbol{\psi} \ \mathrm{d}V \tag{1.25}$$

where $\boldsymbol{\psi}$ is an arbitrary vector and the operator, ∇, is the scalar divergence of the vector, $\boldsymbol{\psi}$. Thus, in Cartesian coordinates:

$$\nabla \cdot \boldsymbol{\psi} = \frac{\partial \boldsymbol{\psi}}{\partial x} + \frac{\partial \boldsymbol{\psi}}{\partial y} + \frac{\partial \boldsymbol{\psi}}{\partial z} \tag{1.26}$$

and Equation (1.24) becomes:

$$\iiint_V \nabla \cdot (\rho_{\text{tot}} \boldsymbol{U}_{\text{tot}}) \, \mathrm{d}V = -\frac{\mathrm{d}}{\mathrm{d}t} \iiint_V \rho_{\text{tot}} \, \mathrm{d}V = -\iiint_V \frac{\partial \rho_{\text{tot}}}{\partial t} \, \mathrm{d}V \tag{1.27}$$

Rearranging gives:

$$\iiint_V \left[\nabla \cdot (\rho_{\text{tot}} \boldsymbol{U}_{\text{tot}}) + \frac{\partial \rho_{\text{tot}}}{\partial t} \right] \, \mathrm{d}V = 0 \tag{1.28}$$

or:

$$\nabla \cdot (\rho_{\text{tot}} \boldsymbol{U}_{\text{tot}}) = -\frac{\partial \rho_{\text{tot}}}{\partial t} \tag{1.29}$$

Equation (1.29) is the continuity equation.

1.4.2 Euler's Equation

In 1775 Euler derived his well-known equation of motion for a fluid, based on Newton's first law of motion. That is, the mass of a fluid particle multiplied by its acceleration is equal to the sum of the external forces acting on it.

Consider the fluid particle of dimensions Δx, Δy and Δz shown in Figure 1.5.

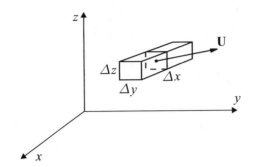

FIGURE 1.5 Particle of fluid.

The external forces, $\boldsymbol{F}$, acting on this particle at time instant, t, are equal to the sum of the pressure differentials across each of the three pairs of parallel forces. Thus:

$$\boldsymbol{F} = \mathbf{i} \cdot \frac{\partial P_{\mathrm{tot}}}{\partial x} + \mathbf{j} \cdot \frac{\partial P_{\mathrm{tot}}}{\partial y} + \mathbf{k} \cdot \frac{\partial P_{\mathrm{tot}}}{\partial z} = \nabla P_{\mathrm{tot}} \tag{1.30}$$

where P_{tot} is the total pressure (mean plus varying component) in the arbitrary volume at time instant, t, $\mathbf{i}$, $\mathbf{j}$ and $\mathbf{k}$ are the unit vectors in the x-, y- and z-directions, and where the operator, ∇, is the grad operator, which is the vector gradient of a scalar quantity.

The inertia force of the fluid particle is its mass multiplied by its acceleration and is equal to:

$$m\dot{\boldsymbol{U}}_{\mathrm{tot}} = m\frac{\mathrm{d}\boldsymbol{U}_{\mathrm{tot}}}{\mathrm{d}t} = \rho_{\mathrm{tot}}V\frac{\mathrm{d}\boldsymbol{U}_{\mathrm{tot}}}{\mathrm{d}t} \tag{1.31}$$

Assume that the fluid particle is accelerating in the positive x-, y- and z-directions. Then the pressure across the particle must be decreasing as x, y and z increase, and the external force must be negative. Thus:

$$\boldsymbol{F} = -\nabla P_{\mathrm{tot}}V = \rho_{\mathrm{tot}}V\frac{\mathrm{d}\boldsymbol{U}_{\mathrm{tot}}}{\mathrm{d}t} \tag{1.32}$$

This is the Euler equation of motion for a fluid.

If sound propagation through porous acoustic media were of interest, then it would be necessary to add the term, $A\boldsymbol{U}_{\mathrm{tot}}$, to the right-hand side of Equation (1.32), where A is a constant dependent on the properties of the fluid. The term $\mathrm{d}\boldsymbol{U}_{\mathrm{tot}}/\mathrm{d}t$ on the right side of Equation (1.32) can be expressed in partial derivative form as:

$$\frac{\mathrm{d}\boldsymbol{U}_{\mathrm{tot}}}{\mathrm{d}t} = \frac{\partial \boldsymbol{U}_{\mathrm{tot}}}{\partial t} + (\boldsymbol{U}_{\mathrm{tot}} \cdot \nabla)\boldsymbol{U}_{\mathrm{tot}} \tag{1.33}$$

where:

$$(\boldsymbol{U}_{\mathrm{tot}} \cdot \nabla)\boldsymbol{U}_{\mathrm{tot}} = \frac{\partial \boldsymbol{U}_{\mathrm{tot}}}{\partial x} \cdot \frac{\partial x}{\partial t} + \frac{\partial \boldsymbol{U}_{\mathrm{tot}}}{\partial y} \cdot \frac{\partial y}{\partial t} + \frac{\partial \boldsymbol{U}_{\mathrm{tot}}}{\partial z} \cdot \frac{\partial z}{\partial t} \tag{1.34}$$

1.4.3   Equation of State

As sound propagation is associated with only very small perturbations to the ambient state of a fluid, it may be regarded as adiabatic. Thus the total pressure P_{tot} will be functionally related to the total density, ρ_{tot}, as:

$$P_{\mathrm{tot}} = \mathrm{f}(\rho_{\mathrm{tot}}) \tag{1.35}$$

Since the acoustic perturbations are small, and P_s and ρ are constant, $\mathrm{d}p = \mathrm{d}P_{\mathrm{tot}}$, $\mathrm{d}\sigma = \mathrm{d}\rho$ and Equation (1.35) can be expanded into a Taylor series as:

$$\mathrm{d}p = \frac{\partial \mathrm{f}}{\partial \rho}\mathrm{d}\sigma + \frac{1}{2}\frac{\partial \mathrm{f}}{\partial \rho}(\mathrm{d}\sigma)^2 + \text{higher order terms} \tag{1.36}$$

The equation of state is derived by using Equation (1.36) and ignoring all of the higher order terms on the right-hand side. This approximation is adequate for moderate sound pressure levels, but becomes less and less satisfactory as the sound pressure level exceeds 130 dB (60 Pa). Thus, for moderate sound pressure levels:

$$\mathrm{d}p = c^2 \, \mathrm{d}\sigma \tag{1.37}$$

where $c^2 = \partial \mathrm{f}/\partial \rho$ is assumed to be constant. Integrating Equation (1.37) gives:

$$p = c^2 \sigma + \text{const} \tag{1.38}$$

which is the linearised equation of state.

Thus the curve $\mathrm{f}(\rho_{\text{tot}})$ of Equation (1.35) has been replaced by its tangent at P_{tot}, ρ_{tot}. The constant may be eliminated by differentiating Equation (1.38) with respect to time. Thus:

$$\frac{\partial p}{\partial t} = c^2 \frac{\partial \sigma}{\partial t} \tag{1.39}$$

Equation (1.39) will be used to eliminate $\partial \sigma/\partial t$ in the wave equation to follow.

1.4.4   Wave Equation (Linearised)

The wave equation may be derived from Equations (1.29), (1.32) and (1.39) by making the linearising approximations listed below. These assume that the acoustic pressure, p, is small compared with the ambient pressure, P_s, and that P_s is constant over time and space. It is also assumed that the mean velocity $\boldsymbol{U} = 0$. Thus:

$$P_{\text{tot}} = P_s + p \approx P_s \tag{1.40}$$

$$\rho_{\text{tot}} = \rho + \sigma \approx \rho \tag{1.41}$$

$$\boldsymbol{U}_{\text{tot}} = \boldsymbol{u} \tag{1.42}$$

$$\frac{\partial P_{\text{tot}}}{\partial t} = \frac{\partial p}{\partial t} \tag{1.43}$$

$$\frac{\partial \rho_{\text{tot}}}{\partial t} = \frac{\partial \sigma}{\partial t} \tag{1.44}$$

$$\nabla P_{\text{tot}} = \nabla p \tag{1.45}$$

Using Equation (1.33), the Euler equation (1.32) may be written as:

$$-\nabla P_{\text{tot}} = \rho_{\text{tot}} \left[\frac{\partial \boldsymbol{U}_{\text{tot}}}{\partial t} + (\boldsymbol{U}_{\text{tot}} \cdot \nabla)\boldsymbol{U}_{\text{tot}} \right] \tag{1.46}$$

Using Equations (1.40), (1.41) and (1.42), Equation (1.46) may be written as:

$$-\nabla p = \rho \left[\frac{\partial \boldsymbol{u}}{\partial t} + \boldsymbol{u} \cdot \nabla \boldsymbol{u} \right] \tag{1.47}$$

As $\boldsymbol{u}$ is small and $\nabla \boldsymbol{u}$ is approximately the same order of magnitude as $\boldsymbol{u}$, the quantity $\boldsymbol{u} \cdot \nabla \boldsymbol{u}$ may be neglected and Equation (1.47) written as:

$$-\nabla p = \rho \frac{\partial \boldsymbol{u}}{\partial t} \tag{1.48}$$

Using Equations (1.41), (1.42) and (1.44), the continuity equation, Equation (1.29), may be written as:

$$\nabla \cdot (\rho \boldsymbol{u} + \sigma \boldsymbol{u}) = -\frac{\partial \sigma}{\partial t} \tag{1.49}$$

As $\sigma \boldsymbol{u}$ is so much smaller than ρ, the equality in Equation (1.49) can be approximated as:

$$\nabla \cdot (\rho \boldsymbol{u}) = -\frac{\partial \sigma}{\partial t} \tag{1.50}$$

Using Equation (1.39), Equation (1.50) may be written as:

$$\nabla \cdot (\rho \boldsymbol{u}) = -\frac{1}{c^2}\frac{\partial p}{\partial t} \tag{1.51}$$

Taking the time derivative of Equation (1.51) gives:

$$\nabla \cdot \rho \frac{\partial \boldsymbol{u}}{\partial t} = -\frac{1}{c^2}\frac{\partial^2 p}{\partial t^2} \tag{1.52}$$

Substituting Equation (1.48) into the left side of Equation (1.52) gives:

$$-\nabla \cdot \nabla p = -\frac{1}{c^2}\frac{\partial^2 p}{\partial t^2} \tag{1.53}$$

or:

$$\nabla^2 p = \frac{1}{c^2}\frac{\partial^2 p}{\partial t^2} \tag{1.54}$$

The operator ∇^2 is the (div grad) or the Laplacian operator, and Equation (1.54) is known as the linearised wave equation or the Helmholtz equation.

The wave equation can be expressed in terms of the particle velocity by taking the gradient of the linearised continuity equation, Equation (1.51). Thus:

$$\nabla(\nabla \cdot \rho \boldsymbol{u}) = -\nabla\left(\frac{1}{c^2}\frac{\partial p}{\partial t}\right) \tag{1.55}$$

Differentiating the Euler equation (1.48) with respect to time gives:

$$-\nabla\frac{\partial p}{\partial t} = \rho\frac{\partial^2 \boldsymbol{u}}{\partial t^2} \tag{1.56}$$

Substituting Equation (1.56) into (1.55) gives:

$$\nabla(\nabla \cdot \boldsymbol{u}) = \frac{1}{c^2}\frac{\partial^2 \boldsymbol{u}}{\partial t^2} \tag{1.57}$$

However, it may be shown that grad div = div grad + curl curl, or:

$$\nabla(\nabla \cdot \boldsymbol{u}) = \nabla^2\boldsymbol{u} + \nabla \times (\nabla \times \boldsymbol{u}) \tag{1.58}$$

Thus Equation (1.57) may be written as:

$$\nabla^2\boldsymbol{u} + \nabla \times (\nabla \times \boldsymbol{u}) = \frac{1}{c^2}\frac{\partial^2 \boldsymbol{u}}{\partial t^2} \tag{1.59}$$

which is the wave equation for the acoustic particle velocity.

1.4.5 Acoustic Potential Function

Situations may arise in which the simplifications of linear acoustics are inappropriate; the associated phenomena are then referred to as nonlinear. For example, a sound wave incident on a perforated plate may incur large energy dissipation due to nonlinear effects. Convection of sound through or across a small hole, due either to a superimposed steady flow or to relatively large amplitudes associated with the sound field, may convert the cyclic flow of the sound field into local fluid streaming. Such nonlinear effects take energy from the sound field, thus reducing the sound level to produce local streaming of the fluid medium, which produces no sound. Similar nonlinear effects also may be associated with acoustic energy dissipation at high sound pressure levels, in excess of 130 dB re 20 μPa.

In general, except for special cases, such as those mentioned, which may be dealt with separately, the losses associated with an acoustic field are quite small, and consequently, the acoustic field may be treated as conservative, meaning that energy dissipation is insignificant and may be neglected. Under such circumstances, it is possible to define a potential function, ϕ, which is a solution to the wave equation (Pierce, 1981) and may be real or complex. It is a scalar quantity and has the advantage that it provides a means for determining both the acoustic pressure and the particle velocity by simple differentiation. That is:

$$\boldsymbol{u} = -\nabla\phi \tag{1.60}$$

However, if Equation (1.60) is substituted into Equation (1.59), then by the Stokes theorem (curl grad = 0 or $\nabla \times \nabla\phi = 0$) the second term on the left-hand side of the equation vanishes. This effectively means that postulating a velocity potential solution to the wave equation causes some loss of generality and restricts the solutions to those which do not involve fluid rotation. Fortunately, acoustic motion in liquids and gases is nearly always rotationless.

Introducing Equation (1.60) for the velocity potential into Euler's equation (1.32), an expression for the acoustic pressure gradient can be derived as:

$$-\nabla p = -\rho\frac{\partial\nabla\phi}{\partial t} = -\rho\nabla\frac{\partial\phi}{\partial t} \tag{1.61}$$

Integrating gives:

$$p = \rho\frac{\partial\phi}{\partial t} + \text{const} \tag{1.62}$$

Introducing Equation (1.62) into the wave equation (1.54) for acoustic pressure, integrating with respect to time and dropping the integration constant gives:

$$\nabla^2\phi = \frac{1}{c^2}\frac{\partial^2\phi}{\partial t^2} \tag{1.63}$$

This is the preferred form of the Helmholtz equation as both acoustic pressure and particle velocity can be derived from the velocity potential solution by simple differentiation using Equation (1.62) with the constant removed and Equation (1.60), respectively.

At high sound pressure levels, or in cases where the particle velocity is large (as in the case when intense sound induces streaming through a small hole or many small holes in parallel), Equation (1.62) takes the form (Morse and Ingard, 1968):

$$p = \rho\left[\partial\phi/\partial t - \frac{1}{2}(\partial\phi/\partial x)^2\right] \quad \text{(Pa)} \tag{1.64}$$

where the coordinate, x, is along the centre line (axis) of a hole. In writing Equation (1.64) a third term on the right side of the equation given in the reference has been omitted as it is inversely proportional to the square of the phase speed, and thus in the cases considered here, it is negligible. Alternatively, if a convection velocity, $\boldsymbol{U}$, is present and large, and the particle velocity, $\boldsymbol{u}$, is small, Equation (1.62) takes the form:

$$p = \rho\left[\partial\phi/\partial t - \boldsymbol{U}\partial\phi/\partial x\right] \quad \text{(Pa)} \tag{1.65}$$

Taking the gradient of Equation (1.62), interchanging the order of differentiation on the right-hand side of the equation and introducing Equation (1.60) gives Euler's famous equation of motion for a unit volume of fluid acted on by a pressure gradient:

$$\rho \frac{\partial \boldsymbol{u}}{\partial t} = -\nabla p \tag{1.66}$$

1.4.6 Inhomogeneous Wave Equation (Medium Containing Acoustic Sources)

Assume that the acoustic medium contains acoustic sources with a net volume velocity output of q units per unit volume per unit time (note that by this definition, the quantity q has the dimensions of T^{-1}). The mass introduced by these sources must be added to the left-hand side of the continuity equation, Equation (1.51). Thus:

$$\rho q - \nabla \cdot (\rho \boldsymbol{u}) = \frac{1}{c^2} \frac{\partial p}{\partial t} \tag{1.67}$$

which reduces to:

$$\nabla \cdot \boldsymbol{u} = -\frac{1}{\rho c^2} \frac{\partial p}{\partial t} + q \tag{1.68}$$

Substituting Equations (1.60) and (1.65) into Equation (1.68) gives:

$$\nabla^2 \phi = \frac{1}{c^2} \frac{\partial^2 \phi}{\partial t^2} - q \tag{1.69}$$

which is often referred to as the inhomogeneous wave equation. Note that the inhomogeneous wave equation describes the response of a forced acoustic system, whereas the homogeneous wave equation describes the response of an acoustic medium that contains no sources or sinks.

The quantity, q, is referred to as a source distribution or source strength per unit volume. For a point source of strength, q, at location, $\boldsymbol{r}_0$, in the acoustic medium, the quantity, q, in Equation (1.69) would be replaced by $q\delta(\boldsymbol{r} - \boldsymbol{r}_0)$. For a source of unit strength at location $\boldsymbol{r}_0$, the quantity q in Equation (1.69) would be replaced by $\delta(\boldsymbol{r} - \boldsymbol{r}_0)$. Note that $\boldsymbol{r}$ is the location at which the velocity potential, ϕ, is to be evaluated. The use of the Dirac delta function, $\delta(\boldsymbol{r} - \boldsymbol{r}_0)$, is discussed in more detail in Chapter 3. An identical equation to Equation (1.69) applies for the acoustic pressure, p. It is obtained simply by replacing ϕ with p and replacing q with $j\omega\rho q$, where $j = \sqrt{-1}$.

1.4.7 Wave Equation for One-Dimensional Mean Flow

Consider a medium with a mean flow of U_z along the z-axis in the positive direction. Consider also a reference frame, X, Y, Z and T_f moving along with the fluid. Then the wave equation in this moving reference frame is:

$$\frac{\partial^2 \phi}{\partial X^2} + \frac{\partial^2 \phi}{\partial Y^2} + \frac{\partial^2 \phi}{\partial Z^2} = \frac{1}{c^2} \frac{\partial^2 \phi}{\partial T_f^2} \tag{1.70}$$

Introducing a second, stationary reference frame, x, y, z and t, results in the following relationships between the two sets of coordinates:

$$X = x \text{ and } Y = y \tag{1.71}$$

$$Z = z - U_z t \tag{1.72}$$

and:

$$T_f = t \tag{1.73}$$

In terms of the stationary reference frame, the quantity, $\dfrac{\partial \phi}{\partial T_f}$, can be written as:

$$\frac{\partial \phi}{\partial T_f} = \frac{\partial \phi}{\partial t}\frac{\partial t}{\partial T_f} + \frac{\partial \phi}{\partial z}\frac{\partial z}{\partial T_f} \tag{1.74}$$

From Equation (1.73), $\dfrac{\partial T_f}{\partial t} = 1$ and from Equation (1.72)

$$\frac{\partial z}{\partial T_f} = \frac{\partial(z + U_z t)}{\partial T_f} = U_z \tag{1.75}$$

Thus Equation (1.74) becomes:

$$\frac{\partial \phi}{\partial T_f} = \frac{\partial \phi}{\partial t} + U_z \frac{\partial \phi}{\partial z} \tag{1.76}$$

Differentiating a second time:

$$\frac{\partial}{\partial T_f}\left(\frac{\partial \phi}{\partial T_f}\right) = \frac{\partial}{\partial T_f}\left(\frac{\partial \phi}{\partial t} + U_z \frac{\partial \phi}{\partial z}\right) \tag{1.77}$$

Expanding gives:

$$\frac{\partial^2 \phi}{\partial T_f^2} = \frac{\partial^2 \phi}{\partial t^2}\cdot\frac{\partial t}{\partial T_f} + \frac{\partial^2 \phi}{\partial z\partial t}\cdot\frac{\partial z}{\partial T_f} + U_z\frac{\partial^2 \phi}{\partial t\partial z}\cdot\frac{\partial t}{\partial T_f} + U_z\frac{\partial^2 \phi}{\partial z^2}\cdot\frac{\partial z}{\partial t}$$

$$= \frac{\partial^2 \phi}{\partial t^2} + 2U_z\frac{\partial^2 \phi}{\partial z\partial t} + U_z^2\frac{\partial^2 \phi}{\partial z^2} \tag{1.78}$$

Thus:

$$\frac{\partial^2 \phi}{\partial T_f^2} = \left(\frac{\partial}{\partial t} + U_z\frac{\partial}{\partial z}\right)^2 \phi \tag{1.79}$$

Expressing the left-hand side of Equation (1.70) in terms of the stationary reference frame gives:

$$\frac{\partial^2 \phi}{\partial X^2} + \frac{\partial^2 \phi}{\partial Y^2} + \frac{\partial^2 \phi}{\partial Z^2} = \frac{\partial^2 \phi}{\partial x^2} + \frac{\partial^2 \phi}{\partial y^2} + \frac{\partial^2 \phi}{\partial z^2} \tag{1.80}$$

Substituting Equations (1.79) and (1.80) into Equation (1.70) gives the following wave equation for a fluid with a mean velocity of U_z along the positive z-direction in terms of a stationary reference frame, x, y, z and t:

$$\nabla^2 \phi = \frac{1}{c^2}\left(\frac{\partial}{\partial t} + U_z\frac{\partial}{\partial z}\right)^2 \phi \tag{1.81}$$

1.4.8 Wave Equation in Cartesian, Cylindrical and Spherical Coordinates

1.4.8.1 Cartesian Coordinates

$$\nabla^2 \phi = \frac{\partial^2 \phi}{\partial x^2} + \frac{\partial^2 \phi}{\partial y^2} + \frac{\partial^2 \phi}{\partial z^2} \tag{1.82}$$

Thus the wave equation in Cartesian coordinates is:

$$\frac{\partial^2 \phi}{\partial x^2} + \frac{\partial^2 \phi}{\partial y^2} + \frac{\partial^2 \phi}{\partial z^2} = \frac{1}{c^2}\frac{\partial^2 \phi}{\partial t^2} \tag{1.83}$$

1.4.8.2 Cylindrical Coordinates

The cylindrical coordinate system is illustrated in Figure 1.6 and the gradient of the velocity potential in cylindrical coordinates is:

$$\nabla \phi = \frac{\partial \phi}{\partial r} + \frac{1}{r} \frac{\partial \phi}{\partial \theta} + \frac{\partial \phi}{\partial z} \tag{1.84}$$

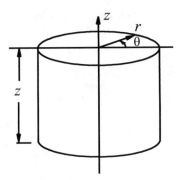

FIGURE 1.6 Cylindrical coordinate system.

and:

$$\nabla^2 \phi = \frac{\partial^2 \phi}{\partial r^2} + \frac{1}{r} \frac{\partial \phi}{\partial r} + \frac{1}{r^2} \frac{\partial^2 \phi}{\partial \theta^2} + \frac{\partial^2 \phi}{\partial z^2} \tag{1.85}$$

Thus the wave equation in cylindrical coordinates is:

$$\frac{\partial^2 \phi}{\partial r^2} + \frac{1}{r} \frac{\partial \phi}{\partial r} + \frac{1}{r^2} \frac{\partial^2 \phi}{\partial \theta^2} + \frac{\partial^2 \phi}{\partial z^2} = \frac{1}{c^2} \frac{\partial^2 \phi}{\partial t^2} \tag{1.86}$$

1.4.8.3 Spherical Coordinates

The spherical coordinate system is illustrated in Figure 1.7 and the velocity potential gradient in spherical coordinates is:

$$\nabla \phi = \frac{\partial \phi}{\partial r} + \frac{1}{r} \frac{\partial \phi}{\partial \theta} + \frac{1}{r \sin \theta} \frac{\partial \phi}{\partial \vartheta} \tag{1.87}$$

and:

$$\nabla^2 \phi = \frac{\partial^2 \phi}{\partial r^2} + \frac{2}{r} \frac{\partial \phi}{\partial r} + \frac{1}{r^2} \frac{\partial^2 \phi}{\partial \theta^2} + \frac{1}{r^2 \sin \theta} \frac{\partial \phi}{\partial \theta} + \left(\frac{1}{r \sin \theta} \right)^2 \frac{\partial^2 \phi}{\partial \vartheta^2} \tag{1.88}$$

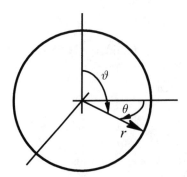

FIGURE 1.7 Spherical coordinate system.

Thus, the wave equation in spherical coordinates is:

$$\frac{\partial^2 \phi}{\partial r^2} + \frac{2}{r}\frac{\partial \phi}{\partial r} + \frac{1}{r^2}\frac{\partial^2 \phi}{\partial \theta^2} + \frac{1}{r^2 \sin\theta}\frac{\partial \phi}{\partial \theta} + \left(\frac{1}{r\sin\theta}\right)^2 \frac{\partial^2 \phi}{\partial \vartheta^2} = \frac{1}{c^2}\frac{\partial^2 \phi}{\partial t^2} \tag{1.89}$$

1.4.9   Plane and Spherical Waves

In general, sound wave propagation is quite complicated and not amenable to simple analysis. However, sound wave propagation can often be described in terms of the propagation properties of plane and spherical waves. Plane and spherical waves, in turn, have the convenient property that they can be described in terms of one dimension. Thus, the investigation of plane and spherical waves, although by no means exhaustive, is useful as a means of greatly simplifying and rendering tractable what, in general, may be a very complicated problem. The investigation of plane and spherical wave propagation is the subject of Sections 1.4.10 and 1.4.11, respectively.

1.4.10   Plane Wave Propagation

For the case of plane wave propagation, only one spatial dimension, x, the direction of propagation, is required to describe the acoustic field. An example of plane wave propagation is sound propagating along the centre line of a tube with rigid walls. In this case, Equation (1.63) reduces to:

$$\frac{\partial^2 \phi}{\partial x^2} = \frac{1}{c^2}\frac{\partial^2 \phi}{\partial t^2} \tag{1.90}$$

A solution of Equation (1.90), which may be verified by direct substitution, is:

$$\phi = \mathrm{f}(ct \pm x) \tag{1.91}$$

The function, f, in Equation (1.91) describes a distribution along the x-axis at any fixed time, t, as well as the variation with time at any fixed place, x, along the direction of propagation. If the argument, $(ct \pm x)$, is fixed and the positive sign is chosen, then with increasing time, t, x must decrease with speed, c. Alternatively, if the argument, $(ct \pm x)$, is fixed and the negative sign is chosen, then with increasing time, t, x must increase with speed, c. Consequently, a wave travelling in the positive x-direction is represented by taking the negative sign and a wave travelling in the negative x-direction is represented by taking the positive sign in the argument of Equation (1.91). A very important relationship between acoustic pressure and particle velocity will now be determined. A prime sign, $'$, will indicate differentiation of a function by its argument as, for example, $\mathrm{df}(w)/\mathrm{d}w = \mathrm{f}'(w)$. Substitution of Equation (1.91) in Equation (1.60) gives:

$$u = \mp \mathrm{f}'(ct \pm x) \tag{1.92}$$

and substitution of Equation (1.91) in Equation (1.62) gives:

$$p = \rho c \mathrm{f}'(ct \pm x) \tag{1.93}$$

Division of Equation (1.93) by Equation (1.92) gives:

$$p/u = \pm \rho c \tag{1.94}$$

which is a very important result — the characteristic impedance, ρc, of a plane wave.

In Equation (1.94), the positive sign is taken for waves travelling in the positive x-direction, while the negative sign is taken for waves travelling in the negative x-direction. The characteristic impedance is one of three kinds of impedance used in acoustics (see Section 4.2 for a more detailed discussion of impedance types). It provides a very useful relationship between acoustic pressure and particle velocity in a plane wave. It also has the property that a duct terminated in its characteristic impedance will respond as an infinite duct, as no wave will be reflected at its termination. Fourier analysis enables the representation of any function, $f(ct \pm x)$, as a sum or integral of harmonic functions. Thus, it will be useful, for consideration of the wave equation, to investigate the special properties of harmonic solutions. Consideration will begin with the following harmonic solution for the acoustic potential function:

$$\phi = A\cos(k_a(ct \pm x) + \beta) \tag{1.95}$$

where k_a is a constant, which will be investigated, and β represents an arbitrary relative phase. As β is arbitrary in Equation (1.95), for fixed time, t, β may be chosen so that:

$$k_a ct + \beta = 0 \tag{1.96}$$

In this case, Equation (1.91) reduces to the following representation of the spatial distribution:

$$\phi = A\cos(k_a x) = A\cos(2\pi x/\lambda) \tag{1.97}$$

From Equations (1.97) it may be concluded that the unit of length, λ, defined as the wavelength of the propagating wave and the constant, k_a, defined as the wavenumber are related as:

$$2\pi/\lambda = k_a \tag{1.98}$$

An example of harmonic (single frequency) plane wave propagation in a tube is illustrated in Figure 1.3. The type of wave generated is longitudinal, as shown in Figure 1.3(a) and the corresponding pressure fluctuations as a function of time are shown in Figure 1.3(b).

The distribution in space has been considered and now the distribution in time for a fixed point in space will be considered. The arbitrary phase constant, β, of Equation (1.95) will be chosen so that, for fixed position, x:

$$\beta \pm k_a x = 0 \tag{1.99}$$

Equation (1.95) then reduces to the following representation for the temporal distribution:

$$\phi = A\cos(k_a ct) = A\cos\frac{2\pi}{T_p}t \tag{1.100}$$

The period, T_p, of the propagating wave is:

$$2\pi/k_a c = T_p \quad (\text{s}) \tag{1.101}$$

Its reciprocal is the more familiar frequency, f. Since the angular frequency, ω (radians/s), is quite often used as well, the following relation should be noted:

$$2\pi/T_p = 2\pi f = \omega \quad (\text{rad/s}) \tag{1.102}$$

and from Equations (1.101) and (1.102):

$$k_a = \omega/c \quad (\text{rad/m}) \tag{1.103}$$

and from Equations (1.98), (1.102), and (1.103):

$$f\lambda = c \quad (\text{m/s}) \tag{1.104}$$

The wavenumber, k_a, may be thought of as a spatial frequency, where k_a is the analog of frequency, f, and wavelength, λ, is the analog of the period, T.

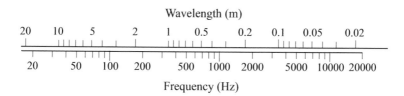

FIGURE 1.8 Wavelength in air versus frequency under normal conditions.

The relationship between wavelength and frequency, for sound propagating in air, is illustrated in Figure 1.8.

The wavelength of generally audible sound varies by a factor of about one thousand. The shortest audible wavelength is 17 mm (corresponding to 20 000 Hz) and the longest is 17 m (corresponding to 20 Hz), although humans can detect sound via their vestibular system (which the ear is part of) at much lower frequencies if it is sufficiently loud. Letting $A = B/\rho\omega$ in Equation (1.95) and use of Equations (1.91) to (1.93) and (1.103) gives the following useful expressions for the particle velocity and the acoustic pressure, respectively, for a plane wave:

$$p = B\sin(\omega t \mp k_a x + \beta) \quad \text{(Pa)} \qquad (1.105)$$

$$u = \pm \frac{B}{\rho c} \sin(\omega t \mp k_a x + \beta) \quad \text{(m/s)} \qquad (1.106)$$

It may be mentioned in passing that the group speed, briefly introduced in Section 1.3.8, has the following form:

$$c_g = \mathrm{d}\omega/\mathrm{d}k_a \quad \text{(m/s)} \qquad (1.107)$$

By differentiating Equation (1.52) with respect to wavenumber, k_a, it may be concluded that for non-dispersive wave propagation, where the wave speed is independent of frequency, as for longitudinal compressional waves in unbounded media, the phase and group speeds are equal. Thus, in the case of longitudinal waves propagating in unbounded media, the rate of acoustic energy transport is the same as the speed of sound, as earlier stated.

A convenient form of harmonic solution for the wave equation is the complex solution, written in either of the following equivalent forms:

$$\phi = \hat{\phi}\mathrm{e}^{\mathrm{j}\omega t} = A\mathrm{e}^{\mathrm{j}(\omega t \pm k_a x + \beta)} = A\cos(\omega t \pm k_a x + \beta) + \mathrm{j}A\sin(\omega t \pm k_a x + \beta) \qquad (1.108)$$

where $\mathrm{j} = \sqrt{-1}$, A is a real amplitude, and $\hat{\phi}$ is the complex amplitude of the velocity potential. In either form, the negative sign represents a wave travelling in the positive x-direction, while the positive sign represents a wave travelling in the negative x-direction. The real parts of Equations (1.108) are just the solutions given by Equation (1.95). The imaginary parts of Equations (1.108) are also solutions, but in quadrature (90° out of phase) with the former solutions. By convention, the complex notation is defined so that what is measured with an instrument corresponds to the real part; the imaginary part is then inferred from the real part. The complex exponential form of the harmonic solution to the wave equation is used as a mathematical convenience, as it greatly simplifies mathematical manipulations, allows waves with different phases to be added together easily and allows graphical representation of the solution as a rotating vector in the complex plane. Setting $\beta = 0$ and $x = 0$ allows Equation (1.108) to be rewritten as:

$$\phi = \hat{\phi}\mathrm{e}^{\mathrm{j}\omega t} = A\mathrm{e}^{\mathrm{j}\omega t} = A(\cos\omega t + \mathrm{j}\sin\omega t) \qquad (1.109)$$

Equation (1.109) represents harmonic motion that may be represented at any time, t, as a rotating vector of constant amplitude, A, and constant angular velocity, ω, as illustrated in Figure 1.9. Referring to the figure, the projection of the rotating vector on the abscissa, x-axis,

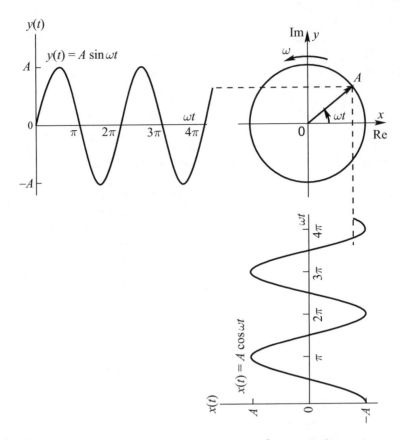

FIGURE 1.9 Harmonic motions represented as a rotating vector.

is given by the real term on the right-hand side of Equation (1.109), and the projection of the rotating vector on the ordinate, y-axis, is given by the imaginary term.

For the special case of single frequency sound propagating in 3D, complex notation may be introduced. For example, the acoustic pressure of real amplitude, A, and the particle velocity of real amplitude, B, may then be written in the following general form where the wavenumber, k_a, is given by Equation (1.103):

$$p(\boldsymbol{r}, t) = A\mathrm{e}^{\mathrm{j}k_a(ct+|\boldsymbol{r}|+\theta/k_a)} = A\mathrm{e}^{\mathrm{j}(\omega t+\theta_p(\boldsymbol{r}))} = \hat{p}(\boldsymbol{r})\mathrm{e}^{\mathrm{j}\omega t} \tag{1.110}$$

and:

$$\boldsymbol{u}(\boldsymbol{r}, t) = B\mathrm{e}^{\mathrm{j}(\omega t+\theta_u(\boldsymbol{r}))} = \hat{\boldsymbol{u}}(\boldsymbol{r})\mathrm{e}^{\mathrm{j}\omega t} \tag{1.111}$$

where $\hat{p}(\boldsymbol{r})$ and $\hat{\boldsymbol{u}}(\boldsymbol{r})$ are complex amplitudes.

In writing Equations (1.110) and (1.111) it has been assumed that the origin of vector, $\boldsymbol{r}$, is at the source centre and θ_p and θ_u are, respectively, the phases of the pressure and particle velocity, both of which are also functions of location $\boldsymbol{r}$. Use of the complex form of the solution makes integration and differentiation particularly simple. Also, impedances are conveniently described using this notation. For these reasons, the complex notation will be used throughout this book. However, care must be taken in the use of the complex notation when multiplying one function by another. In the calculation of products of quantities expressed in complex notation, it is important to remember that the product implies that in general, the real parts of the quantities are multiplied. This is important, for example, in the calculation of intensity associated with single frequency sound fields expressed in complex notation; for example, the product of $p(\boldsymbol{r}, t)$

and $\boldsymbol{u}(\boldsymbol{r},t)$ in Equations (1.110) and (1.111) is:

$$\begin{aligned} p(\boldsymbol{r},t) \times \boldsymbol{u}(\boldsymbol{r},t) &= A\cos(\omega t + \theta_p(\boldsymbol{r})) \times B\cos(\omega t + \theta_p(\boldsymbol{r})) \\ &= \hat{p}(\boldsymbol{r})\cos(\omega t) \times \hat{\boldsymbol{u}}(\boldsymbol{r})\cos(\omega t) \end{aligned} \tag{1.112}$$

1.4.11 Spherical Wave Propagation

A second important case is that of spherical wave propagation; an example is the propagation of sound waves from a small source in free space with no boundaries nearby. In this case, the wave Equation (1.38) may be written in spherical coordinates in terms of a radial term only, since no angular dependence is implied. Thus Equation (1.38) becomes (Morse and Ingard (1968), p. 309):

$$\frac{1}{r^2}\frac{\partial}{\partial r}\left[r^2\frac{\partial\phi}{\partial r}\right] = \frac{1}{c^2}\frac{\partial^2\phi}{\partial t^2} \tag{1.113}$$

which can be written as:

$$\frac{1}{r^2}\frac{\partial}{\partial r}\left[r^2\frac{\partial\phi}{\partial r}\right] = \frac{2}{r}\frac{\partial\phi}{\partial r} + \frac{\partial^2\phi}{\partial r^2} = \frac{1}{r}\frac{\partial}{\partial r}\left[\phi + r\frac{\partial\phi}{\partial r}\right] = \frac{1}{r}\frac{\partial^2(r\phi)}{\partial r^2} \tag{1.114}$$

Thus, the wave equation may be rewritten as:

$$\frac{\partial^2(r\phi)}{\partial r^2} = \frac{1}{c^2}\frac{\partial^2(r\phi)}{\partial t^2} \tag{1.115}$$

The difference between, and similarity of, Equations (1.90) and (1.115) should be noted. Evidently, $r\phi = \mathrm{f}(ct \mp r)$ is a solution of Equation (1.115) where the source is located at the origin. Thus:

$$\phi = \frac{\mathrm{f}(ct \mp r)}{r} \tag{1.116}$$

The implications of the latter solution will now be investigated. To proceed, Equations (1.62) and (1.60) are used to write expressions for the acoustic pressure and particle velocity in terms of the potential function given by Equation (1.116). The expression for the acoustic pressure is:

$$p = \rho c\frac{\mathrm{f}'(ct \mp r)}{r} \quad \text{(Pa)} \tag{1.117}$$

and the expression for the acoustic particle velocity in the direction normal to the wavefront is:

$$u = \frac{\mathrm{f}(ct \mp r)}{r^2} \pm \frac{\mathrm{f}'(ct \mp r)}{r} \quad \text{(m/s)} \tag{1.118}$$

In Equations (1.116), (1.117) and (1.118) the upper sign describes a spherical wave that decreases in amplitude as it diverges outward from the origin, where the source is located. Alternatively, the lower sign describes a converging spherical wave, which increases in amplitude as it converges towards the origin. The characteristic impedance of the spherical wave may be computed, as was done earlier for the plane wave, by dividing Equation (1.117) by Equation (1.118) to obtain the expression:

$$\frac{p}{u} = \rho c\frac{r\mathrm{f}'(ct \mp r)}{\mathrm{f}(ct \mp r) \pm r\mathrm{f}'(ct \mp r)} \tag{1.119}$$

If the distance, r, from the origin is very large, the quantity, $r\mathrm{f}'$, will be sufficiently large compared to the quantity, f, for the latter to be neglected; in this case, for outward-going waves the characteristic impedance becomes ρc, while for inward-going waves it becomes $-\rho c$. In summary, at large enough distance from the origin of a spherical wave, the curvature of any part of the wave finally becomes negligible, and the characteristic impedance becomes that of a plane wave, as given by Equation (1.94). See the discussion following Equation (1.91) in Section 1.4.10 for a definition of the use of the prime, $'$. A moment's reflection, however, immediately raises the question: how large is a large distance? The answer concerns the curvature of the wavefront; a large distance must be where the curvature or radius of the wavefront as measured in wavelengths is large. For example, referring to Equation (1.98), a large distance must be where:

$$k_a r \gg 1 \qquad\qquad (1.120)$$

For harmonic waves, the solution given by Equation (1.116) can also be written as:

$$\phi = \frac{\mathrm{f}(k_a(ct \pm r))}{r} = \frac{\mathrm{f}(\omega t \pm k_a r)}{r} = \frac{A}{r}\mathrm{e}^{\mathrm{j}(\omega t \pm k_a r)} \qquad\qquad (1.121)$$

Substitution of Equation (1.121) into Equation (1.62) gives an expression for the acoustic pressure for outwardly travelling waves (corresponding to the negative sign in Equation (1.121)), which can be written as:

$$p = \frac{\mathrm{j}\omega A\rho}{r}\mathrm{e}^{\mathrm{j}(\omega t - k_a r)} = \frac{\mathrm{j}k_a \rho c A}{r}\mathrm{e}^{\mathrm{j}(\omega t - k_a r)} \qquad\qquad (1.122)$$

Substitution of Equation (1.121) into Equation (1.60) gives an expression for the acoustic particle velocity, as:

$$u = \frac{A}{r^2}\mathrm{e}^{\mathrm{j}(\omega t - k_a r)} + \frac{\mathrm{j}k_a A}{r}\mathrm{e}^{\mathrm{j}(\omega t - k_a r)} \qquad\qquad (1.123)$$

Dividing Equation (1.122) by Equation (1.123) gives:

$$\frac{p}{u} = \rho c\frac{\mathrm{j}k_a r}{1 + \mathrm{j}k_a r} \qquad\qquad (1.124)$$

which holds for a harmonic wave characterised by a wavenumber k_a, and also for a narrow band of noise characterised by a narrow range of wavenumbers around k_a. For inward-travelling waves, the sign of k_a is negative.

Consideration of Equation (1.124) now gives explicit meaning to large distance, as according to Equations (1.98) and (1.120), large distance means that the distance measured in wavelengths is large; for example, $r > 3\lambda$. Note that when Equation (1.120) is satisfied, Equation (1.124) reduces to the positive, outward-travelling form of Equation (1.94), which is a plane wave. For the case of a narrow band of noise, for example an octave band, the wavelength is conveniently taken as the wavelength associated with the centre frequency of the band.spherical wave

1.4.12 Wave Summation

It will be shown that any number of harmonic waves, of the same frequency travelling in one particular direction, combine to produce one wave travelling in the same direction. For example, a wave that is reflected back and forth between two terminations many times may be treated as a single wave travelling in each direction.

Assume that many waves, all of the same frequency, travel together in the same direction. The waves may each be represented by rotating vectors as shown in Figure 1.10. The wave vectors in the figure will all rotate together with the passage of time and thus they will add vectorially as illustrated in the figure for the simple case of two waves separated in phase by β.

Consider any two waves, with real amplitudes, A_1 and A_2 travelling in one direction, which may be described respectively, as $p_1 = A_1\mathrm{e}^{\mathrm{j}\omega t}$ and $p_2 = A_2\mathrm{e}^{\mathrm{j}(\omega t + \beta)}$, where β is the phase

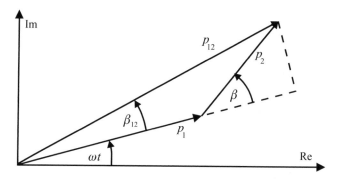

FIGURE 1.10 Graphical representation of the addition of two vectors in the complex plane, which represents the sum of two complex-valued harmonic pressure waves.

difference between the two waves. The cosine rule gives for the real amplitude, A_{12}, and the relative phase, β_{12}, of the combined wave:

$$A_{12}^2 = A_1^2 + A_2^2 + 2A_1 A_2 \cos \beta \tag{1.125}$$

$$\beta_{12} = \tan^{-1} \frac{A_2 \sin \beta}{A_1 + A_2 \cos \beta} \tag{1.126}$$

Equations (1.125) and (1.126) define the vector sum of the two complex-valued harmonic pressure waves, as:

$$p_{12} = A_{12} e^{j(\omega t + \beta_{12})} \tag{1.127}$$

The process is then repeated, each time adding to the cumulative sum of the previous waves, a new wave not already considered, until the sum of all waves travelling in the same direction has been obtained. It can be observed that the sum will always be like any of its constituent parts; thus it may be concluded that the cumulative sum may be represented by a single wave travelling in the same direction as all of its constituent parts.

1.4.13 Plane Standing Waves

If a loudspeaker emitting a tone is placed at one end of a closed tube, there will be multiple reflections of waves from each end of the tube. As has been shown, all of the reflections in the tube result in two waves, one propagating in each direction. These two waves will also combine, and form a 'standing wave'. This effect can be illustrated by writing the following expression for sound pressure at any location in the tube as a result of the two waves of amplitudes A and B, respectively, travelling in the two opposite directions, where $A \geq B$:

$$p = A e^{j(\omega t + k_a x)} + B e^{j(\omega t - k_a x + \beta)} \tag{1.128}$$

Equation (1.128) can be rewritten making use of the identity:

$$0 = -B e^{j(k_a x + \beta)} + B e^{j(k_a x + \beta)} \tag{1.129}$$

Thus:

$$p = (A - B e^{j\beta}) e^{j(\omega t + k_a x)} + 2B e^{j(\omega t + \beta)} \cos k_a x \tag{1.130}$$

Equation (1.130) consists of two terms. The first term on the right-hand side is a left travelling wave of amplitude $(A - B e^{j\beta})$ and the second term on the right-hand side is a standing wave of amplitude $2B e^{j\beta}$. In the latter case, the wave is described by a cosine, which varies in amplitude with time, but remains stationary in space.

1.4.14 Spherical Standing Waves

Standing waves are most easily demonstrated using plane waves, but any harmonic wave motion may produce standing waves. An example of spherical standing waves is provided by the sun, which may be considered as a fluid sphere of radius, r, in a vacuum. At the outer edge, the acoustic pressure may be assumed to be effectively zero. Using Equation (1.122), the sum of the outward travelling wave and the reflected inward travelling wave gives the following relation for the acoustic pressure, p, at the surface of the sphere:

$$p = \mathrm{j}k_a\rho c\frac{2A\mathrm{e}^{\mathrm{j}\omega t}}{r}\cos k_a r = 0 \tag{1.131}$$

where the identity, $\mathrm{e}^{-\mathrm{j}k_a r} + \mathrm{e}^{\mathrm{j}k_a r} = 2\cos k_a r$ has been used. Evidently, the simplest solution for Equation (1.131) is $k_a r = (2N-1)/2$ where N is an integer. If it is assumed that there are no losses experienced by the wave travelling through the media making up the sun, the first half of the equation is valid everywhere except at the centre, where $r = 0$ and the solution is singular. Inspection of Equation (1.131) shows that it describes a standing wave. Note that the largest difference between maximum and minimum pressures occurs in the standing wave when $p = 0$ at the boundary. However, standing waves (with smaller differences between the maximum and minimum pressures) will be also be generated for conditions where the pressure at the outer boundary is not equal to 0.

1.5 Application of the Wave Equation to Analysis of Acoustic Enclosures

The wave equation can be used to find mode shapes and resonance frequencies associated with an acoustic enclosure. As with structural analysis considered in Chapter 2, where analysis is restricted to analytically tractable problems, such as beams, plates and cylinders, here the analysis will be restricted to rectangular and cylindrical enclosures, thus allowing principles to be illustrated without an unwarranted amount of complexity.

In the low-frequency range, an enclosure sound field is dominated by standing waves at certain characteristic frequencies. Large spatial variations in the reverberant field are observed if the enclosure is excited with pure tone sound, and the sound field in the enclosure is said to be dominated by resonant or modal response.

When a source of sound in an enclosure is turned on, the resulting sound waves spread out in all directions from the source. When the advancing sound waves reach the walls of the enclosure they are reflected, generally with a small loss of energy, eventually resulting in waves travelling around the enclosure in all directions. If each path that a wave takes is traced around the enclosure, there will be certain paths of travel that repeat upon themselves to form normal modes of vibration, and at certain frequencies, waves travelling around such paths will arrive back at any point along the path in phase. Amplification of the wave disturbance will result and the normal mode will be resonant. When the frequency of the source equals one of the resonance frequencies of a normal mode, resonance occurs and the interior space of the enclosure responds strongly, being only limited by the absorption present within the enclosure.

A normal mode has been associated with paths of travel that repeat upon themselves. Evidently, waves may travel in either direction along such paths so that, in general, normal modes are characterised by waves travelling in opposite directions along any repeating path. As waves travelling along the same path, but in opposite directions, produce standing waves, a normal mode may be characterised as a system of standing waves, which in turn, is characterised by nodes (locations of minimum response) and antinodes (locations of maximum response). At locations where the oppositely travelling waves arrive, for example in pressure anti-phase, pressure cancellation will occur, resulting in a sound pressure minimum called a node. Similarly, at locations where the oppositely travelling waves arrive in pressure phase, pressure amplification will occur, resulting in a sound pressure maximum called an antinode.

In an enclosure at low frequencies, the number of resonance frequencies within a band, such as an octave or 1/3-octave, will be small. Thus, at low frequencies, the response of a room as a function of frequency and location will be quite irregular; that is, the spatial distribution in the reverberant field will be characterised by sound pressure nodes and antinodes.

1.5.1 Rectangular Enclosures

If the source in the rectangular room illustrated in Figure 1.11 is arranged to produce a single frequency, which is slowly increased, the sound level at any location (other than at a node in the room for that frequency) will at first rapidly increase, momentarily reach a maximum at resonance, then rapidly decrease. The process repeats with each room resonance. The measured frequency response of a 180 m^3 rectangular reverberation room is shown in Figure 1.12 for illustration. The sound pressure was measured in a corner of the room (where there are no pressure nodes) while the frequency of the source (placed at an opposite corner) was very slowly increased.

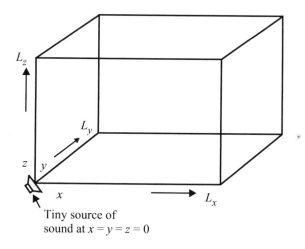

FIGURE 1.11 Rectangular enclosure.

Consideration of a rectangular room provides a convenient model for understanding modal response and the placement of sound absorbents for sound control. The mathematical description of the modal response of the rectangular room, illustrated in Figure 1.11, takes on a particularly simple form; thus it will be advantageous to use the rectangular room as a model for the following discussion of modal response. It is emphasised that modal response is by no means peculiar to rectangular or even regular-shaped rooms. Modal response characterises enclosures of all shapes. Splayed, irregular or odd numbers of walls will not prevent resonances and accompanying pressure nodes and antinodes in an enclosure constructed of reasonably reflective walls; nor will such peculiar construction necessarily result in a more uniform distribution in frequency of the resonances of an enclosure than would occur in a rectangular room of appropriate dimensions. However, it is simpler to calculate the resonance frequencies and mode shapes for rectangular rooms.

For sound in a rectangular enclosure, a standing wave solution for the acoustic potential function takes the following simple form (see Section 1.4.5):

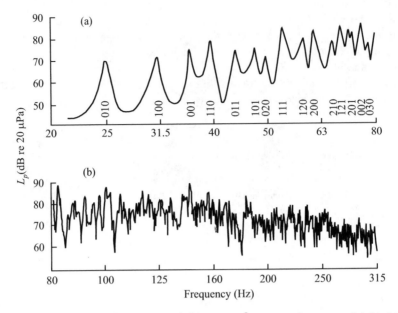

FIGURE 1.12 Measured frequency response of a 180 m^3 rectangular room. (a) In this frequency range, room resonances are identified by mode numbers. (b) In this frequency range, peaks in the room response cannot be associated with room resonances identified by mode numbers.

$$\phi = X(x)Y(y)Z(z)\mathrm{e}^{\mathrm{j}\omega t} \tag{1.132}$$

Substitution of Equation (1.132) into the wave equation (Equation (1.63)), use of $k_a^2 = \omega^2/c^2$, and rearrangement gives:

$$\frac{X''}{X} + \frac{Y''}{Y} + \frac{Z''}{Z} = -k_a^2 \tag{1.133}$$

Each term of Equation (1.133) on the left-hand side is a function of a different independent variable, whereas the right-hand side of the equation is a constant. It may be concluded that each term on the left must also equal a constant; that is, Equation (1.133) takes the form:

$$k_x^2 + k_y^2 + k_z^2 = k_a^2 \tag{1.134}$$

This implies the following:

$$X'' + k_x^2 X = 0 \tag{1.135}$$

$$Y'' + k_y^2 Y = 0 \tag{1.136}$$

$$Z'' + k_z^2 Z = 0 \tag{1.137}$$

Solutions of Equations (1.135), (1.136) and (1.137) are as:

$$X = A_x \mathrm{e}^{\mathrm{j}k_x x} + B_x \mathrm{e}^{-\mathrm{j}k_x x} \tag{1.138}$$

$$Y = A_y \mathrm{e}^{\mathrm{j}k_y y} + B_y \mathrm{e}^{-\mathrm{j}k_y y} \tag{1.139}$$

$$Z = A_z \mathrm{e}^{\mathrm{j}k_z z} + B_z \mathrm{e}^{-\mathrm{j}k_z z} \tag{1.140}$$

Boundary conditions will determine the values of the constants. For example, if it is assumed that the walls are essentially rigid, so that the normal particle velocity, u_x, at the walls is zero, then, using Equations (1.60), (1.132) and (1.138), the following is obtained:

$$u_x = -\partial\phi/\partial x \tag{1.141}$$

and:

$$-jk_x YZ e^{j\omega t}[A_x e^{jk_x x} - B_x e^{-jk_x x}]_{x=0,L_x} = 0 \qquad (1.142)$$

Since:

$$jk_x YZ e^{j\omega t} \neq 0 \qquad (1.143)$$

then:

$$[A_x e^{jk_x x} - B_x e^{-jk_x x}]_{x=0,L_x} = 0 \qquad (1.144)$$

First consider the boundary condition at $x = 0$. This condition leads to the conclusion that $A_x = B_x$. Similarly, it may be shown that:

$$A_i = B_i; \qquad i = x,\ y,\ z \qquad (1.145)$$

Next consider the boundary condition at $x = L_x$. This second condition leads to the following equation:

$$e^{jk_x L_x} - e^{-jk_x L_x} = 2j\sin(k_x L_x) = 0 \qquad (1.146)$$

Similar expressions follow for the boundary conditions at $y = L_y$ and $z = L_z$. From these considerations it may be concluded that the k_i are defined as:

$$k_i = n_i \frac{\pi}{L_i}; \qquad n_i = 0,\ \pm 1,\ \pm 2,\ ...; \qquad i = x,\ y,\ z \qquad (1.147)$$

Substitution of Equation (1.147) into Equation (1.134) and use of $k_a^2 = \omega^2/c^2$ leads to the following useful result:

$$f_n = \frac{c}{2}\sqrt{\left[\frac{n_x}{L_x}\right]^2 + \left[\frac{n_y}{L_y}\right]^2 + \left[\frac{n_z}{L_z}\right]^2}\ (\text{Hz}) \qquad (1.148)$$

In this equation, the subscript, n, on the frequency variable, f, indicates that the particular solutions or 'eigen' frequencies of the equation are functions of the particular mode numbers, n_x, n_y and n_z.

Following Section 1.4.5 and using Equation (1.34), the following expression for the acoustic pressure is obtained:

$$p = \rho\frac{\partial\phi}{\partial t} = j\omega\rho X(x)Y(y)Z(z)e^{j\omega t} \qquad (1.149)$$

Substitution of Equations (1.145) and (1.147) into Equations (1.138), (1.139) and (1.140) and, in turn, substituting these altered equations into Equation (1.149) gives the following expression for the acoustic pressure for mode number, n_x, n_y, n_z in a rectangular room with rigid walls:

$$p = \hat{p}\cos\left[\frac{\pi n_x x}{L_x}\right]\cos\left[\frac{\pi n_y y}{L_y}\right]\cos\left[\frac{\pi n_z z}{L_z}\right]e^{j\omega t} \qquad (1.150)$$

where $\hat{p}$ is the acoustic pressure amplitude of the standing wave.

In Equations (1.148) and (1.150), the mode numbers n_x, n_y and n_z have been introduced. These numbers take on all positive integer values including zero. There are three types of normal modes of vibration in a rectangular room, which have their analogues in enclosures of other shapes. They may readily be understood as:

1. axial modes for which only one mode number is not zero;
2. tangential modes for which one mode number is zero; and
3. oblique modes for which no mode number is zero.

Axial modes correspond to wave travel back and forth parallel to an axis of the room. For example, the $(n_x, 0, 0)$ mode in the rectangular room of Figure 1.11 corresponds to a wave travelling back and forth parallel to the x-axis. Such a system of waves forms a standing wave having n_x nodal planes normal to the x-axis and parallel to the end walls. This may be verified by using Equation (1.150). The significance for noise control is that only sound absorption on the walls normal to the axis of sound propagation, where the sound is multiply reflected, will significantly affect the energy stored in the mode. The significance for sound coupling is that a speaker placed in the nodal plane of any mode will couple at best very poorly to that mode. Thus, the best place to drive an axial mode is to place the sound source on the end wall where the axial wave is multiply reflected; that is, at a pressure antinode.

Tangential modes correspond to waves travelling essentially parallel to two opposite walls of an enclosure while successively reflecting from the other four walls. For example, the $(n_x, n_y, 0)$ mode of the rectangular enclosure of Figure 1.11 corresponds to a wave travelling around the room parallel to the floor and ceiling. In this case, the wave impinges on all four vertical walls and absorptive material on any of these walls would be most effective in attenuating this mode. Absorptive material on the floor or ceiling would be less effective.

Oblique modes correspond to wave travel oblique to all room surfaces. For example, the $(n_x, n_y, \text{ and } n_z)$ mode in the rectangular room of Figure 1.11 would successively impinge on all six walls of the enclosure. Consequently, absorptive treatment on the floor, ceiling or any wall would be equally effective in attenuating an oblique mode.

For the placement of a speaker to drive a room, it is of interest to note that every mode of vibration has a pressure antinode at the corners of a room. This may be verified by using Equation (1.150). A corner is a good place to drive a rectangular room when it is desirable to introduce sound. It is also a good location to place absorbents to attenuate sound and to sample the sound field for the purpose of determining room frequency response.

In Figure 1.12, the first 15 room resonant modes have been identified using Equation (1.148). Reference to the figure shows that of the first 15 lowest order modes, seven are axial modes, six are tangential modes and two are oblique modes. Reference to the figure also shows that as the frequency increases, the resonances become too numerous to identify individually and in this range, the number of axial and tangential modes will become negligible compared to the number of oblique modes. It may be useful to note that the frequency at which this occurs is about 80 Hz in the reverberation room described in Figure 1.12 and this corresponds to a room volume of about 2.25 cubic wavelengths. As the latter description is non-dimensional, it is probably general; however, a more precise boundary between low- and high-frequency behaviour is given in Section 6.5.

In a rectangular room, for every mode of vibration for which one of the mode numbers is odd, the sound pressure is zero at the centre of the room, as shown by consideration of Equation (1.150); that is, when one of the mode numbers is odd the corresponding term in Equation (1.150) is zero at the centre of the corresponding coordinate (room dimension). Consequently, the centre of the room is a very poor place to couple, either with a speaker or an absorber, into the modes of the room. Consideration of all the possible combinations of odd and even in a group of three mode numbers shows that only one-eighth of the modes of a rectangular room will not have nodes at the centre of the room. At the centre of the junction of two walls, only one-quarter of the modes of a rectangular room will not have nodes, and at the centre of any wall only half of the modes will not have nodes.

1.5.2   Cylindrical Rooms

The analysis of cylindrical rooms follows the same procedure as for rectangular rooms except that the cylindrical coordinate system is used instead of the cartesian system. The result of this analysis is the following expression for the resonance frequencies of a cylindrical room.

$$f(n_z, m, n) = \frac{c}{2}\sqrt{\left(\frac{n_z}{L}\right)^2 + \left(\frac{\psi_{m,n}}{a}\right)^2} \tag{1.151}$$

where n_z is the number (varying from 0 to ∞) of nodal planes normal to the axis of the cylinder, L is the length of the cylinder and a is its radius. The characteristic values, ψ_{mn}, are functions of the mode numbers m, n, where m is the number of diametral pressure nodes and n is the number of circumferential pressure nodes. Values of ψ_{mn} for the first few modes are given in Table 1.2.

TABLE 1.2 Values of $\psi_{m,n}$

$m \backslash n$	0	1	2	3	4
0	0.0000	1.2197	2.2331	3.2383	4.2411
1	0.5861	1.6971	2.7172	3.7261	4.7312
2	0.9722	2.1346	3.1734	4.1923	5.2036
3	1.3373	2.5513	3.6115	4.6428	5.6623
4	1.6926	2.9547	4.0368	5.0815	6.1103
5	2.0421	3.3486	4.4523	5.5108	6.5494
6	2.3877	3.7353	4.8600	5.9325	6.9811
7	2.7034	4.1165	5.2615	6.3477	7.4065

1.5.3 Boundary between Low-Frequency and High-Frequency Behaviour in Acoustic Spaces

Referring to Figure 1.12, where the frequency response of a rectangular enclosure is shown, it can be observed that the number of peaks in response increases rapidly with increasing frequency. At low frequencies, the peaks in response are well separated and can be readily identified with resonant modes of the room. However, at high frequencies, so many modes may be driven in strong response at once that they tend to interfere, so that at high frequencies individual peaks in response cannot be associated uniquely with individual resonances. In this range, statistical analysis is appropriate.

Clearly, a need exists for a frequency bound that defines the crossover from the low-frequency range, where modal analysis, as discussed in this chapter and in Chapter 5, is appropriate, to the high-frequency range where statistical analysis, as discussed in Chapter 6, is appropriate. Reference to Figure 1.12 provides no clear indication of a possible bound, as a continuum of gradual change is observed. However, analysis does provide a bound, referred to as the crossover frequency, which is discussed in Section 6.5.

1.6   Sound Propagation in Porous Media

Before considering sound propagation in porous media, the definition and measurement of an important parameter, flow resistance, will be discussed. Flow resistance is a useful parameter for characterising sound propagation in porous materials as well as their sound absorbing properties.

1.6.1   Flow Resistance

A solid material that contains many voids is said to be porous. The voids may or may not be interconnected; however, for acoustical purposes, it is the interconnected voids that are important; the voids that are not connected are generally of little importance. One easily measured property of porous materials that can be used to determine their effectiveness in absorbing sound in various applications is the resistance of the material to induced flow through it, as a result of a pressure gradient. Flow resistance is a measure of this property, and is defined according to the following simple experiment. A uniform layer of porous material of thickness, ℓ (m), and area, S (m^2), is subjected to an induced mean volume flow, V_0 (m^3/s), through the material, and the pressure drop, ΔP_s (Pa), across the layer is measured. Very low pressures and mean volume velocities are assumed (of the order of the particle velocity amplitude of a sound wave having a sound pressure level between 80 and 100 dB). The flow resistance of the material, R_f, is defined as the induced pressure drop across the layer of material divided by the resulting mean volume velocity, V_0, per unit area, S, of the material:

$$R_f = \Delta P_s S / V_0 \tag{1.152}$$

The units of flow resistance are the same as for specific acoustic impedance, ρc; thus it is sometimes convenient to specify flow resistance in dimensionless form in terms of numbers of ρc units. The flow resistance of unit thickness of material is defined as the flow resistivity, R_1, which has the units Pa s m^{-2}, often referred to as MKS rayls per metre. Experimental investigation shows that porous materials of generally uniform composition may be characterised by a unique flow resistivity. Thus, for such materials, the flow resistance is proportional to the material thickness, ℓ, as:

$$R_f = R_1 \ell \tag{1.153}$$

Flow resistance characterises a layer of specified thickness, whereas flow resistivity characterises a bulk material in terms of resistance per unit thickness. For fibreglass and rockwool fibrous porous materials, which may be characterised by a mean fibre diameter, d, the following relation holds (Allard and Atalla, 2009):

$$\frac{R_1 \ell}{\rho c} = 27.3 \left(\frac{\rho_B}{\rho_f} \right)^{1.53} \left(\frac{\mu}{d \rho c} \right) \left(\frac{\ell}{d} \right) \tag{1.154}$$

In the above equation, in addition to the quantities already defined, the gas density, ρ ($=$ 1.206 kg/m^2 for air at 20°C), the porous material bulk density, ρ_B, and the fibre material density, ρ_f have been introduced. The remaining variables are the speed of sound, c, of the gas and the dynamic gas viscosity, μ (1.84×10^{-5} Pa s for air at 20°C). The dependence of flow resistance on bulk density, ρ_B, and fibre diameter, d, of the porous material is to be noted. A decrease in fibre diameter results in an increase of flow resistivity and an increase in sound absorption, so a useful fibrous material will have very fine fibres. Values of flow resistivity for some fibreglass and rockwool products have been measured and published (Bies and Hansen, 1979, 1980; Tarnow, 2002; Wang and Torng, 2001).

The flow resistance of a sample of porous material may be measured using an apparatus that meets the requirements of ASTM C522-03 (2016), such as illustrated in Figure 1.13. Flow rates between 5×10^{-4} and 5×10^{-2} m/s are easily realisable, and yield good results.

Higher flow rates should be avoided due to the possible introduction of nonlinear effects. To ensure that the pressure within the flow meter 7 is the same as that measured by the manometer 8, valve 6 must be adjusted so that flow through it is choked. The flow resistivity of the specimen shown in Figure 1.13 is calculated from the measured quantities as:

$$R_1 = \rho \Delta P_s S / (\dot{m} \ell) = \Delta P_s S / (V_0 \ell) \tag{1.155}$$

where ρ is the density of the gas (kg/m^3), ΔP_s is the differential static pressure (N/m^2), S is the specimen cross-sectional area (m^2), $\dot{m}$ is the air mass flow rate (kg/s), V_0 is the volume flow rate through the sample (m^3/s) and ℓ is the specimen thickness (m).

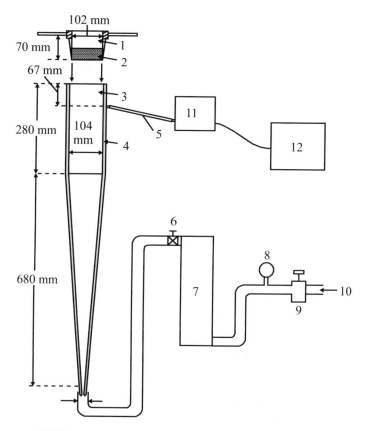

FIGURE 1.13 Flow resistance measuring apparatus.

Key

1 sample holder and cutter
2 porous material
3 O-ring seal
4 conical tube to ensure
   uniform air flow through sample

5 tube
6 valve
7 flow meter
8 manometer

9 pressure regulator
10 air supply
11 barocell
12 electronic manometer

Alternatively, acoustic flow resistance may be measured using a closed end tube, a sound source and any inexpensive microphones arranged as shown in Figure 1.14 (Ingard and Dear, 1985). To make a measurement, the sound source is driven with a pure tone signal, preferably below 100 Hz, at a frequency chosen to produce an odd number of quarter wavelengths over the distance $w + \ell$ from the closed end to the sample under test. The first step is to satisfy the latter requirement by adjusting the chosen frequency to achieve a minimum sound pressure level at microphone 1 in the absence of the sample. The sample is then inserted and the sound pressure level is measured at locations 1 and 3. The normalised flow impedance is a complex quantity made up of a real term (flow resistance) and an imaginary term (flow reactance) and is:

$$\frac{Z}{\rho c} = \frac{p_1 - p_2}{\rho c u_1} \tag{1.156}$$

As the tube is rigidly terminated and the losses along it are assumed small, the amplitude of the reflected wave will be the same as the incident wave and there will be zero phase shift between the incident and reflected waves at the rigid termination. If the coordinate system is chosen so that $x = 0$ corresponds to the rigid end and $x = -L$ (where $L = w + \ell$), at microphone

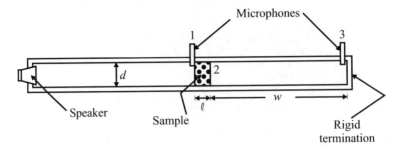

FIGURE 1.14 An alternative arrangement for measuring flow resistance.

location 1, then the particle velocity at location 1 can be shown to be:

$$\boldsymbol{u}_1 = -\frac{\mathrm{j}p_3}{\rho c}\sin(k_a L) \tag{1.157}$$

The acoustic pressure at location 2 is very similar to the pressure at location 1 in the absence of the sample and is given as:

$$p_2 = p_3 \cos(k_a L) \tag{1.158}$$

Thus, Equation (1.156) can be written as:

$$\frac{Z}{\rho c} = \frac{\mathrm{j}p_1}{p_3 \sin(k_a L)} + \mathrm{j}\cot(k_a L) \tag{1.159}$$

If L is chosen to be an odd number of quarter wavelengths such that $L = (2n-1)\lambda/4$ where n is an integer (preferably $n = 1$), the normalised flow impedance becomes:

$$\frac{Z}{\rho c} = \mathrm{j}(-1)^n (p_1/p_3) \tag{1.160}$$

so the flow resistance is the imaginary part of p_1/p_3. Taking into account that the flow reactance is small at low frequencies, the flow resistance is given by the magnitude of the ratio of the acoustic pressure at location 1 to that at location 3. Thus the flow resistivity (flow resistance divided by sample thickness) is:

$$R_1 = \frac{\rho c}{\ell} 10^{(L_{p1}-L_{p3})/20} \tag{1.161}$$

As only the sound pressure level difference between locations 1 and 3 is required, then prior to taking the measurement, microphones 1 and 3 are placed together near the closed end in the absence of the sample so that they measure the same sound pressure and the gain of either one is adjusted so that they read the same level. The closed end location is chosen for this as here the sound pressure level varies only slowly with location. The final step is to place the calibrated microphones at positions 1 and 3 as shown in Figure 1.13 and measure L_{p1} and L_{p3} and then use Equation (1.161) to calculate the flow resistivity, R_1.

Measured values of flow resistivity for various commercially available sound-absorbing materials are available in the literature (Bies and Hansen, 1980), and can sometimes be obtained from material manufacturers. For fibrous materials with a reasonably uniform fibre diameter and with only a small quantity of binder (such that the flow resistance is minimally affected), Figure 1.15 may be used to obtain an estimate of flow resistivity. Figure 1.15 is derived from the empirical equation (Bies and Hansen, 1980):

$$R_1 = K_2 d^{-2} \rho_B^{K_1} \tag{1.162}$$

where ρ_B is the bulk density of material, d is the fibre diameter, $K_1 = 1.53$ and $K_2 = 3.18 \times 10^{-9}$.

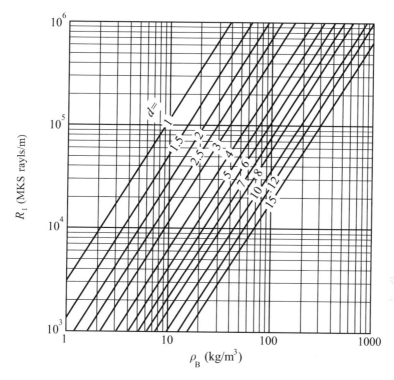

FIGURE 1.15 Flow resistivity, R_1, as a function of material bulk density, ρ_B (kg/m^3), and fibre diameter, d (μm).

For fibres with diameters larger than those in Figure 1.15, such as polyester, the flow resistivity may be estimated using Equation (1.162), but with different values for K_1 and K_2. In this case, Garai and Pompoli (2005) found that $K_1 = 1.404$ and $K_2 = 25.989d^2$, with d the fibre diameter in metres.

1.6.2 Calculation of Parameters Used for Characterising Sound Propagation in Porous Media

As mentioned at the end of Section 1.4.2, viscous effects due to internal friction in an acoustic medium, such as a porous acoustic material (e.g., rockwool), can be taken into account by adding the term, AU_{tot}, to the right-hand side of the Euler equation of motion. Alternatively, these effects could be taken into account by replacing the speed of sound c with a complex quantity, $c(1 + j\eta)^{1/2}$, as outlined elsewhere (Skudrzyk, 1971, p. 283).

For the purpose of the analysis, the porous, gas-filled medium is replaced by an effective medium, which is characterised in dimensionless variables by a complex density, ρ_m, and complex compressibility, κ. In terms of these quantities, a complex characteristic impedance and propagation coefficient are defined, analogous to the similar quantities for the contained gas in the medium. Implicit in the formulation of the following expressions is the assumption that time dependence has the positive form, $e^{j\omega t}$, consistent with the practice adopted throughout the text.

The characteristic impedance of porous material may be written in terms of the gas density, ρ, the gas speed of sound, c, the gas compressibility, κ_0, the porous material complex density, ρ_m, and the porous material complex compressibility, κ, as:

$$Z_m = \rho c \sqrt{\frac{\rho_m \kappa}{\rho \kappa_0}} \tag{1.163}$$

Similarly, a complex propagation coefficient, k_m, may be defined as:

$$k_m = \frac{2\pi}{\lambda}(1 - j\alpha_m) = (\omega/c)\sqrt{\frac{\rho_m \kappa_0}{\rho \kappa}} \tag{1.164}$$

where $\omega = 2\pi f$ is the angular frequency (rad/s) of the sound wave, and the complex density is then:

$$\rho_m = \frac{Z_m k_m}{\omega} \tag{1.165}$$

The quantities, ρ_m/ρ and κ/κ_0, may be calculated using the following procedure (Bies, 1981). This procedure gives results for fibrous porous materials within 4% of the mean of published data (Delany and Bazley, 1969, 1970), and unlike the Delaney and Bazley model, it tends to the correct limits at both high and low values of the dimensionless frequency, $\rho f/R_1$. However, this model and the Delaney and Bazley model have only been verified for fibreglass and rockwool materials with a small amount of binder and having short fibres smaller than 15 µm in diameter, which excludes such materials as polyester and acoustic foam. The normalised compressibility and normalised density of a porous material can be calculated from a knowledge of the material flow resistivity, R_1, using the following Equations (1.166) to (1.176).

$$\kappa/\kappa_0 = [1 + (1 - \gamma)\tau]^{-1} \tag{1.166}$$

$$\rho_m/\rho = [1 + \sigma]^{-1} \tag{1.167}$$

where γ is the ratio of specific heats for the gas (=1.40 for air), ρ is the density of gas (=1.205 kg/m^3 for air at 20°C), f is the frequency (Hz), R_1 is the flow resistivity of the porous material (MKS rayls/m) and:

$$\tau = 0.592a(X_1) + jb(X_1) \tag{1.168}$$

$$\sigma = a(X) + jb(X) \tag{1.169}$$

$$a(X) = \frac{T_3(T_1 - T_3)T_2^2 - T_4^2 T_1^2}{T_3^2 T_2^2 + T_4^2 T_1^2} \tag{1.170}$$

$$b(X) = \frac{T_1^2 T_2 T_4}{T_3^2 T_2^2 + T_4^2 T_1^2} \tag{1.171}$$

$$T_1 = 1 + 9.66X \tag{1.172}$$

$$T_2 = X(1 + 0.0966X) \tag{1.173}$$

$$T_3 = 2.537 + 9.66X \tag{1.174}$$

$$T_4 = 0.159(1 + 0.7024X) \tag{1.175}$$

$$X = \rho f/R_1 \tag{1.176}$$

The quantities, $a(X_1)$ and $b(X_1)$, are calculated by substituting $X_1 = 0.856X$ for the quantity, X, in Equations (1.163)–(1.169).

The relationships that have been generally accepted in the past (Delany and Bazley, 1969, 1970), and which are accurate in the flow resistivity range $R_1 = 10^3$ to 5×10^4 MKS rayls/m, are:

$$Z_m = \rho c[1 + C_1 X^{-C_2} - jC_3 X^{-C_4}] \tag{1.177}$$

$$k_m = (\omega/c)[1 + C_5 X^{-C_6} - jC_7 X^{-C_8}] \tag{1.178}$$

The quantities X, Z_m, k_m, c and ρ have all been defined previously. Values of the coefficients $C_1 - C_8$ are given in Table 1.3 for various materials from various references.

TABLE 1.3 Values of the coefficients C_1–C_8 for various materials

Material type reference	C_1	C_2	C_3	C_4	C_5	C_6	C_7	C_8
Rockwool/fibreglass Delany and Bazley (1970)	0.0571	0.754	0.087	0.732	0.0978	0.700	0.189	0.595
Polyester Garai and Pompoli (2005)	0.078	0.623	0.074	0.660	0.159	0.571	0.121	0.530
Polyurethane foam of low flow resistivity Dunn and Davern (1986)	0.114	0.369	0.0985	0.758	0.168	0.715	0.136	0.491
Porous plastic foams of medium flow resistivity Wu (1988)	0.212	0.455	0.105	0.607	0.163	0.592	0.188	0.544

1.6.3 Sound Reduction Due to Propagation through a Porous Material

For the purpose of the calculation, three frequency ranges are defined: low, middle and high, as indicated in Figure 1.16. The quantities in the parameters $\rho f/R_1$ and $f\ell/c$ are defined in Sections 1.6.1 and 1.6.

In the low-frequency range, the inertia of the porous material is small enough for the material to move with the particle velocity associated with the sound wave passing through it. The transmission loss (TL) to be expected in this frequency range can be obtained from Figure 1.17. If the material is used as pipe wrapping, the noise reduction will be approximately equal to the transmission loss. In the high-frequency range, the porous material is many wavelengths thick and, in this case, reflection at both surfaces of the layer, as well as propagation losses through the layer, must be taken into account when estimating noise reduction. The reflection loss at an air/porous medium interface may be calculated using Figure 1.18 and the transmission loss (TL) may be estimated using Figure 1.19.

In the middle-frequency range it is generally sufficient to estimate the transmission loss graphically, with a faired, smooth curve connecting plotted estimates of the low- and high-frequency transmission loss versus log frequency.

1.6.4 Measurement of Absorption Coefficients of Porous Materials

1.6.4.1 Measurement Using the Moving Microphone Method

Absorption coefficients may be determined using impedance tube measurements of the normal incidence absorption coefficient, as an alternative to the measurement of the Sabine absorption coefficient using a reverberant test chamber. This same measuring apparatus may be used to obtain the normal specific acoustic impedance looking into the sample, which can be used to estimate the statistical absorption coefficient.

When a tonal (single frequency) sound field is set up in a tube terminated in an impedance, Z, a pattern of regularly spaced maxima and minima along the tube will result, which is uniquely determined by the driving frequency and the terminating impedance. The absorption coefficients are related to the terminating impedance and the characteristic impedance, ρc, of air.

An impedance tube is relatively easily constructed and therein lies its appeal. Any heavy walled tube may be used for its construction. A source of sound should be placed at one end

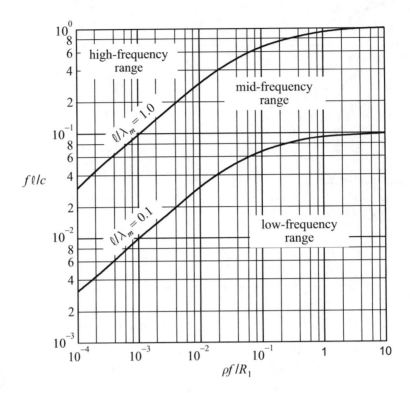

$f\ell/c$

$\rho f/R_1$

FIGURE 1.16 Limits showing when low- and high-frequency models should be used for estimating the transmission loss through a porous layer. The low-frequency model should be used when the design point lies below the $\ell/\lambda_m = 0.1$ curve, and the high-frequency model should be used when the design point lies above the $\ell/\lambda_m = 1.0$ curve. The quantity, λ_m, is the wavelength of sound in the porous material, ρ is the gas density, c is the speed of sound in the gas, f is frequency, ℓ is the material thickness and R_1 is the material flow resistivity.

of the tube and the material to be tested should be mounted at the other end. Means must be provided for probing the standing wave within the tube. An example of a possible configuration is shown in Figure 1.20.

The older and simpler method by which the sound field in the impedance tube is explored, using a moveable microphone which traverses the length of the tube, will be described first (see ASTM C384-04 (2016)). In this case, the impedance of the test sample is determined from measurements of the sound field in the tube. This method is slow but easily implemented. A much quicker method, which makes use of two fixed microphones and a digital frequency analysis system, is described in Section 1.6.4.2 (see ASTM E1050-12 (2012)).

It is also possible to measure the specific acoustic surface impedance and absorption coefficient of materials in situ, without using an impedance tube, by using two microphones close to the surface of the material (Dutilleaux et al., 2001). However, the procedure is quite complex.

Implicit in the use of the impedance tube is the assumption that only plane waves propagate back and forth in the tube. This assumption puts upper and lower frequency bounds on the use of the impedance tube. Let d be the tube diameter if it is circular in cross section, or the larger dimension if it is rectangular in cross section. Then the upper frequency limit (or cut-on frequency) is:

$$f_u = \begin{cases} 0.586c/d; & \text{for circular section ducts} \\ 0.5c/d; & \text{for rectangular section ducts} \end{cases} \quad (1.179)$$

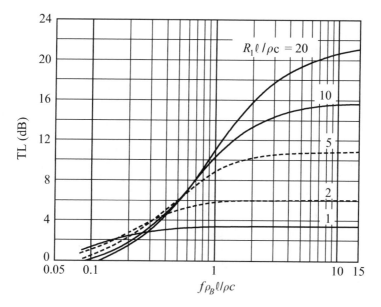

FIGURE 1.17 Transmission loss (TL) through a porous layer for a design point lying in the low-frequency range of Figure 1.16. The quantity, ρ, is the gas density, c is the speed of sound in the gas, ρ_B is the material bulk density, ℓ is the material thickness, f is frequency and R_1 is the material flow resistivity.

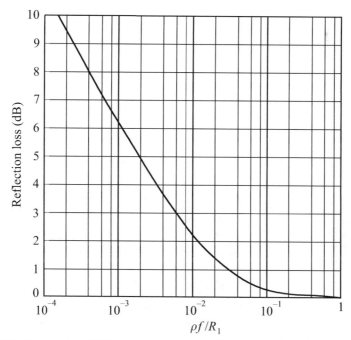

FIGURE 1.18 Reflection loss (dB) at a porous material–air interface for a design point in the high-frequency range of Figure 1.16. The quantity ρ is the gas density, f is the frequency and R_1 is the material flow resistivity.

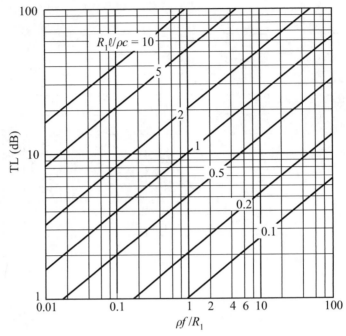

FIGURE 1.19 Transmission loss of a porous lining of thickness ℓ and normalised flow resistance, $R_1\ell/\rho c$, for a design point lying in the high-frequency region of Figure 1.16. The quantity, ρ, is the gas density, c is the speed of sound in the gas and R_1 is the material flow resistivity.

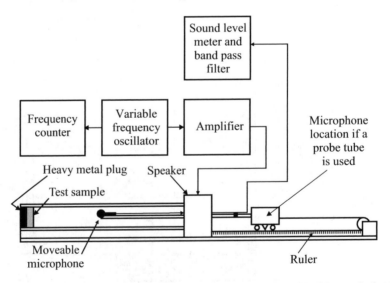

FIGURE 1.20 Equipment for impedance tube measurement using the older and simpler method.

Here c is the speed of sound and the frequency limit is given in Hz. The required length, ℓ, of the tube is a function of the lowest frequency, f_ℓ, to be tested and is:

$$\ell = d + \frac{3c}{4f_\ell} \tag{1.180}$$

In general, the frequency response of the apparatus will be very much dependent on the material under test. To reduce this dependence and to ensure a more uniform response, some sound absorptive material may be placed permanently in the tube at the source end. Furthermore, as energy losses due to sound propagation along the length of the tube are undesirable, the following equations may be used as a guide during design to estimate and minimise such losses:

$$L_{D1} - L_{D2} = a\lambda/2 \quad \text{(dB)} \tag{1.181}$$

$$a = 0.19137\sqrt{f}/cd \quad \text{(dB/m)} \tag{1.182}$$

In the above equations, a is the loss (in dB per metre of tube length) due to propagation of the sound wave down the tube, and L_{D1} and L_{D2} are the sound pressure levels at the first and second minima relative to the surface of the test sample. The other quantities are the frequency, f (Hz), the corresponding wavelength, λ (m), and the speed of sound, c (m/s). For tubes of any cross section, d (m) is defined as $d = 4S/P_D$, where S is the cross-sectional area and P_D is the cross-sectional perimeter.

The propagation loss in dB is related to the attenuation coefficient, ζ, in nepers/m for a propagating wave, expressed as follows for the sound pressure as a function of distance, x, along a tube from the sound source:

$$p(x) = p_0 e^{-\zeta x} \tag{1.183}$$

where the sound wave has an RMS sound pressure of p_0 at location x_0 and an RMS sound pressure of $p(x)$ at a location that is a distance, $x + x_0$, from the sound source. One neper $= 20/(\log_e 10) = 8.686$ dB.

The sound field within the impedance tube may be explored either with a small microphone placed at the end of a probe and immersed in the sound field, as illustrated in Figure 1.20, or with a probe tube attached to a microphone placed externally to the field. Equation (1.182) may be useful in selecting a suitable tube for exploring the field; the smaller the probe tube diameter, the greater will be the energy loss suffered by a sound wave travelling down it, but the smaller will be the disturbance to the acoustic field being sampled. An external linear scale should be provided for locating the probe. As the sound field will be distorted slightly by the presence of the probe, it is recommended that the actual location of the sound field point sampled be determined by replacing the specimen with a heavy solid metal face located at the specimen surface. The first minimum will then be $\lambda/4$ away from the solid metal face. Subsequent minima will always be spaced at intervals of $\lambda/2$. Thus, this experiment will allow the determination of how far in front of the end of the probe tube that the sound field is effectively being sampled. It will be found that the minima will be subject to contamination by acoustic and electronic noise; thus, it is recommended that a narrow band filter, for example one which is an octave or 1/3-octave wide, be used in the detection system.

When the material to be tested is in place and the sound source is excited with a single frequency, a series of maximum and minimum sound pressures in the impedance tube will be observed. The maxima will be effectively constant in level but the minima will increase in level, according to Equation (1.181), as one moves away from the surface of the test material. For best results, it is recommended that losses in the tube be taken into account by extrapolating successive minima back to the surface of the sample by drawing a straight line joining the first two minima to the location corresponding to the surface of the sample on the plot of sound pressure level in dB vs distance along the tube (see Figure 1.21). The standing wave ratio, L_0, is then determined as the difference between the maximum level, $L_{\max}$, and the extrapolated minimum level, $L_{\min}$.

It is of interest to derive an expression for the normal incidence absorption coefficient of a sample of acoustic material in an impedance tube, as shown in Figure 1.21. For the following analysis, the impedance tube contains the material sample at the right end and the loudspeaker sound source at the left end.

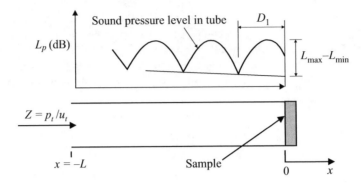

FIGURE 1.21 Schematic of an impedance tube.

For simplicity, it is assumed that there are no losses due to dissipation in the tube; that is, the quantity, a, in Equation (1.182) is assumed to be zero. The loudspeaker is not shown in the figure and the origin is set at the right end of the tube at the face of the test sample.

Reference should be made to Section 1.4.12, where it is shown that multiple waves travelling in a single direction may be summed together to give a single wave travelling in the same direction. For the case considered here, the multiple waves travelling in each direction are a result of multiple reflections from each end of the tube.

As the origin is at the right end of the tube, the resultant incident wave will be travelling in the positive x-direction. Assuming a phase shift between the incident and reflected waves of β at $x = 0$, the incident wave and reflected wave pressures may be written as:

$$p_i = A\mathrm{e}^{\mathrm{j}(\omega t - k_a x)} \quad \text{and} \quad p_r = B\mathrm{e}^{\mathrm{j}(\omega t + k_a x + \beta)} \tag{1.184}$$

The total sound pressure is thus:

$$p_t = A\mathrm{e}^{\mathrm{j}(\omega t - k_a x)} + B\mathrm{e}^{\mathrm{j}(\omega t + k_a x + \beta)} \tag{1.185}$$

The first maximum pressure (closest to the sample) will occur when:

$$\beta = -2k_a x \tag{1.186}$$

and the first minimum will occur when:

$$\beta = -2k_a x + \pi \tag{1.187}$$

Thus:

$$p_{\max} = \mathrm{e}^{-\mathrm{j}k_a x}\left(A + B\right) \quad \text{and} \quad p_{\min} = \mathrm{e}^{-\mathrm{j}k_a x}\left(A - B\right) \tag{1.188}$$

and the ratio of maximum to minimum pressures is $(A+B)/(A-B)$. The standing wave ratio, L_0, is the difference in decibels between the maximum and minimum sound pressures in the standing wave and is defined as:

$$10^{L_0/20} = \frac{A + B}{A - B} \tag{1.189}$$

Thus, the ratio (B/A) is:

$$\frac{B}{A} = \left[\frac{10^{L_0/20} - 1}{10^{L_0/20} + 1}\right] \tag{1.190}$$

The amplitude of the pressure reflection coefficient squared is defined as $|R_p|^2 = (B/A)^2$, which can be written in terms of L_0 as:

$$|R_p|^2 = \left[\frac{10^{L_0/20} - 1}{10^{L_0/20} + 1}\right]^2 \tag{1.191}$$

The normal incidence absorption coefficient is defined as:

$$\alpha_n = 1 - |R_p|^2 \tag{1.192}$$

and it can also be determined from Table 1.4.

TABLE 1.4 Normal incidence sound absorption coefficient, α_n, vs standing wave ratio, L_0 (dB)

L_0	α_n	L_0	α_n	L_0	α_n	L_0	α_n
0	1.000	10	0.730	20	0.331	30	0.119
1	0.997	11	0.686	21	0.301	31	0.107
2	0.987	12	0.642	22	0.273	32	0.096
3	0.971	13	0.598	23	0.247	33	0.086
4	0.949	14	0.555	24	0.223	34	0.077
5	0.922	15	0.513	25	0.202	35	0.069
6	0.890	16	0.472	26	0.182	36	0.061
7	0.854	17	0.434	27	0.164	37	0.055
8	0.815	18	0.397	28	0.147	38	0.049
9	0.773	19	0.363	29	0.132	39	0.04

It is also of interest to continue the analysis to determine the normal specific acoustic impedance of the surface of the sample. This can then be used to determine the statistical absorption coefficient of the sample, which is the absorption coefficient averaged over all possible angles of an incident wave.

The total particle velocity can be calculated using Equations (1.60), (1.62) and (1.184) to give:

$$u_t = \frac{1}{\rho c}(p_i - p_r) \tag{1.193}$$

Thus:

$$u_t = \frac{1}{\rho c}\left(A\mathrm{e}^{\mathrm{j}(\omega t - k_a x)} - B\mathrm{e}^{\mathrm{j}(\omega t + k_a x + \beta)}\right) \tag{1.194}$$

The specific acoustic impedance (or characteristic impedance) at any point in the tube may be written as:

$$Z_s = \frac{p_t}{u_t} = \rho c\frac{A\mathrm{e}^{-\mathrm{j}k_a x} + B\mathrm{e}^{\mathrm{j}k_a x + \mathrm{j}\beta}}{A\mathrm{e}^{-\mathrm{j}k_a x} - B\mathrm{e}^{\mathrm{j}k_a x + \mathrm{j}\beta}} = \rho c\frac{A + B\mathrm{e}^{\mathrm{j}(2k_a x + \beta)}}{A - B\mathrm{e}^{\mathrm{j}(2k_a x + \beta)}} \tag{1.195}$$

At $x = 0$, the impedance is the normal specific acoustic impedance, Z_N, of the surface of the sample. Thus:

$$\frac{Z_N}{\rho c}\frac{p_t}{\rho c u_t} = \frac{A + B\mathrm{e}^{\mathrm{j}\beta}}{A - B\mathrm{e}^{\mathrm{j}\beta}} \tag{1.196}$$

The above impedance equation may be expanded to give:

$$\frac{Z_N}{\rho c} = \frac{A/B + \cos\beta + \mathrm{j}\sin\beta}{A/B - \cos\beta - \mathrm{j}\sin\beta} = \frac{(A/B)^2 - 1 + (2A/B)\mathrm{j}\sin\beta}{(A/B)^2 + 1 - (2A/B)\cos\beta} \tag{1.197}$$

In practice, the phase angle, β, is evaluated by measuring the distance, D_1, of the first sound pressure minimum in the impedance tube from the sample surface, and the tube that corresponds to Figure 1.21 is known as an impedance tube. Referring to Equation (1.187) and Figure 1.21, the phase angle, β, may be expressed in terms of D_1 (which is a positive number) as:

$$\beta = 2k_a D_1 + \pi = 2\pi\left(\frac{2D_1}{\lambda} + \frac{1}{2}\right) \tag{1.198}$$

Equation (1.197) may be rewritten in terms of a real and imaginary components as:

$$\frac{Z_N}{\rho c} = R + \mathrm{j}X = \frac{(A/B)^2 - 1}{(A/B)^2 + 1 - (2A/B)\cos\beta} + \mathrm{j}\frac{(2A/B)\sin\beta}{(A/B)^2 + 1 - (2A/B)\cos\beta} \tag{1.199}$$

where β is defined by Equation (1.198) and the ratio, A/B, is defined by the reciprocal of Equation (1.190) where L_0 is the difference in dB between the maximum and minimum sound pressure levels in the tube.

In terms of an amplitude and phase, the normal specific acoustic impedance may also be written as:

$$Z_N/(\rho c) = \xi e^{\mathrm{j}\psi} \tag{1.200}$$

where:

$$\xi = \sqrt{R^2 + X^2} \tag{1.201}$$

and:

$$\psi = \tan^{-1}(X/R) \tag{1.202}$$

The statistical absorption coefficient, α_{st}, may be calculated as:

$$\alpha_{st} = \frac{1}{\pi} \int_0^{2\pi} \mathrm{d}\varphi \int_0^{\pi/2} \alpha(\theta)\cos\theta\sin\theta\,\mathrm{d}\theta \tag{1.203}$$

where θ is the angle of incidence of the sound wave with respect to the normal to the material surface.

Rewriting the absorption coefficient in terms of the reflection coefficient gives:

$$|R(\theta)|^2 = 1 - \alpha(\theta) \tag{1.204}$$

then Equation (1.204) can be written as:

$$\alpha_{st} = 1 - 2\int_0^{\pi/2} |R(\theta)|^2 \cos\theta\sin\theta\,\mathrm{d}\theta \tag{1.205}$$

For bulk reacting materials (in which there is limited sound propagation occurring parallel to the surface of the material), Equations (1.206) and (1.207) may be used to calculate the complex amplitude reflection coefficient, $R(\theta)$, for waves incident on the material surface at an angle, θ, from the normal to the surface and refracted into the material at an angle, ψ, from the normal to the surface. The expression is given by (Bies et al., 2017):

$$R(\theta) = \frac{A_R}{A_I} = \frac{Z_N\cos\theta - \rho c\cos\psi}{Z_N\cos\theta + \rho c\cos\psi} = |R(\theta)|e^{\mathrm{j}\alpha_p} \tag{1.206}$$

where the phase of the reflected wave relative to the incident wave is $\alpha_p = \tan^{-1}[\mathrm{Im}\{R(\theta)\}/\mathrm{Re}\{R(\theta)\}]$ and the angle, ψ, is defined as:

$$\cos\psi = \sqrt{1 - \left(\frac{k_a}{k_m}\right)^2 \sin^2\theta} \tag{1.207}$$

where k_m is the complex propagation coefficient in the material, defined by Equation (1.164). If the porous material is infinite in extent, or sufficiently thick that waves reflected from any termination back towards the interface have negligible amplitude on arrival at the interface, then Z_N may be replaced with the material characteristic impedance, Z_m, defined in Equation (1.163). In fact, this is what is done when the material is an outdoor ground surface.

Reference to Equation (1.207) shows that when $k_m = k_2 \gg k_1 = k_a$, the angle ψ tends to zero and Equation (1.206) reduces to the following form:

$$R(\theta) = \frac{Z_N \cos\theta - \rho c}{Z_N \cos\theta + \rho c} \tag{1.208}$$

which is the equation for a locally reactive surface, for which sound propagation in the material is only in a direction normal to the surface.

Use of Equations (1.206) and (1.208) requires a knowledge of the normal specific acoustic impedance, Z_N, of the material. This can be measured using an impedance tube in which a sample of the material, in the same configuration as to be used in practice, in terms of how it is backed, is tested. Alternatively, the normal specific acoustic impedance for any configuration may be calculated using the methods described in Section 1.6.5.

Using Equations (1.205) and (1.208), Morse and Bolt (1944) derive the following expression for the statistical absorption coefficient for a locally reactive surface of normal specific acoustic impedance, Z_N, given by Equation (1.200):

$$\alpha_{st} = \left\{ \frac{8\cos\psi}{\xi} \right\} \left\{ 1 - \left[\frac{\cos\psi}{\xi} \right] \log_e(1 + 2\xi\cos\psi + \xi^2) \right.$$
$$\left. + \left[\frac{\cos(2\psi)}{\xi\sin\psi} \right] \tan^{-1} \left[\frac{\xi\sin\psi}{1 + \xi\cos\psi} \right] \right\} \tag{1.209}$$

Alternatively Figure 1.22 may be used to determine the statistical absorption coefficient. The statistical absorption coefficient can be used with the Norris-Eyring equation (Equation (1.210)) to calculate the reverberation time of an acoustic space and this can be used with Equation (6.38) to calculate the damping loss factor of the space for use in the calculations outlined in Chapter 6.

$$T_{60} = -\frac{55.25V}{Sc \log_e(1 - \bar{\alpha}_{st})} \tag{1.210}$$

where V and S are, respectively, the volume and surface area of the acoustic space.

Note that Equation (1.209) and Figure 1.22 are based on the explicit assumption that sound propagation within the sample is always normal to the surface. However, calculations indicate that the error in ignoring propagation in the porous material in other directions is negligible.

The incoherent reflection coefficient, $\bar{\Re}_S$, is defined as:

$$\bar{\Re}_S = 1 - \alpha_{st} \tag{1.211}$$

1.6.4.2 Measurement Using the 2-Microphone Method

The advantage of the 2-microphone method over the moving microphone method discussed in the previous subsection is that it considerably reduces the time required to determine the normal specific acoustic impedance and normal incidence absorption coefficient of a sample, as it allows the sample to be evaluated over the frequency range of interest in one or two measurements. The upper frequency limit for each tube is the same as for the moving microphone method. The lower frequency limit is a function of the spacing, ℓ_s, of the two microphones (which should exceed 1% of the wavelength at the lowest frequency of interest) and the accuracy of the analysis system. However, the microphone spacing must not be larger than 40% of the wavelength at the highest frequency of interest. In some cases it may be necessary to repeat measurements with two different microphone spacings and two tubes with different diameters. The microphones are mounted through the wall of the tube and flush with its inside surface, as illustrated in Figure 1.23. The microphone closest to the sample surface should be located at least one tube diameter from the surface for a rough surface and half a tube diameter for a smooth surface.

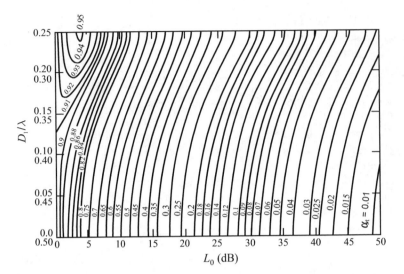

FIGURE 1.22 A chart for determining the statistical absorption coefficient, α_{st}, from measurements in an impedance tube of the standing wave ratio, L_0, and position, D_1/λ, of the first minimum sound pressure level. α_{st} is shown parametrically in the chart.

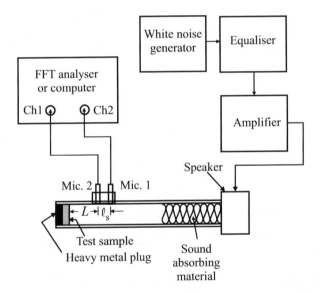

FIGURE 1.23 Arrangement for measuring the normal incidence complex reflection coefficient, normal specific acoustic impedance and absorption coefficient of a sample of acoustic material using the 2-microphone method.

The sample is mounted in a holder attached to the end of the impedance tube as illustrated in Figure 1.23 and described in ASTM E1050-12 (2012). An FFT analyser is used to determine the transfer function between the two microphones as a function of frequency. The loudspeaker at the end of the tube is excited with white noise (see Section 1.3.9) and the resulting transfer function (or frequency response function), $H^M(f_n)$, between the two microphones is measured using the FFT analyser (see Section 7.3.14). The transfer function is written in terms of its real, $H_R^M(f_n)$, and imaginary, $H_I^M(f_n)$, parts as:

$$H^M(f_n) = H_R^M(f_n) + \mathrm{j}H_I^M(f_n) = H_R^M + \mathrm{j}H_I^M = |H^M|\mathrm{e}^{\mathrm{j}\phi^M} \qquad (1.212)$$

It is important that in Equation (1.212) and in the context of Equation (7.75), microphone 1 is the input microphone, X, and microphone 2 is the response microphone, Y.

The transfer function must be corrected to account for any phase and amplitude mismatch between the microphones. The correction factor is obtained by first measuring the complex transfer function as described above with any or no test sample, given by:

$$H^{\mathrm{I}} = H_R^{\mathrm{I}} + jH_I^{\mathrm{I}} = |H^{\mathrm{I}}|e^{j\phi^{\mathrm{I}}} \qquad (1.213)$$

and then repeating the measurement with the positions of the two microphones in the tube wall physically swapped, without changing any cable connections, to obtain the complex transfer function:

$$H^{\mathrm{II}} = H_R^{\mathrm{II}} + jH_I^{\mathrm{II}} = |H^{\mathrm{II}}|e^{j\phi^{\mathrm{II}}} \qquad (1.214)$$

Care must be taken to ensure the microphone positions are exchanged as accurately as possible, or there will be errors in the calibration factor. The calibration factor, H^c, is calculated as:

$$H^c = (H^{\mathrm{I}} \times H^{\mathrm{II}})^{1/2} = |H^c|e^{j\phi^c} \qquad (1.215)$$

where:

$$|H^c| = \left(|H^I| \times |H^{II}|\right)^{1/2} = \left\{ \left[\left(H_R^I\right)^2 + \left(H_I^I\right)^2 \right] \times \left[\left(H_R^{II}\right)^2 + \left(H_I^{II}\right)^2 \right] \right\}^{1/4} \qquad (1.216)$$

and:

$$\phi^c = \frac{1}{2}(\phi^I + \phi^{II}) = \frac{1}{2}\arctan\left[\frac{H_I^I H_R^{II} + H_R^I H_I^{II}}{H_R^I H_R^{II} - H_I^I H_I^{II}} \right] \qquad (1.217)$$

Once the calibration factor has been determined with an arbitrary or no test sample, the transfer function, H^M, between microphones 1 and 2 is measured with the required sample in place and the corrected complex transfer function, H, is then calculated as:

$$H = H^M/H^c = |H|e^{j\phi} = H_R + jH_I \qquad (1.218)$$

where:

$$H_R = \frac{1}{|H^c|}(H_R\cos\phi^c + H_I\sin\phi^c) \qquad (1.219)$$

$$H_I = \frac{1}{|H^c|}(H_I\cos\phi^c - H_R\sin\phi^c) \qquad (1.220)$$

To calculate the normal incidence absorption coefficient and normal surface impedance for the test material, it is first necessary to calculate the plane wave, complex amplitude reflection coefficient, R_p (using the corrected transfer function, H). Thus for each frequency component in the transfer function spectrum:

$$R_p = |R_p|e^{j\phi_R} = R_{pR} + jR_{pI} \qquad (1.221)$$

where:

$$R_{pR} = \frac{2H_R\cos[k_a(2\ell+\ell_s)] - \cos(2k_a\ell) - [(H_R)^2 + (H_I)^2]\cos[2k_a(\ell+\ell_s)]}{1 + (H_R)^2 + (H_I)^2 - 2[H_R\cos k_a\ell_s + H_I\sin(k_a\ell_s)]} \qquad (1.222)$$

$$R_{pI} = \frac{2H_R\cos[k_a(2\ell+\ell_s)] - \sin(2k_a\ell) - [(H_R)^2 + (H_I)^2]\sin[2k_a(\ell+\ell_s)]}{1 + (H_R)^2 + (H_I)^2 - 2[H_R\cos k_a\ell_s + H_I\sin(k_a\ell_s)]} \qquad (1.223)$$

The normal incidence absorption coefficient, α_n is then calculated using:

$$\alpha_n = 1 - |R_p|^2 \qquad (1.224)$$

The normal specific acoustic impedance at the face of the sample is then:

$$Z_N = Z_{NR} + jZ_{NI} = \rho c \frac{1 + R_p}{1 - R_p} \qquad (1.225)$$

where:

$$Z_{NR} = \rho c \frac{\alpha_n}{2(1 - R_{pR}) - \alpha_n} \qquad (1.226)$$

and:

$$Z_{NI} = 2\rho c R_{pI}[2(1 - R_{pR}) - \alpha_n] \qquad (1.227)$$

Measurement errors are discussed in ASTM E1050-12 (2012).

1.6.4.3   Measurement Using the 4-Microphone Method

The advantage of the 4-microphone method over the 2-microphone method is that the 4-microphone method allows evaluation of the complex wavenumber of the porous material in addition to the complex characteristic impedance (only the same as the normal specific acoustic impedance for materials of sufficient extent that waves reflected upstream from the downstream face are insignificant in amplitude when they arrive back at the upstream face), complex normal incidence reflection coefficient and normal incidence absorption coefficient. The 4-microphone method also allows the transmission loss of the porous material to be determined. The measurement method is described in detail in ASTM E2611-09 (2009). The experimental arrangement for the 4-microphone method is illustrated in Figure 1.24. The analysis of the 4-microphone method also serves as an excellent introduction to the transfer matrix method. As for both previous methods involving measurements in a tube, the tube diameter must be sufficiently small so that only plane waves are propagating. That is, the test frequencies must lie below the cut-on frequency of the first higher order mode (see Equation (1.179)).

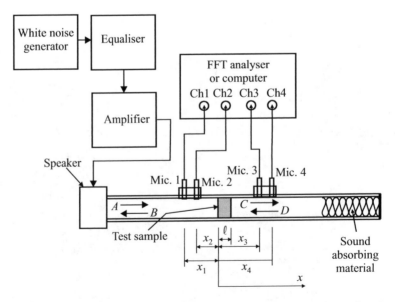

FIGURE 1.24 Arrangement for measuring the normal incidence complex reflection coefficient, characteristic impedance, wavenumber and absorption coefficient of a sample of acoustic material using the 4-microphone method.

Referring to Figure 1.24, the sound pressure, p, at frequency, ω, recorded by any microphone, normalised by some complex reference signal, p_{ref}, is:

$$\frac{p(\omega)}{p_{\text{ref}}(\omega)} = \frac{\hat{p}(\omega)}{\hat{p}_{\text{ref}}(\omega)} e^{j\omega t} \tag{1.228}$$

where $\dfrac{\hat{p}(\omega)}{\hat{p}_{\text{ref}}(\omega)}$ is the normalised complex pressure amplitude given in terms of coefficients, A, B, C and D, corresponding to waves A, B, C and D shown in Figure 1.24. Thus, the normalised pressure amplitudes at the 4 microphones at frequency, ω, are:

$$\begin{cases} \dfrac{\hat{p}_1}{\hat{p}_{\text{ref}}} = Ae^{-jk_a x_1} + Be^{jk_a x_1} \\[2ex] \dfrac{\hat{p}_2}{\hat{p}_{\text{ref}}} = Ae^{-jk_a x_2} + Be^{jk_a x_2} \\[2ex] \dfrac{\hat{p}_3}{\hat{p}_{\text{ref}}} = Ce^{-jk_a x_3} + De^{jk_a x_3} \\[2ex] \dfrac{\hat{p}_4}{\hat{p}_{\text{ref}}} = Ce^{-jk_a x_4} + De^{jk_a x_4} \end{cases} \tag{1.229}$$

where the ω in brackets is not included in the previous or following equations to simplify the notation. All equations and unknowns apply to a single frequency, with a different set of A to D for each frequency. If the system is excited with random noise, the normalised sound pressure amplitude and relative phase at frequency, ω, can be found by taking a transfer function in the frequency domain, which is derived from the cross-spectrum between the microphone signal and the reference signal as well as the power spectra of both the microphone signal and the reference signal (see Section 7.3.11). The complex transfer function, $H_{n,\text{ref}} = |H_{n,\text{ref}}| \exp[j\phi_{n,\text{ref}}]$, represented by an amplitude and phase, can then replace the ratio, $\dfrac{\hat{p}(\omega)}{\hat{p}_{\text{ref}}(\omega)}$, in Equations (1.229), so that Equations (1.229) become:

$$\begin{cases} H_{1,\text{ref}} = Ae^{-jk_a x_1} + Be^{jk_a x_1} \\[2ex] H_{2,\text{ref}} = Ae^{-jk_a x_2} + Be^{jk_a x_2} \\[2ex] H_{3,\text{ref}} = Ce^{-jk_a x_3} + De^{jk_a x_3} \\[2ex] H_{4,\text{ref}} = Ce^{-jk_a x_4} + De^{jk_a x_4} \end{cases} \tag{1.230}$$

Equations (1.230) can be rearranged to provide expressions for A, B, C and D as follows:

$$\begin{cases} A = \dfrac{j \left(H_{1,\text{ref}} e^{jk_a x_2} - H_{2,\text{ref}} e^{jk_a x_1} \right)}{2 \sin[k_a(x_1 - x_2)]} \\[3ex] B = \dfrac{j \left(H_{2,\text{ref}} e^{-jk_a x_1} - H_{1,\text{ref}} e^{-jk_a x_2} \right)}{2 \sin[k_a(x_1 - x_2)]} \\[3ex] C = \dfrac{j \left(H_{3,\text{ref}} e^{jk_a x_4} - H_{4,\text{ref}} e^{jk_a x_3} \right)}{2 \sin[k_a(x_3 - x_4)]} \\[3ex] D = \dfrac{j \left(H_{4,\text{ref}} e^{-jk_a x_3} - H_{3,\text{ref}} e^{-jk_a x_4} \right)}{2 \sin[k_a(x_3 - x_4)]} \end{cases} \tag{1.231}$$

Note that x_1 and x_2 will both be negative.

In practice, the transfer functions between the acoustic pressures at the microphone locations and the reference signal can be obtained by moving the same microphone to each location or by using different microphones at each location. If the same microphone is used, then it is important that any unused holes in the test tube are plugged when measurements are taken. If four different microphones are used, then they need to be calibrated relative to one another so that any differences between the microphone amplitude and phase responses are accounted for. When a single microphone is used, a convenient reference signal is the input to the loudspeaker that generates the sound in the tube. The accuracy of this method is limited by nonlinearities (or harmonic distortion) in the response of the loudspeaker. Thus, when more than one microphone is used to obtain measurements at the four microphone locations, it is better to choose a microphone at one of the four measurement locations to provide the reference signal. Of course, this will result in a transfer function of unity amplitude and zero phase for the microphone position used as the reference.

Using Equation (1.229), the complex pressure and particle velocity amplitudes (relative to the reference signal) in the acoustic medium adjacent to the two surfaces of the porous material sample may be expressed in terms of the positive and negative going wave amplitudes, A, B, C and D, as:

$$\begin{cases} \dfrac{\hat{p}_{(x=0)}}{\hat{p}_{\mathrm{ref}}} = A + B \\[2mm] \dfrac{\hat{u}_{(x=0)}}{\hat{p}_{\mathrm{ref}}} = \dfrac{A - B}{\rho c} \\[2mm] \dfrac{\hat{p}_{(x=\ell)}}{\hat{p}_{\mathrm{ref}}} = C\mathrm{e}^{-\mathrm{j}k_a\ell} + D\mathrm{e}^{\mathrm{j}k_a\ell} \\[2mm] \dfrac{\hat{u}_{(x=\ell)}}{\hat{p}_{\mathrm{ref}}} = \dfrac{C\mathrm{e}^{-\mathrm{j}k_a\ell} - D\mathrm{e}^{\mathrm{j}k_a\ell}}{\rho c} \end{cases} \tag{1.232}$$

where A, B, C and D are calculated from the transfer function measurements and Equation (1.231).

Equation (1.232) can be rewritten in terms of the normal incidence complex reflection coefficient, $R_n = B/A$, and the normal incidence complex transmission coefficient, $\tau_n = C/A$, to give:

$$\begin{cases} \hat{p}_{(x=0)} = A[1 + R_n] \\[2mm] \hat{u}_{(x=0)} = \dfrac{A(1 - R_n)}{\rho c} \\[2mm] \hat{p}_{(x=\ell)} = A\tau_n\mathrm{e}^{-\mathrm{j}k_a\ell} \\[2mm] \hat{u}_{(x=\ell)} = \dfrac{A\tau_n\mathrm{e}^{-\mathrm{j}k_a\ell}}{\rho c} \end{cases} \tag{1.233}$$

where the equations containing the transmission coefficient are only valid if the amplitude, D, of the wave reflected from the end of the tube is small compared to the amplitude, C, of the incident wave, implying an almost anechoic termination.

A transfer matrix can be used to relate the acoustic pressure and particle velocity at location $x = 0$ to the acoustic pressure and particle velocity at location $x = \ell$. The appropriate transfer matrix contains elements T_{11}, T_{12}, T_{21} and T_{22}, so that:

$$\left[\begin{array}{c} \hat{p} \\ \hat{u} \end{array} \right]_{x=0} = \left[\begin{array}{cc} T_{11} & T_{12} \\ T_{21} & T_{22} \end{array} \right] \left[\begin{array}{c} \hat{p} \\ \hat{u} \end{array} \right]_{x=\ell} \tag{1.234}$$

where the variables on the RHS of Equation (1.234) are calculated using Equation (1.232) for termination conditions, a and b. Note that in Equation (1.234), p_{ref} cancels out as it appears in all variables on the numerators and denominators.

If the two surfaces of the sample are not the same, it is necessary to measure all of the four transfer functions between the microphone locations and the reference for two different end conditions at the downstream end of the tube. The first condition (condition, a) can be an almost anechoic one, while the second condition (condition, b) can be an open or rigidly closed end. In this case the elements of the transfer matrix are:

$$\begin{cases} T_{11} = \dfrac{\hat{p}_{a(x=0)}\hat{u}_{b(x=\ell)} - \hat{p}_{b(x=0)}\hat{u}_{a(x=\ell)}}{\hat{p}_{a(x=\ell)}\hat{u}_{b(x=\ell)} - \hat{p}_{b(x=\ell)}\hat{u}_{a(x=\ell)}} \\[12pt] T_{12} = \dfrac{\hat{p}_{b(x=0)}\hat{p}_{a(x=\ell)} - \hat{p}_{a(x=0)}\hat{p}_{b(x=\ell)}}{\hat{p}_{a(x=\ell)}\hat{u}_{b(x=\ell)} - \hat{p}_{b(x=\ell)}\hat{u}_{a(x=\ell)}} \\[12pt] T_{21} = \dfrac{\hat{u}_{a(x=0)}\hat{u}_{b(x=\ell)} - \hat{u}_{b(x=0)}\hat{u}_{a(x=\ell)}}{\hat{p}_{a(x=\ell)}\hat{u}_{b(x=\ell)} - \hat{p}_{b(x=\ell)}\hat{u}_{a(x=\ell)}} \\[12pt] T_{22} = \dfrac{\hat{p}_{a(x=\ell)}\hat{u}_{b(x=0)} - \hat{p}_{b(x=\ell)}\hat{u}_{a(x=0)}}{\hat{p}_{a(x=\ell)}\hat{u}_{b(x=\ell)} - \hat{p}_{b(x=\ell)}\hat{u}_{a(x=\ell)}} \end{cases} \tag{1.235}$$

If the two surfaces of the sample are the same, $T_{11} = T_{22}$ and reciprocity requires that the determinant of the transfer matrix is unity so that $T_{11}T_{22} - T_{12}T_{21} = 1$. Thus, Equation (1.235) simplifies to:

$$\begin{cases} T_{11} = T_{22} = \dfrac{\hat{p}_{(x=\ell)}\hat{u}_{(x=\ell)} + \hat{p}_{(x=0)}\hat{u}_{(x=0)}}{\hat{p}_{(x=0)}\hat{u}_{(x=\ell)} + \hat{p}_{(x=\ell)}\hat{u}_{(x=0)}} \\[12pt] T_{12} = \dfrac{\hat{p}_{(x=0)}^2 - \hat{p}_{(x=\ell)}^2}{\hat{p}_{(x=0)}\hat{u}_{(x=\ell)} + \hat{p}_{(x=\ell)}\hat{u}_{(x=0)}} \\[12pt] T_{21} = \dfrac{\hat{u}_{(x=0)}^2 - \hat{u}_{(x=\ell)}^2}{\hat{p}_{(x=0)}\hat{u}_{(x=\ell)} + \hat{p}_{(x=\ell)}\hat{u}_{(x=0)}} \end{cases} \tag{1.236}$$

Transmission Coefficient, Anechoic Termination

For an anechoic termination, the transmission coefficient, τ_a, can be expressed as (Song and Bolton, 2000):

$$\tau_a = \frac{2\mathrm{e}^{\mathrm{j}k_a\ell}}{T_{11} + T_{12}/(\rho c) + \rho c T_{21} + T_{22}} \tag{1.237}$$

This transmission coefficient also applies to the measurement of transmission loss of a muffler, in which case the sample of porous material is replaced with a muffler. However, the upper limiting frequency for this analysis is related to the diameter of the inlet and discharge ducts servicing the muffler, as indicated by Equation (1.179). The transmission loss, TL, is:

$$\mathrm{TL}_a = -10\log_{10}(\tau_a) \tag{1.238}$$

Absorption Coefficient, Anechoic Termination

For an anechoic termination, the normal incidence absorption coefficient, α_n, can be written in terms of the complex reflection coefficient, R_a, as:

$$\alpha_n = 1 - |R_a|^2 \tag{1.239}$$

where R_a is (Song and Bolton, 2000):

$$R_a = \frac{T_{11} + T_{12}/(\rho c) - \rho c T_{21} - T_{22}}{T_{11} + T_{12}/(\rho c) + \rho c T_{21} + T_{22}} \tag{1.240}$$

If face $(x = \ell)$ of the porous material is mounted against a rigid backing, then $\hat{u}_{(x=\ell)} = 0$, so that terms involving T_{22} and T_{12} are zero, in which case the reflection coefficient is (Song and Bolton, 2000):

$$R_h = \frac{T_{11} - \rho c T_{21}}{T_{11} + \rho c T_{21}} \tag{1.241}$$

Complex Wavenumber, Impedance and Density of the Test Sample

The transfer matrix for sound normally incident on a finite thickness sample of isotropic, homogeneous, porous acoustic material is (Allard and Atalla, 2009):

$$\begin{bmatrix} T_{11} & T_{12} \\ T_{21} & T_{22} \end{bmatrix} = \begin{bmatrix} \cos(k_m \ell) & \dfrac{\mathrm{j}\rho_m \omega}{k_m} \sin(k_m \ell) \\ \dfrac{\mathrm{j}k_m}{\omega \rho_m} \sin(k_m \ell) & \cos(k_m \ell) \end{bmatrix} \tag{1.242}$$

The four transfer matrix elements in Equation (1.242) may be directly linked to various properties of such a test sample. Note that as both the surfaces of the material are the same, the only end condition needing to be tested is the nearly anechoic one. From Equation (1.242), the wavenumber, k_m, of the test material is (Song and Bolton, 2000):

$$k_m = \frac{1}{\ell} \cos^{-1} (T_{11}) = \frac{1}{\ell} \sin^{-1} \sqrt{-(T_{12} T_{21})} \tag{1.243}$$

the characteristic impedance is (Song and Bolton, 2000):

$$Z_m = \rho_m c_m = \sqrt{\frac{T_{12}}{T_{21}}} \tag{1.244}$$

The complex normal specific acoustic impedance, Z_N, is then calculated using Equation (1.225), with R_p replaced with R_a or R_h of Equations (1.240) or (1.241), respectively.

The phase speed of sound within the material is:

$$c_m = \omega / k_m \tag{1.245}$$

Thus, the complex density of the material is:

$$\rho_m = Z_m k_m / \omega \tag{1.246}$$

Correction of the Measured Transfer Functions to Account for Differences in the Microphones

When more than one test microphone is used, corrections to the measured transfer functions must be made to account for differences in the phase and amplitude responses of the microphones (ASTM E2611-09, 2009). The transfer function, H, used in the calculations is then calculated from the measured transfer function, H^M, using a correction transfer function, H^c, the calculation of which is discussed in the following paragraphs. Thus, for microphone, n, in Figure 1.24, where $n = 1$, 2, 3 or 4:

$$H_{n,\mathrm{ref}} = \frac{H^M_{n,\mathrm{ref}}}{H^c_{n,\mathrm{ref}}} \tag{1.247}$$

where one of the four microphones shown in Figure 1.24 may be used as the reference one, and $H_{n,\mathrm{ref}}$, $H^c_{n,\mathrm{ref}}$ and $H^M_{n,\mathrm{ref}}$ are all complex numbers, so that, for example:

$$H_{n,\mathrm{ref}} = |H_{n,\mathrm{ref}}| e^{\mathrm{j}\phi_{n,\mathrm{ref}}} = H_R + \mathrm{j}H_I \tag{1.248}$$

The value of the correction transfer function is different for each microphone, but the procedure is the same for all microphones. Of course, the correction transfer function for the microphone chosen as the reference is unity with a phase shift of zero, so it is only necessary to use the following procedure to determine the correction transfer function for the three microphones that are not the reference microphone.

1. The transfer function is first measured between the reference microphone and microphone, n, where n is the microphone number for which the correction transfer function is to be determined. This transfer function is denoted $H_{n,\mathrm{ref}}^{\mathrm{I}}$.

2. Next, the microphone positions are interchanged; that is, the reference microphone is moved to the physical position occupied by microphone, n, in the previous step and microphone, n, is moved to the physical position previously occupied by the reference microphone. However, the connections to the measuring system are not disturbed or changed.

3. The transfer function between the reference microphone and microphone, n, in their new positions is then measured and denoted, $H_{n,\mathrm{ref}}^{\mathrm{II}}$.

4. The correction transfer function, $H_{n,\mathrm{ref}}^{c}$, for microphone, n, at frequency, ω, is then:

$$H_{n,\mathrm{ref}}^{c}(\omega) = |H_{n,\mathrm{ref}}^{c}|e^{\mathrm{j}\phi_{n,\mathrm{ref}}^{c}} = \left[H_{n,\mathrm{ref}}^{\mathrm{I}}(\omega) \cdot H_{n,\mathrm{ref}}^{\mathrm{II}}(\omega)\right]^{1/2} \qquad (1.249)$$

The procedure for the case of the speaker driving signal being the reference when only one physical microphone exists (not recommended) is outlined in ASTM E2611-09 (2009).

1.6.5 Calculation of Absorption Coefficients of Porous Materials

1.6.5.1 Porous Materials with a Backing Cavity

For porous acoustic materials, such as rockwool or fibreglass, the normal specific acoustic impedance of Equation (1.200) may also be calculated from the material characteristic impedance and propagation coefficient of Equations (1.163) and (1.164). For a material of infinite depth (or sufficiently thick that waves transmitted through the material from one face and reflected from the opposite face are of insignificant amplitude by the time they arrive back at the first face), the normal specific acoustic impedance is equal to the characteristic impedance of Equation (1.163). For a porous blanket of thickness, ℓ, backed by a cavity of any depth, L (including $L = 0$), with a rigid back, the normal specific acoustic impedance (in the absence of flow past the cavity) may be calculated using an electrical transmission line analogy (Magnusson, 1965) and is:

$$Z_N = Z_m \frac{Z_L + \mathrm{j}Z_m \tan(k_m\ell)}{Z_m + \mathrm{j}Z_L \tan(k_m\ell)} \qquad (1.250)$$

The quantities, Z_m and k_m, in Equation (1.250) are defined in Equations (1.163) and (1.164). The normal specific acoustic impedance, Z_L, of a rigidly terminated, partitioned backing cavity is:

$$Z_L = -\mathrm{j}\rho c / \tan(2\pi f L / c) \qquad (1.251)$$

and for a rigidly terminated, non-partitioned backing cavity, the impedance, Z_L, for a wave incident at angle, θ, is:

$$Z_L = -\mathrm{j}\rho c \cos\theta / \tan(2\pi f L / c) \qquad (1.252)$$

where θ is the angle of incidence of the sound wave measured from the normal to the surface.

A partitioned cavity is one that is divided into compartments by partitions that permit propagation normal to the surface, while inhibiting propagation parallel to the surface of the liner. The depth of each compartment is equal to the overall cavity depth.

If the porous material is rigidly backed so that $L = 0$ or, equivalently, L is an integer multiple of half wavelengths, Equation (1.250) reduces to:

$$Z_N = -jZ_m/\tan(k_m\ell) \tag{1.253}$$

1.6.5.2 Multiple Layers of Porous Liner Backed by an Impedance

Equation (1.250) can be easily extended to cover the case of multiple layers of porous material by applying it to each layer successively, beginning with the layer closest to the termination (rigid wall or cavity backed by a rigid wall) with impedance Z_L. The normal specific acoustic impedance looking into the ith layer surface that is furthest from the termination is:

$$Z_{N,i} = Z_{m,i}\frac{Z_{N,i-1} + jZ_{m,i}\tan(k_{m,i}\ell_i)}{Z_{m,i} + jZ_{N,i-1}\tan(k_{m,i}\ell_i)} \tag{1.254}$$

The variables in the above equation have the same definitions as those in Equation (1.250), with the added subscript, i, which refers to the ith layer or the added subscript, $i - 1$, which refers to the $(i - 1)$th layer.

Equation (1.254) could also be used for materials whose density was smoothly varying, by dividing the material into a number of very thin layers, with each layer assumed to have uniform properties.

1.6.5.3 Porous Liner Covered with a Limp Impervious Layer

If the porous material is protected by covering or enclosing it in an impervious blanket of thickness, h, and mass per unit area, σ', the effective normal specific acoustic impedance, Z_{NB}, at the outer surface of the blanket, which can be used together with Equations (1.200) and (1.209) to find the statistical absorption coefficient of the construction, is:

$$Z_{NB} = Z_N + j2\pi f\sigma' \tag{1.255}$$

where f is the frequency of the incident tonal sound or tone, or, alternatively, the centre frequency of a narrow band of noise. Typical values for σ' and c_L are included in Table 1.5 for commonly used covering materials. Guidelines for the selection of suitable protective coverings are given by Andersson (1981).

TABLE 1.5 Properties of commonly used limp impervious wrappings for environmental protection of porous materials

Material	Density (kg/m^3)	Typical thickness (microns = 10^{-6} m)	σ' (kg/m^2)[a]	c_L (approx.) (m/s)
Polyethylene (LD)	930	6–35	0.0055–0.033	460
Polyurethane	900	6–35	0.005–0.033	1330
Aluminium	2700	2–12	0.0055–0.033	5150
PVC	1400	4–28	0.005–0.033	1310
Melinex (polyester)	1390	15–30	0.021–0.042	1600
Metalised polyester	1400	12	0.017	1600

[a] σ' and c_L are, respectively, the surface density and speed of sound in the wrapping material.

1.6.5.4 Porous Liner Covered with a Perforated Sheet

If the porous liner were covered with a perforated sheet, the effective normal specific acoustic impedance (locally reactive rather than extended reactive) at the outer surface of the perforated sheet is (Bolt, 1947):

$$Z_{NP} = Z_N + \frac{\dfrac{100}{P}\left\{\mathrm{j}\rho c \tan\left[k_a \ell_e (1 - M)\right] + R_a S_h\right\}}{1 + \dfrac{100}{\mathrm{j}\omega m P}\left\{\mathrm{j}\rho c \tan\left[k_a \ell_e (1 - M)\right] + R_a S_h\right\}} \tag{1.256}$$

where Z_N is the normal specific acoustic impedance of the porous acoustic material with or without a cavity backing (and in the absence of flow), ω is the radian frequency, P is the % open area of the holes, R_a is the acoustic resistance of each hole, S_h is the area of each hole, M is the Mach number of the flow past the holes and m is the mass per unit area of the perforated sheet, all in consistent SI units. The effective length, ℓ_e, of each of the holes in the perforated sheet is Bies et al. (2017):

$$\ell_e = w + \left[\frac{16a}{3\pi}\left(1 - 0.43\frac{a}{q}\right)\right](1 - M)^2 \tag{1.257}$$

where a is the radius of the holes in the perforated sheet, w is the thickness of the perforated sheet, $M = U/c$ is the Mach number of flow through or across the holes, U is the flow speed through or across the holes, c is the speed of sound and q is the separation distance between hole centres.

1.6.5.5 Porous Liner Covered with a Limp Impervious Layer and a Perforated Sheet

In this case, the impedance of the perforated sheet and impervious layer are both added to the normal specific acoustic impedance of the porous acoustic material, so that:

$$Z_{NBP} = Z_N + \frac{\dfrac{100}{P}\left\{\mathrm{j}\rho c \tan\left[k_a \ell_e (1 - M)\right] + R_a S_h\right\}}{1 + \dfrac{100}{\mathrm{j}\omega m P}\left\{\mathrm{j}\rho c \tan\left[k_a \ell_e (1 - M)\right] + R_a S_h\right\}} + \mathrm{j}2\pi f \sigma' \tag{1.258}$$

It is important that the impervious layer and the perforated sheet are separated, using something like a mesh spacer (with a grid size of at least 2 cm); otherwise the performance of the construction as an absorber will be severely degraded, as the impervious layer will no longer be acting as a limp blanket.

2

Structural Mechanics Fundamentals

2.1 Introduction

This chapter begins with explanations of the basic principles underpinning Newtonian and analytical mechanics, including generalised coordinates, the principle of virtual work, D'Alembert's principle, Hamilton's principle, Lagrange's equations of motion and influence coefficients. This is followed by a discussion of the vibration of and equations of motion for continuous systems, including beams, plates and cylinders.

2.2 Vibration of Discrete Systems

In this section, the analysis of systems that consist of discrete units will be undertaken, as it forms the basis of the analysis of continuous systems such as beams, plates and shells discussed in Section 2.3.

2.2.1 Summary of Newtonian Mechanics

One way of analysing a mechanical system is to extend Newton's laws for a single particle to systems of particles and use the concepts of force and momentum, both of which are vector quantities. An approach such as this is referred to as Newtonian mechanics or vectorial mechanics. For more detailed treatment of this subject the reader is advised to consult Meirovitch (1970) or McCuskey (1959).

A second approach known as analytical mechanics is attributed principally to Lagrange, and later Hamilton, and considers the system as a whole rather than as a number of individual components. Using this approach, problems are formulated in terms of two scalar quantities: energy and work. This second approach will be discussed in detail in Section 2.2.2.

The three fundamental physical laws describing the dynamics of particles are those enunciated by Isaac Newton in the Principia (1686). The first law states that a particle of constant mass will remain at rest or move in a straight line unless acted upon by a force. The most useful concept used in the solution of dynamics problems is embodied in Newton's second law, which effectively states that the time rate of change of the linear momentum vector for a particle is equal to the force vector acting on the particle. That is:

$$\boldsymbol{F} = \frac{\mathrm{d}(m\boldsymbol{u})}{\mathrm{d}t} \tag{2.1}$$

For the systems that are discussed in this book, the mass remains constant and thus Equation (2.1) may be written as:

$$F = m\frac{\mathrm{d}\boldsymbol{u}}{\mathrm{d}t} \tag{2.2}$$

Newton's third law of motion states that the force exerted on one particle by a second particle is equal and opposite to the force exerted by the first particle on the second. That is:

$$\boldsymbol{F}_{12} = -\boldsymbol{F}_{21} \tag{2.3}$$

A particle is an idealisation of a body whose dimensions are very small in comparison with the distance to other bodies and whose internal motion does not affect the motion of the body as a whole. Mathematically, it is represented by a mass point of infinitesimally small size.

All three of Newton's laws apply equally well to particles experiencing angular motion. In this case, the force vector is replaced by a torque vector and the linear momentum vector is replaced by an angular momentum vector equal to $J_I\dot{\boldsymbol{\theta}}$, where J_I is the moment of inertia of the particle (not to be confused with the second moment of area, J, of a structural section discussed in Section 2.3.3.3 and Chapter 4) and $\dot{\boldsymbol{\theta}}$ is the angular velocity vector.

It can be shown (McCuskey, 1959) that Newton's laws lead to three very important conservation laws.

1. If the vector sum of the linear forces acting on a particle are zero, then the linear momentum of the particle will remain unchanged.
2. If the vector sum of the torques acting on a particle are zero, then the angular momentum of the particle will remain unchanged.
3. If a particle is acted on only by conservative forces, then its total energy (kinetic plus potential) will remain unchanged.

A non-conservative force is one for which the work done on the particle (or the change in total energy of the particle) is dependent on the path taken by the particle from its initial to final position. Friction and air drag are two examples of non-conservative forces, as the work done on a particle will depend on the path taken from its initial to final position. A longer path will result in more energy dissipation. Non-conservative forces do not have any potential energy associated with them and do not conserve the total energy in the system. If a particle is acted on by a non-conservative force, $\boldsymbol{F}_1$, then the rate of change of total energy of the particle moving with velocity $\boldsymbol{u}$ is:

$$\frac{\mathrm{d}E_{\mathrm{tot}}}{\mathrm{d}t} = \boldsymbol{F}_1 \cdot \boldsymbol{u} \tag{2.4}$$

A conservative force is one that is not associated with any dissipation. Thus, for a particle acted on by a conservative force, $E_p + E_k = \mathrm{const}$, where E_k is the kinetic energy and E_p is the potential energy. Examples of a conservative force are gravity and the force exerted on a particle by a spring, so if a particle moves from an initial to a final position as a result of being acted on by a conservative force, the work done on the particle is independent of the path taken from the initial to final position. The kinetic energy, E_k, of a particle is:

$$E_k = \frac{1}{2}mu^2 \tag{2.5}$$

where m is its mass and $u = |\boldsymbol{u}|$ is its speed.

The potential energy, E_p, of a particle is:

$$E_p = mgh \tag{2.6}$$

where g is the acceleration due to gravity and h is the height above some reference datum.

For a particle attached to a horizontal massless spring, the potential energy is:

$$E_p = -kx \tag{2.7}$$

where k is the spring stiffness and x is its extension.

2.2.1.1 Systems of Particles

Newton's laws and the conservation laws also apply to bodies and systems of particles as well as to single particles. Thus, as shown by Meirovitch (1970), for a system of particles acted on by no external forces, the linear and angular momentum of the system will remain unchanged and will be equal to the vector sum of the linear and angular momentum of each particle. In addition, the total system energy will remain unchanged, provided that all of the external forces acting on the system are conservative. Thus, the motion of the centre of mass of a system of particles or a body is the same as if all the system mass were concentrated at that point and were acted upon by the resultant of all the external forces.

In other words, for a linear conservative system, the mass of a body multiplied by its acceleration is equal to the sum of the external forces acting upon it. This principle will be used in later sections of this text to derive the equations of motion for both discrete and continuous systems.

2.2.2 Summary of Analytical Mechanics

The following topics, which are relevant to the analysis in later parts of this book, will be considered briefly: generalised coordinates; principle of virtual work; d'Alembert's principle; Hamilton's principle; and Lagrange's equations.

2.2.2.1 Generalised Coordinates

In many physical problems, the bodies of interest are not completely free, but are subject to some kinematic constraints. As an example, consider two point masses, m_1 and m_2, connected by a rigid link of length, L, as shown in Figure 2.1.

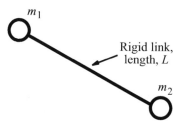

FIGURE 2.1 Rigid link.

The motions of m_1 and m_2 are completely defined in Cartesian coordinates using x_1, y_1, z_1, x_2, y_2 and z_2. However, the six coordinates are not independent of each other. In fact:

$$(x_1 - x_2)^2 + (y_1 - y_2)^2 + (z_1 - z_2)^2 = L^2 \qquad (2.8)$$

Thus, in this case, the problem is solved if only five of the coordinates are known, as the sixth one may be determined using Equation (2.8). The motion of this system may be completely defined by the three coordinates of either m_1 or m_2 and two angles, which define the orientation of the rigid link.

In general, the motion of N particles, subject to a number, a, of kinematic restraints, can be described uniquely by n independent coordinates, q_k ($k = 1, 2, \ldots, n$), where:

$$n = 3N - a \qquad (2.9)$$

The n coordinates, $q_1, \ldots, q_n$, are referred to as generalised coordinates — each is independent of all of the others. The generalised coordinates may not be physically identifiable and they are not necessarily unique. Thus, there may be several sets of generalised coordinates that can describe a particular physical system. They must be finite, single-valued, differentiable with respect to time and continuous.

The generalised coordinates for a given system represent the least number of variables required to specify the positions of the elements of the system at any particular time. A system that can be described in terms of generalised independent coordinates is referred to as holonomic.

2.2.2.2 Principle of Virtual Work

This principle was first enunciated by Johann Bernoulli in 1717, and is essentially a statement of the static equilibrium of a mechanical system.

Consider a single particle at a vector location, $\boldsymbol{r}$, in space, acted upon by a number of forces, N. Defining $\delta \boldsymbol{r}$ as any small virtual (or imaginary) displacement arbitrarily imposed on the particle, the virtual work done by the forces is:

$$\delta W = \boldsymbol{F}_1 \cdot \delta \boldsymbol{r} + \boldsymbol{F}_2 \cdot \delta \boldsymbol{r} + \ldots + \boldsymbol{F}_N \cdot \delta \boldsymbol{r} \tag{2.10}$$

or:

$$\delta W = \sum_{i=1}^{N} \boldsymbol{F}_i \cdot \delta \boldsymbol{r} \tag{2.11}$$

or:

$$\delta W = \boldsymbol{F} \cdot \delta \boldsymbol{r} \tag{2.12}$$

where $\boldsymbol{F}$ is the resultant of all the forces acting on the particle.

The principle of virtual work states that if and only if, for any arbitrary virtual displacement, $\delta \boldsymbol{r}$, the virtual work, $\delta W = 0$, under the action of the forces, $\boldsymbol{F}_i$, the particle is in equilibrium. For non-zero $\delta \boldsymbol{r}$, Equation (2.11) shows that either $\delta \boldsymbol{r}$ is perpendicular to $\Sigma \boldsymbol{F}_i$ or $\Sigma \boldsymbol{F}_i = 0$. Since Equation (2.11) must hold for any $\delta \boldsymbol{r}$, the first possibility is ruled out and $\Sigma \boldsymbol{F}_i = 0$. Thus, for a particle to be in equilibrium:

$$\delta W = \boldsymbol{F} \cdot \delta \boldsymbol{r} = 0 \tag{2.13}$$

where $\boldsymbol{F}$ is the resultant of all forces acting on the particle.

For a system subject to constraint forces, distinguishing between the applied forces, $\boldsymbol{F}_i$, and the constraint forces, $\boldsymbol{F}_i'$, and using Equation (2.13), the following is obtained:

$$\delta W = \sum_{i=1}^{n} \boldsymbol{F}_i \cdot \delta \boldsymbol{r}_i + \sum_{i=1}^{m} \boldsymbol{F}_i' \cdot \delta \boldsymbol{r}_i = 0 \tag{2.14}$$

An example of a constraint force would be the vertical reaction force on a particle resting on a smooth surface.

As the work of the constraint forces through virtual displacements compatible with the system constraints is zero, Equation (2.14) becomes:

$$\delta W = \sum_{i=1}^{n} \boldsymbol{F}_i \cdot \delta \boldsymbol{r}_i = 0 \tag{2.15}$$

However, for systems with constraints, all of the virtual displacements, $\delta \boldsymbol{r}_i$ are not independent and so Equation (2.15) cannot be interpreted simply to imply that $\boldsymbol{F}_i = 0$; $(i = 1, 2, \ldots, n)$.

If the problem is described by a set of independent generalised coordinates, Equation (2.15) may be written as:

$$\delta W = \sum_{k=1}^{m} \boldsymbol{Q}_k \delta \boldsymbol{q}_k = 0 \tag{2.16}$$

where Q_k; $(k = 1, \ldots, m)$ are known as generalised forces. Since the number of generalised coordinates is now the same as the number of degrees of freedom of the system, the virtual displacements, δq_k, are all independent so that $Q_k = 0$; $(k = 1, \ldots, m)$.

Example 2.1

Use the principle of virtual work to calculate the angle, θ, for the link of length, L, shown in Figure 2.2 to be in static equilibrium. Here, x = elongation of the spring, and y = the extent of lowering of one end of the link from the horizontal position.

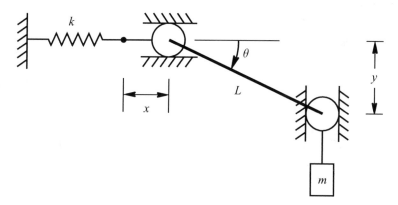

FIGURE 2.2 Example 2.1 link mechanism.

Solution 2.1

The position of the ends of the link at equilibrium may be described by:

$$x = L(1 - \cos\theta); \qquad y = L\sin\theta \tag{a}$$

The virtual work can be written as:

$$\delta W = -kx\,\delta x + mg\,\delta y = 0 \tag{b}$$

where the virtual displacements, δx and δy, are obtained from Equation (a) and are equal to:

$$\delta x = L\sin\theta\,\delta\theta; \qquad \delta y = L\cos\theta\,\delta\theta \tag{c}$$

Substituting the first of the Equations (a) and both Equations (c) into (b):

$$\delta W = -kL(1 - \cos\theta)L\sin\theta\,\delta\theta + mgL\cos\theta\,\delta\theta = 0 \tag{d}$$

from which the following is obtained:

$$(1 - \cos\theta)\tan\theta = \frac{mg}{kL} \tag{e}$$

Solving Equation (e) gives the value of θ corresponding to the equilibrium position of the system shown in the figure.

2.2.2.3   D'Alembert's Principle

In 1743 in his Traité de Dynamique, D'Alembert proposed a principle to reduce a dynamics problem to an equivalent statics problem. By introducing 'inertial forces' he applied Bernoulli's principle of virtual work to systems in which motion resulted from the applied forces. The inertial force acting upon the ith particle of mass, m_i, in a system is $-m_i\ddot{r}$ or $-m_i\dot{u}_i$, where the dot above a variable represents differentiation with respect to time and a double dot represents a double differentiation. If the resultant force acting on the ith particle of a system of n particles is $\boldsymbol{F}_i$, then D'Alembert's principle states that the system is in equilibrium if the total virtual work performed by the inertia forces and applied forces is zero. That is,

$$\sum_{i=1}^{n}(\boldsymbol{F}_i - m_i\ddot{\boldsymbol{r}}_i)\cdot\delta\boldsymbol{r}_i = 0 \qquad (2.17)$$

D'Alembert's principle represents the most general formulation of dynamics problems and all the various principles of mechanics (including Hamilton's principle to follow) are derived from it.

    The problem with D'Alembert's principle is that it is not very convenient for deriving equations of motion, as problems are formulated in terms of position coordinates which, in contrast with generalised coordinates, may not all be independent. A different formulation, Hamilton's principle, which avoids this difficulty, will now be discussed.

2.2.2.4   Hamilton's Principle

Hamilton's principle is a consideration of the motion of an entire system between two times, t_1 and t_2. It is an integral principle and reduces dynamics problems to a scalar definite integral.

    Consider a system of n particles of masses, m_i, located at points, $\boldsymbol{r}_i$, and acted upon by resultant external forces, $\boldsymbol{F}_i$. By D'Alembert's principle the following can be written:

$$\sum_{i=1}^{n}(m_i\ddot{\boldsymbol{r}}_i - \boldsymbol{F}_i)\cdot\delta\boldsymbol{r}_i = 0 \qquad (2.18)$$

This is the dynamic condition to be satisfied by the applied forces and inertia forces for arbitrary $\delta\boldsymbol{r}_i$, consistent with the constraints on the system. The virtual work done by the applied forces is:

$$\delta W = \sum_{i=1}^{n}\boldsymbol{F}_i\cdot\delta\boldsymbol{r}_i \qquad (2.19)$$

Furthermore:

$$\frac{\mathrm{d}}{\mathrm{d}t}(\dot{\boldsymbol{r}}_i\cdot\delta\boldsymbol{r}_i) = \dot{\boldsymbol{r}}_i\frac{\mathrm{d}}{\mathrm{d}t}(\delta\boldsymbol{r}_i) + \ddot{\boldsymbol{r}}_i\cdot\delta\boldsymbol{r}_i \qquad (2.20)$$

Interchanging the derivative and variational operators gives:

$$\frac{\mathrm{d}}{\mathrm{d}t}\delta\boldsymbol{r}_i = \delta\dot{\boldsymbol{r}}_i \qquad (2.21)$$

Thus:

$$\dot{\boldsymbol{r}}_i\cdot\delta\dot{\boldsymbol{r}}_i = \delta\left(\frac{1}{2}\dot{r}_i^2\right) = \delta\left(\frac{1}{2}u_i^2\right) \qquad (2.22)$$

where u_i is the speed of the ith particle. When the mass, m_i, is included as a factor, then Equation (2.22) represents the variation in kinetic energy, δE_k, of the particle.

Substituting Equations (2.21) and (2.22) into Equation (2.20) and rearranging gives:

$$\ddot{\boldsymbol{r}}_i \cdot \delta\boldsymbol{r}_i = \frac{\mathrm{d}}{\mathrm{d}t}\left(\dot{\boldsymbol{r}}_i \cdot \delta\boldsymbol{r}_i\right) - \delta\left(\frac{1}{2}u_i^2\right) \tag{2.23}$$

Multiplying Equation (2.23) by m_i and summing over all particles in the system gives:

$$\sum_{i=1}^{n} m_i \ddot{\boldsymbol{r}}_i \cdot \delta\boldsymbol{r}_i = \frac{\mathrm{d}}{\mathrm{d}t}\left(\dot{\boldsymbol{r}}_i \cdot \delta\boldsymbol{r}_i\right) - \delta E_k \tag{2.24}$$

Substituting Equations (2.24) and (2.19) into Equation (2.18) gives:

$$\delta E_k + \delta W = \sum_{i=1}^{n} m_i \frac{\mathrm{d}}{\mathrm{d}t}\left(\dot{\boldsymbol{r}}_i \cdot \delta\boldsymbol{r}_i\right) \tag{2.25}$$

The instantaneous configuration of a system is given by the values of the n generalised coordinates defining a representative point in the n-dimensional configuration space.

$$\delta\boldsymbol{r}_i(t_1) = \delta\boldsymbol{r}_i(t_2) = 0 \tag{2.26}$$

The system configuration changes with time, tracing a true path (or dynamic path) in the configuration space. In addition to this true path, there will be an infinite number of imagined variations of this path. Consider two times, t_1 and t_2, at which it is assumed that $\delta r_i = 0$; that is, two times at which the dynamic path and all the imagined variations of this path coincide (see Figure 2.3).

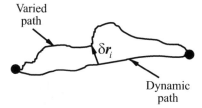

Varied path

δr_i

Dynamic path

FIGURE 2.3 Dynamic path.

Integrating Equation (2.25) between t_1 and t_2 gives:

$$\int_{t_1}^{t_2} (\delta E_k + \delta W)\mathrm{d}t = \sum_{i=1}^{n} m_i(\dot{\boldsymbol{r}}_i \cdot \delta\boldsymbol{r}_i)\bigg|_{t_1}^{t_2} \tag{2.27}$$

which, given the conditions of Equation (2.26), reduces to:

$$\int_{t_1}^{t_2} (\delta E_k + \delta W)\mathrm{d}t = 0 \tag{2.28}$$

If it is assumed that δW is a work function arising from the potential energy so that $\delta W = -\delta E_p$, then Equation (2.28) may be written for a holonomic system as:

$$\delta \int_{t_1}^{t_2} (E_k - E_p)\mathrm{d}t = 0 \tag{2.29}$$

The function, $E_k - E_p$, is called the Lagrangian, L_A, of the system.

Equation (2.29) is a mathematical statement of Hamilton's principle, which may be stated as follows: 'of all paths in time that the coordinates, r_i, may be imagined to take between two instants t_1 and t_2, the dynamic (or true) path actually taken by the system will be that for which $\int_{t_1}^{t_2}(E_k - E_p)dt$ will have a stationary value, provided that the path variations vanish at the end points, t_1 and t_2.

Hamilton's principle provides a formulation rather than a solution of dynamic problems and as will be shown shortly, it is better used to derive equations of motion which can be used for the solution of dynamic problems.

Example 2.2

Consider the system shown in Figure 2.2 and use Hamilton's principle to derive the equation of motion.

Solution 2.2

Using Equation (a) of Example 2.1, the Lagrangian is:

$$
\begin{aligned}
L_A &= E_k - E_p = \frac{1}{2}m\dot{y}^2 - \frac{1}{2}kx^2 + mgy \\
&= \frac{1}{2}mL^2\left[\dot{\theta}^2\cos^2\theta - \frac{k}{m}(1-\cos\theta)^2 + \frac{2g}{L}\sin\theta\right]
\end{aligned}
\tag{a}
$$

Substituting Equation (a) into Equation (2.29) gives:

$$
\begin{aligned}
\delta\int_{t_1}^{t_2} L_A dt &= \int_{t_1}^{t_2}\left(\frac{\partial L_A}{\partial\theta}\delta\theta + \frac{\partial L_A}{\partial\dot{\theta}}\delta\dot{\theta}\right)dt \\
&= -mL^2\int_{t_1}^{t_2}\left\{\left[\dot{\theta}^2\sin\theta\cos\theta + \frac{k}{m}(1-\cos\theta)\sin\theta - \frac{g}{L}\cos\theta\right]\delta\theta - \dot{\theta}\cos^2\theta\delta\dot{\theta}\right\}dt = 0
\end{aligned}
\tag{b}
$$

The second term in the integrand contains $\delta\dot{\theta}$, making it incompatible with the remaining terms, which are all multiplying $\delta\theta$. But $\delta\dot{\theta} = \mathrm{d}(\delta\theta)/\mathrm{d}t$, so that after an integration by parts of the term containing $\delta\dot{\theta}$, the following is obtained:

$$
\int_{t_1}^{t_2}\left[\frac{\mathrm{d}}{\mathrm{d}t}(\dot{\theta}\cos^2\theta) + \dot{\theta}^2\sin\theta\cos\theta + \frac{k}{m}(1-\cos\theta)\sin\theta - \frac{g}{L}\cos\theta\right]\delta\theta dt
$$

$$
-\dot{\theta}\cos^2\theta\delta\theta\Big|_{t_1}^{t_2} = 0
\tag{c}
$$

where the constant, $-mL^2$, has been ignored.

Invoking the requirement that the variation, $\delta\theta$, vanish at the two instants, t_1 and t_2, the second term in Equation (c) reduces to zero. Moreover, $\delta\theta$ is arbitrary in the time interval between t_1 and t_2, so that the only way for the integral in the first term to be zero is for the coefficient of $\delta\theta$ to vanish for any time, t. Hence, the following must be set:

$$
\frac{\mathrm{d}}{\mathrm{d}t}(\dot{\theta}\cos^2\theta) + \dot{\theta}^2\sin\theta\cos\theta + \frac{k}{m}(1-\cos\theta)\sin\theta - \frac{g}{L}\cos\theta = 0
\tag{d}
$$

which is the desired equation of motion. Letting $\dot{\theta} = \ddot{\theta} = 0$ the same equation for the system equilibrium position as the one derived in Example 2.1 is obtained. In general, it is not necessary to use Hamilton's principle directly for the solution of dynamic problems. Instead, Hamilton's principle is used to derive Lagrange's equations of motion, which can then be used to solve dynamic problems.

2.2.2.5   Lagrange's Equations of Motion

It was mentioned earlier that the generalised D'Alembert's principle, Equation (2.17), is not convenient for the derivation of equations of motion. It is more advantageous to express Equation (2.17) in terms of a set of generalised coordinates, q_k; $(k = 1, 2, \ldots, n)$ in such a way that the virtual displacements, δq_k, are independent and arbitrary. Under these circumstances, the coefficients of δ_k; $(k = 1, 2, \ldots, n)$ can be set equal to zero separately, thus obtaining a set of differential equations in terms of generalised coordinates, known as Lagrange's equations of motion.

Instead of using D'Alembert's principle, Lagrange's equations of motion can be derived using Hamilton's principle which is simply an integrated form of D'Alembert's principle. The derivation of Lagrange's equations of motion using Hamilton's principle is discussed in detail elsewhere (McCuskey, 1959; Mierovitch, 1970; Tse et al., 1978) and only the results will be given here.

For each generalised coordinate, q_k, the Lagrange equation of motion is:

$$\frac{\mathrm{d}}{\mathrm{d}t}\left(\frac{\partial E_k}{\partial \dot{q}_k}\right) - \frac{\partial E_k}{\partial q_k} = Q'_k; \qquad (k = 1, 2, \ldots, n) \tag{2.30}$$

where Q'_k is the kth generalised force, q_k is the kth generalised coordinate, the system has n degrees of freedom, and E_k is the kinetic energy of the total system. The forces, Q'_k, may be made up of potential forces, Q_v, damping forces, Q_D and the applied forces Q_k. The potential forces can be either due to changes in height, spring forces or both. The damping forces are either viscous or hysteretic and are associated with the internal system damping.

The potential energy is a function of the generalised coordinates. Thus:

$$E_p = E_p(q_1, q_2, \ldots, q_n) \tag{2.31}$$

Expanding this about the stable equilibrium position of the system gives:

$$E_p = E_{p0} + \sum_{k=1}^{n}\left(\frac{\partial E_p}{\partial q_k}\right)_0 q_k + \frac{1}{2}\sum_{k=1}^{n}\sum_{i=1}^{n}\left(\frac{\partial^2 E_p}{\partial q_k \partial q_i}\right)_0 q_k q_i +, \ \ldots \tag{2.32}$$

where the subscript 0 denotes the value at the equilibrium position. E_{p0} can be defined as zero if the potential energy is measured with respect to this datum. Since E_p is a minimum at E_{p0}, its first derivative must vanish. If also the third and higher order terms are neglected, Equation (2.32) becomes:

$$E_p = \frac{1}{2}\sum_{k=1}^{n}\sum_{i=1}^{n}\left(\frac{\partial^2 E_p}{\partial q_k \partial q_i}\right)_0 q_k q_i \tag{2.33}$$

For small variations, the second partial derivative term may be assumed constant. Denoting this as an equivalent spring constant, k_{ki}, the following is obtained:

$$E_p = \frac{1}{2}\sum_{k=1}^{n}\sum_{i=1}^{n}k_{ki}q_k q_i \tag{2.34}$$

where $k_{ki} = k_{ik}$. Note that k_{ki} are the elements of an $n \times n$ stiffness matrix.

The spring force (or potential force) associated with coordinate, q_k, is:

$$Q_v = \frac{\partial E_p}{\partial q_k} = -\sum_{k=1}^{n} k_{ki} q_k \qquad (2.35)$$

By analogy with the potential energy, a dissipation function may be defined as:

$$D = \frac{1}{2} \sum_{k=1}^{n} \sum_{i=1}^{n} C_{ki} \dot{q}_k \dot{q}_i \qquad (2.36)$$

and the damping force associated with the velocity, $\dot{q}_k$, is:

$$Q_D = \frac{\partial D}{\partial \dot{q}_k} = -\sum_{k=1}^{n} C_{ki} \dot{q}_k \qquad (2.37)$$

Including the spring force and the damping force in the Lagrangian, Equation (2.30) becomes:

$$\frac{\mathrm{d}}{\mathrm{d}t}\left(\frac{\partial E_k}{\partial \dot{q}_k}\right) - \frac{\partial E_k}{\partial q_k} + \frac{\partial D}{\partial \dot{q}_k} + \frac{\partial E_p}{\partial q_k} = Q_k \qquad (2.38)$$

For the free vibration of a conservative system, the following is obtained:

$$\frac{\mathrm{d}}{\mathrm{d}t}\left(\frac{\partial E_k}{\partial \dot{q}_k}\right) - \frac{\partial E_k}{\partial q_k} + \frac{\partial E_p}{\partial q_k} = 0 \qquad (2.39)$$

where the kinetic energy may be defined as:

$$E_k = \frac{1}{2} \sum_{k=1}^{n} \sum_{i=1}^{n} m_{ki} \dot{q}_k \dot{q}_i \qquad (2.40)$$

Since the potential energy, E_p, is a function of the coordinates only and $\dfrac{\partial E_p}{\partial \dot{q}_k} = 0$, Equation (2.39) may be written in terms of the Lagrangian, $L_A = E_k - E_p$, as:

$$\frac{\mathrm{d}}{\mathrm{d}t}\left(\frac{\partial L_A}{\partial \dot{q}_k}\right) - \frac{\partial L_A}{\partial q_k} = 0; \qquad (k = 1, 2, \ldots, n) \qquad (2.41)$$

Returning to Equations (2.34) and (2.40), they may be expressed using matrix notation, respectively (where matrices and vectors are indicated in bold typeface), as:

$$E_p = \frac{1}{2} \boldsymbol{q}_k^{\mathrm{T}} \boldsymbol{K} \boldsymbol{q}_k \qquad (2.42)$$

and:

$$E_k = \frac{1}{2} \dot{\boldsymbol{q}}_k^{\mathrm{T}} \boldsymbol{M} \dot{\boldsymbol{q}}_k \qquad (2.43)$$

where $\boldsymbol{K}$ and $\boldsymbol{M}$ are referred to, respectively, as the system stiffness and mass matrices.

Substituting Equations (2.42) and (2.43) into Equation (2.39) gives the following equation of motion for the free vibration of a conservative system:

$$\boldsymbol{M}\ddot{\boldsymbol{q}}_k + \boldsymbol{K}\boldsymbol{q}_k = \boldsymbol{0} \qquad (2.44)$$

where:

$$\boldsymbol{M} = \begin{bmatrix} m_{11} & m_{12} & \cdots & m_{1n} \\ m_{21} & m_{22} & \cdots & \vdots \\ \vdots & \vdots & \vdots & \vdots \\ m_{n1} & \cdots & \cdots & m_{nn} \end{bmatrix} \qquad (2.45)$$

$$\boldsymbol{K} = \begin{bmatrix} k_{11} & k_{12} & \cdots & k_{1n} \\ k_{21} & k_{22} & \cdots & \vdots \\ \vdots & \vdots & \vdots & \vdots \\ k_{n1} & \cdots & \cdots & k_{nn} \end{bmatrix} \qquad (2.46)$$

Example 2.3

Use Lagrange's equations to derive the equations of motion for the double pendulum shown in Figure 2.4.

Solution 2.3

This type of system is known as a two-degree-of-freedom system, as it possesses two point masses, each of which is capable of movement in only one coordinate direction (θ_1 and θ_2, respectively).

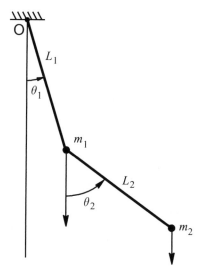

FIGURE 2.4 Double pendulum (Example 2.3).

Because the system is conservative and the constraints of the system are taken into account by choosing θ_1 and θ_2 as generalised coordinates, Equation (2.32) may be used to derive the required equations.

The kinetic energy of the system is:

$$E_k = \frac{1}{2}(m_1 + m_2)L_1^2\dot{\theta}_1^2 + \frac{1}{2}m_2L_2^2\dot{\theta}_2^2 + m_2L_1L_2\dot{\theta}_1\dot{\theta}_2\cos(\theta_1 - \theta_2) \qquad (a)$$

and the potential energy referred to the point of support is:

$$E_p = -(m_1 + m_2)gL_1\cos\theta_1 - m_2gL_2\cos\theta_2 \qquad (b)$$

From Equations (a) and (b) the Lagrangian, L_A, is:

$$L_A = E_k - E_p = (a) - (b) \qquad (c)$$

From Equations (a), (b) and (c):

$$\frac{\partial L_A}{\partial \dot{\theta}_1} = (m_1 + m_2)L_1^2\dot{\theta}_1 + m_2L_1L_2\dot{\theta}_2\cos(\theta_1 - \theta_2) \qquad (d)$$

$$\frac{\partial L_A}{\partial \dot{\theta}_2} = m_2 L_2^2 \dot{\theta}_2 + m_2 L_1 L_2 \dot{\theta}_1 \cos(\theta_1 - \theta_2) \qquad \text{(e)}$$

$$\frac{\partial L_A}{\partial \theta_2} = m_2 L_1 L_2 \dot{\theta}_1 \dot{\theta}_2 \sin(\theta_1 - \theta_2) - m_2 g L_2 \sin \theta_2 \qquad \text{(f)}$$

$$\frac{\partial L_A}{\partial \theta_1} = -m_2 L_1 L_2 \dot{\theta}_1 \dot{\theta}_2 \sin(\theta_1 - \theta_2) - (m_1 + m_2) g L_1 \sin \theta_1 \qquad \text{(g)}$$

Substituting Equations (a), (b) and (c) into Equation (2.32) gives for each of the generalised coordinates, θ_1 and θ_2:

$$(m_1 + m_2) L_1 \ddot{\theta}_1 + m_2 L_2 \ddot{\theta}_2 \cos(\theta_1 - \theta_2) + m_2 L_2 \dot{\theta}_1 \dot{\theta}_2 \sin(\theta_1 - \theta_2) + (m_1 + m_2) g \sin \theta_1 = 0 \quad \text{(h)}$$

$$L_2 \ddot{\theta}_2 + L_1 \ddot{\theta}_1 \cos(\theta_1 - \theta_2) + L_1 \ddot{\theta}_1 \cos(\theta_1 - \theta_2) - L_1 \dot{\theta}_1 \dot{\theta}_2 \sin(\theta_1 - \theta_2) + g \sin \theta_2 = 0 \quad \text{(i)}$$

Equations (h) and (i) are two simultaneous differential equations that may be solved for the coordinates, θ_1 and θ_2, as a function of time. These equations can also be derived with some patience and difficulty using Newton's equations of motion. However, the advantage of the Lagrange approach will soon become obvious to any readers who wish to attempt the derivation using Newton's formulation.

Example 2.4

Use Lagrange's equations to derive the equation of motion for the system in Figure 2.5.

Solution 2.4

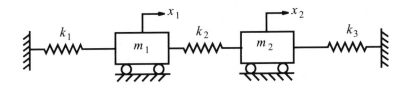

FIGURE 2.5 Simple two-degree-of-freedom, undamped system.

From Figure 2.5, it can be seen that x_1 and x_2 are acceptable generalised coordinates, q_1 and q_2. The spring force acting on m_1 due to displacement x_1 is $-(k_1 + k_2)x_1$. Thus, $k_{11} = k_1 + k_2$. Similarly it can be seen that $k_{22} = k_2 + k_3$. The force acting on m_1 due to a displacement, x_2, is $k_2 x_2$. Thus, $k_{12} = -k_2$ and:

$$E_p = \frac{1}{2} \left[k_{11} x_1^2 + k_{22} x_2^2 + k_{12} x_2 x_1 + k_{21} x_1 x_2 \right] = \frac{1}{2} \left[(k_1 + k_2)x_1^2 + (k_2 + k_3)x_2^2 - 2k_2 x_1 x_2 \right] \quad \text{(a)}$$

The kinetic energy is calculated by first determining the elements of the mass matrix, $\boldsymbol{M}$. In this case, $m_{11} = m_1$, $m_{21} = m_{12} = 0$ and $m_{22} = m_2$ (as x_1 and x_2 refer to the motions of m_1 and m_2 directly). Thus, the kinetic energy, E_k, is:

$$E_k = \frac{1}{2} \left(m_1 \dot{x}_1^2 + m_2 \dot{x}_2^2 \right) \qquad \text{(b)}$$

Substituting Equations (a) and (b) into Equation (2.39) gives the following two equations of motion:

$$\left. \begin{array}{l} m_1 \ddot{x}_1 + (k_1 + k_2)x_1 - k_2 x_2 = 0 \\ m_2 \ddot{x}_2 + (k_2 + k_3)x_2 - k_2 x_1 = 0 \end{array} \right\} \qquad \text{(c)}$$

Alternatively, the equations of motion could have been derived directly from Equation (2.44) by substituting the appropriate quantities for the elements of the mass and stiffness matrices.

2.2.2.6 Influence Coefficients

In addition to the use of Newton's laws (described in Section 2.2.1) or variational techniques leading to Lagrange's equations (described in Section 2.2.2) for the derivation of the equations of motion of a vibrating system, there is a third option, referred to as the influence coefficient method. There are two types of influence coefficient in common use, both of which define the static elastic property of a system: the stiffness influence coefficient, k_{ij}, which is the force acting on mass j (or location j) due to a unit static displacement at mass i (with all other masses stationary); and the flexibility influence coefficient, d_{ij}, which is the static displacement at mass, i, due to a unit force at mass, j, with no other forces acting. Only the stiffness influence coefficients, k_{ij}, will be considered here.

Consider the two-degree-of-freedom system shown in Figure 2.5. The force acting on m_1 due to a unit displacement at m_1 is $k_{11} = k_1 + k_2$. The force acting on m_2 due to a unit displacement of m_1 is $k_{12} = -k_2$. It can be seen easily that the force acting on m_1 due to a unit displacement at m_2 is $k_{21} = -k_2$. Thus, $k_{12} = k_{21}$, which is Maxwell's theory of reciprocity that can be stated in general as: the force produced at any location, j, due to a unit displacement at location, i, in a system is the same as the force produced at location i due to a unit displacement at location j.

Example 2.5

Use the influence coefficient method to derive the equations of motion for the two-degree-of-freedom system shown in Figure 2.5.

Solution 2.5

It has been shown that the equation of motion for a multi-degree-of-freedom, undamped system can be written as:

$$\boldsymbol{M}\ddot{\boldsymbol{q}} + \boldsymbol{K}\boldsymbol{q} = 0 \tag{a}$$

For the two-degree-of-freedom system illustrated in Figure 2.5:

$$\begin{bmatrix} m_{11} & m_{12} \\ m_{21} & m_{22} \end{bmatrix} \begin{bmatrix} \ddot{q}_1 \\ \ddot{q}_2 \end{bmatrix} + \begin{bmatrix} k_{11} & k_{12} \\ k_{21} & k_{22} \end{bmatrix} \begin{bmatrix} q_1 \\ q_2 \end{bmatrix} = 0 \tag{b}$$

In this case, the generalised coordinates used are x_1 and x_2; thus, $m_{12} = m_{21} = 0$ and $m_{11} = m_1$, $m_{22} = m_2$ and $x_1 = \hat{x}_1 e^{j\omega t}$ and $x_2 = \hat{x}_2 e^{j\omega t}$. Thus, Equation (b) becomes:

$$-\omega^2 \begin{bmatrix} m_1 & 0 \\ 0 & m_2 \end{bmatrix} \begin{bmatrix} \hat{x}_1 \\ \hat{x}_2 \end{bmatrix} + \begin{bmatrix} k_{11} & k_{12} \\ k_{21} & k_{22} \end{bmatrix} \begin{bmatrix} \hat{x}_1 \\ \hat{x}_2 \end{bmatrix} = 0 \tag{c}$$

By inspection of Figure 2.5, the influence coefficients are:

$$\begin{cases} k_{11} = k_1 + k_2 \\ k_{22} = k_2 + k_3 \\ k_{12} = k_{21} = -k_2 \end{cases} \tag{d}$$

Thus, the equation of motion becomes:

$$\begin{bmatrix} -m_1\omega^2 & 0 \\ 0 & -m_2\omega^2 \end{bmatrix} \begin{bmatrix} x_1 \\ x_2 \end{bmatrix} + \begin{bmatrix} k_1 + k_2 & -k_2 \\ -k_2 & k_2 + k_3 \end{bmatrix} \begin{bmatrix} x_1 \\ x_2 \end{bmatrix} = 0 \tag{e}$$

2.3   Vibration of Continuous Systems

Vibrating structures are generally characterised by the propagation of waves, although if analyses of the motion are undertaken by use of the discretisation techniques discussed in Section 2.2, this may not always be obvious. The motion of simple structures such as beams, plates and cylinders can be analysed from first principles using wave analysis in much the same way as was done for fluid media in Section 2.2.1. In structures, there will be two wave types in addition to the longitudinal waves present in fluid media and analysed in Section 2.2.1. These are bending waves and shear (or torsional) waves, and all of these wave types must be considered when estimating vibration levels in and sound radiation from vibrating structures. Even though shear and longitudinal waves do not significantly contribute directly to sound radiation, the phenomenon of energy conversion from one wave type to another at structural discontinuities means that these wave types must be taken into account if structural sound radiation is to be estimated accurately for complex structures.

The motion of simple beams, plates and cylinders, analysed using a wave approach, is the subject of Sections 3.2, 3.3 and 3.4. More complex structures must be analysed by dividing them into discrete elements and then using finite element analysis (see Howard and Cazzolato (2015) and Section 5.3).

In the remainder of this chapter, the equations of motion (or wave equations) and corresponding solutions for beams, plates and thin cylinders will be derived and summarised. More detailed treatments are available in books devoted entirely to the topic (Skudrzyk, 1968; Leissa, 1969, 1973; Soedel, 2004). The equations of motion are used in later chapters to estimate structural vibration levels and sound radiation resulting from known structural excitation sources.

2.3.1   Nomenclature and Sign Conventions

The literature on the vibrations of beams, plates and shells is characterised by a wide range of nomenclature and sign conventions. The sign conventions adopted in the past for moments, forces and rotations show little consistency between various authors and are rarely even discussed. Inconsistency in the use of sign conventions can cause problems in the derivation of the equations of motion, generally resulting in one or more sign errors in the final equation. In this book, the right-hand rule for axis labelling in the Cartesian coordinate system and positive external moments and rotations will be followed consistently. This convention is illustrated in Figure 2.6.

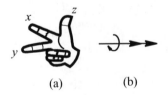

(a)   (b)

FIGURE 2.6 Sign conventions: (a) right-hand axis labelling convention for the Cartesian coordinate system; (b) general positive rotation convention.

Positive external forces and displacements are in the positive directions of the corresponding axes. Thus, in this book, the sign convention adopted for displacements, shear forces, bending moments and twisting moments follows that of Leissa (1969, 1973) for plates and shells, and the convention adopted for beams is the same as that used by Fahy and Gardonio (2007); Cremer et al. (1973). Displacements and applied forces are generally positive upwards and angular deflections and applied moments are positive in the counter-clockwise direction when looking along an axis towards the origin.

For plates and shells, the quantities Q, N and M represent forces and moments per unit width, whereas in beams they simply represent forces and moments. It should be noted that the final equations of motion for each type of structure are independent of the sign conventions used in their derivation. The nomenclature used here is identical to that used by Leissa (1969) except that here, in-plane displacements are denoted by ξ_x, ξ_y and ξ_θ rather than by u and v to avoid confusion with the use of these symbols for acoustic particle velocity and acoustic volume velocity, respectively. Double subscripts are also used here to denote stresses and strains, and the same symbols with different subscripts are used to denote shear stresses and shear strains. For example, σ_{xy} is used to denote shear stress in the $x-y$ plane and σ_{xx} is used for normal stress in the x-direction. Leissa (1969) used τ_{xy} and σ_x to denote these same quantities. Use of the same symbols for normal and shear stresses is preferred, as it allows expressions relating stresses and strains to be more easily generalised. As it will be useful in the derivation of the equations of motion of beams, plates and shells, the general three-dimensional relationship between stresses and strains will now be discussed. It is derived using the generalised three-dimensional Hooke's law (Cremer et al., 1973; Heckl, 1990; Pavic, 1988) and may be written as follows:

$$\sigma_{ik} = \frac{E}{1+\nu} \left[\frac{\nu}{1+2\nu}(\epsilon_{11} + \epsilon_{22} + \epsilon_{33})\delta_{ik} + \epsilon_{ik} \right] \tag{2.47}$$

where $i = 1$, 2, 3 and $k = 1$, 2, 3. For normal stresses $i = k$ and for shear stresses $i \neq k$. The strains are defined as:

$$\epsilon_{ik} = \frac{1}{\delta_{ik}+1} \left[\frac{\partial \xi_i}{\partial k} + \frac{\partial \xi_k}{\partial i} \right] \tag{2.48}$$

and the coefficient δ_{ik} is defined as:

$$\delta_{ik} = \begin{cases} 1; & \text{if } i = k \\ 0; & \text{if } i \neq k \end{cases} \tag{2.49}$$

For the Cartesian coordinate system, $x = 1$, $y = 2$, $z = 3$. For the cylindrical coordinate system $r = 1$, $a\vartheta = 2$, $x = 3$ where a is the radius at which the stresses and strains are evaluated, and ϑ is the cylindrical angular coordinate.

In the remainder of this chapter, the nomenclature and sign conventions used are illustrated at the beginning of each major section and will remain unchanged throughout the remainder of the book. Here, angular rotations are given a subscript to represent the axis about which the rotation is made, that is, θ_x means rotation about the x-axis. This is consistent with generally accepted elasticity conventions. However, in Leissa's work (Leissa, 1969, 1973), the subscript refers to the axis normal to the plane which is rotating. Thus, θ_x means rotation of a plane perpendicular to the x-axis (which is the same as the convention adopted in this book) or rotation of a normal perpendicular to a surface parallel with the x-axis. In both cases, angular displacement is positive in the counter-clockwise direction when viewed along the axis of rotation towards the origin of the coordinate system (right-hand rule — see Figure 2.6). The convention used here is illustrated in Figure 2.7.

The convention used here to define internal moments is consistent with that used by Leissa and is the same for beams, plates and shells. The subscript on a moment refers to a moment acting to rotate the plane perpendicular to the axis denoted by the subscript. In a beam, for example, M_x could refer to a moment about either the y- or z-axes, depending upon the type of bending wave it is characterising. Similarly, M_y could refer to a moment about either the z- or x-axes. However, in a plate or a shell, M_x refers only to a moment about the y-axis or θ-axis, respectively, and not about the axis normal to the surface. A double subscript on the moment refers to a twisting moment about the axis denoted by the first subscript, acting to twist the plane perpendicular to the axis defined by the second subscript. Thus, M_{yx} represents the twisting of a plane perpendicular to the x-axis about the y-axis.

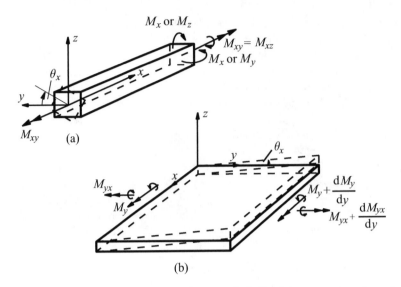

FIGURE 2.7 Sign conventions for beams, plates and shells: (a) moment and angle convention used for beams; (b) moment and angle convention used for plates and shells.

The convention used for externally applied moments is different to that used for internal moments, and again this is consistent with Leissa (1969, 1973). For externally applied moments, the first subscript is e to denote an external moment and the second subscript denotes the axis about which the moment is acting.

Thus, M_{ex} refers to an external moment acting about the x-axis. For beams, M_{ex} has the dimensions of moment per unit length and for plates and shells it has the dimensions of moment per unit area. Similarly, for externally applied forces, q, the dimensions are force per unit length for beams and force per unit area for plates and shells. For applied point forces and line or point moments, it is necessary to use the Dirac delta function to express them in terms of force or moment per unit length or area, to allow them to be used in the equations of motion.

2.3.2 Damping

Damping may be included in any of the following analyses by replacing Young's modulus, E, with the complex modulus, E', given by:

$$E' = E(1 + j\eta) \tag{2.50}$$

and the shear modulus, G, with the complex shear modulus, G', given by:

$$G' = G(1 + j\eta) \tag{2.51}$$

where η is the structural loss factor. Unfortunately, the complex elastic model is not strictly valid in the time dependent forms of the equations of motion, as it can lead to non-causal solutions. However, it is valid if restricted to steady state and simple harmonic (or multiple harmonic) vibration.

When the complex elastic modulus is used, it results in complex wave speeds and complex wavenumbers, as these are calculated using the elastic modulus. For longitudinal waves and torsional (or shear) waves, for which the wave speeds are proportional to $\sqrt{E}$ or $\sqrt{G}$, respectively, the wave speeds and wavenumbers, for small η, are:

$$
\begin{cases}
c'_L \approx c_L \left(1 + \dfrac{\mathrm{j}\eta}{2}\right) \\[2mm]
c'_s \approx c_s \left(1 + \dfrac{\mathrm{j}\eta}{2}\right) \\[2mm]
k'_L \approx k_L \left(1 - \dfrac{\mathrm{j}\eta}{2}\right) \\[2mm]
k'_s \approx k_s \left(1 - \dfrac{\mathrm{j}\eta}{2}\right)
\end{cases}
\tag{2.52}
$$

where $(1 + \mathrm{j}\eta/2) \approx \sqrt{(1 + \mathrm{j}\eta)}$ and $(1 - \mathrm{j}\eta/2) \approx 1/\sqrt{(1 + \mathrm{j}\eta)}$ for small η.

For bending waves, where the wave speed is proportional to $\sqrt[4]{E}$, the bending wavespeed and wavenumber, for small η, are:

$$
\begin{cases}
c'_b \approx c_b \left(1 + \dfrac{\mathrm{j}\eta}{4}\right) \\[2mm]
k'_b \approx k_b \left(1 - \dfrac{\mathrm{j}\eta}{4}\right)
\end{cases}
\tag{2.53}
$$

2.3.3 Waves in Beams

By definition, beams are long in comparison with their width and depth and are sufficiently thin that the cross-sectional dimensions are only a small fraction (less than 10%) of a wavelength at the frequency of interest. Beams that do not satisfy the latter criterion must be analysed by making a correction for the lateral inertia of the section, as will be discussed later.

In a simple, rectangular-section beam it is possible for four different waves to coexist: one longitudinal, one torsional and two bending. Here, the wave equations that describe each wave type of interest will be derived.

The coordinates, x, y and z, displacements, w_z, w_y and ξ_x and angular rotation, θ, of a small beam segment of length, δx, in the axial direction are shown in Figure 2.8.

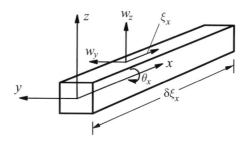

FIGURE 2.8 Coordinates and displacements for wave motion in a beam.

2.3.3.1 Longitudinal Waves

Longitudinal waves in solids that extend many wavelengths in all directions are very similar to acoustic waves in fluid media. However, when the solid medium is a bar (often referred to as a beam) or thin plate and only extends many wavelengths in one (or two) directions, pure longitudinal wave motion cannot occur and the term 'quasi-longitudinal' is used. This is because the lateral surfaces of the beam or the top and bottom surfaces of the plate are free from constraints, allowing the presence of longitudinal stress to produce lateral strains due to the Poisson contraction phenomenon.

The ratio of longitudinal stress to longitudinal strain in a beam is, by definition, equal to Young's modulus, E. Consider a segment of a beam, as shown in Figure 2.9. The x-coordinate is along the axis of the beam and the cross-sectional area is S.

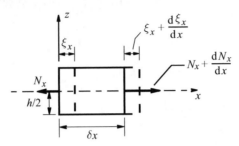

FIGURE 2.9 Axial beam element of infinitely short length, δx.

The strain that the element in Figure 2.9 undergoes in the x-direction along the axis of the beam is:

$$\epsilon_{xx} = \frac{\partial \xi_x}{\partial x} \tag{2.54}$$

By definition, the stress necessary to cause this strain is:

$$\sigma_{xx} = E\epsilon_{xx} = E\frac{\partial \xi_x}{\partial x} \tag{2.55}$$

The force acting in the x-direction is:

$$N_x = b \int_{-h/2}^{h/2} \sigma_{xx} \, dz = bh\sigma_{xx} = S\sigma_{xx} \tag{2.56}$$

where S is the cross-sectional area of the beam of thickness h and width b.

The equation of motion for the element in Figure 2.9 is obtained by equating the force acting on the element with the mass of the element multiplied by its acceleration as follows:

$$(\rho_m S\delta x)\frac{\partial^2 \xi_x}{\partial t^2} = \left[\sigma_{xx} + \frac{\partial \sigma_{xx}}{\partial x}\delta x - \sigma_{xx}\right]S = \frac{\partial \sigma_{xx}}{\partial x}\delta x S \tag{2.57}$$

Combining Equations (2.55) and (2.57) results in the wave equation for longitudinal waves propagating in a beam:

$$\frac{\partial^2 \xi_x}{\partial x^2} = \frac{\rho_m}{E}\frac{\partial^2 \xi_x}{\partial t^2} \tag{2.58}$$

This is analogous to the one-dimensional wave equation in an acoustic medium given in Section 2.2.1.

For a solid beam or bar, the longitudinal wave speed (phase speed) is:

$$c_L = (E/\rho_m)^{1/2} \tag{2.59}$$

As this expression is independent of frequency, longitudinal waves in a beam may be described as non-dispersive, a property also characterising longitudinal waves in any other structure or medium. The longitudinal wavenumber, k_L, at frequency, ω, is:

$$k_L = \omega/c_L = \sqrt{\frac{\rho_m \omega^2}{E}} \tag{2.60}$$

The equation for longitudinal waves in a two dimensional solid such as a thin plate is obtained from Equation (2.58) by replacing E with $E(1 - \nu^2)$ (Cremer et al., 1973). Similarly for waves in a three-dimensional solid, E in Equation (2.58) is replaced by $E(1-\nu)/(1+\nu)(1-2\nu)$, where ν is Poisson's ratio for the material.

Equation (2.58) is based on assumptions which are accurate, provided that the cross-sectional dimensions of the beam are less than one-tenth of a wavelength at the frequency of interest. The assumptions are:

1. The lateral motion due to Poisson's contraction has no effect on the kinetic energy of the beam.
2. The axial displacement is independent of the location on the beam cross section (that is, plane surfaces remain plane).

At higher frequencies when these assumptions are no longer accurate, it is necessary to add correction terms to the equation of motion to account for them (lateral inertia and lateral shear corrections). The most important correction is that of lateral inertia and this can be accounted for by increasing the density, ρ_m, of the beam in the equation of motion by the factor (Skudrzyk, 1968):

$$1 + \frac{\Delta\rho_m}{\rho_m} = 1 + \frac{1}{2}\nu^2 k^2 r^2 \tag{2.61}$$

where ν is Poisson's ratio, r the effective radius of the beam (the actual radius for a circular section beam) and k is the wavenumber $(= 2\pi/\lambda)$ of the longitudinal wave. This correction provides accurate results for beams with cross-sectional dimensions less than two tenths of a wavelength, as shown in Figure 2.10.

Beams with larger cross-sectional dimensions must be analysed using a lateral shear correction term as well as the lateral inertia term. This is discussed in detail elsewhere (Skudrzyk, 1968).

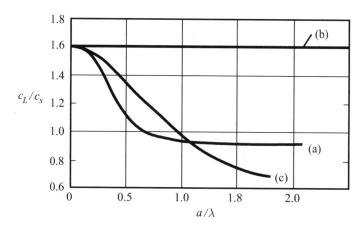

FIGURE 2.10 Longitudinal wave velocities in a beam: (a) exact solution from elasticity theory; (b) classical theory; (c) classical theory with correction for lateral inertia.

Solutions to the equation of motion, Equation (2.58), are dependent upon the boundary conditions of the beam. The general solution to Equation (2.58) is:

$$\xi_x(x,t) = (A_1 e^{-jkx} + A_2 e^{jkx})e^{j\omega t} = \hat{\xi}_x(x)e^{j\omega t} \tag{2.62}$$

where A_1, A_2 and $\hat{\xi}_x(x)$ are complex constants so that Equation (2.62) may also be written as:

$$\xi_x(x,t) = A_3 \cos(kx + \alpha)e^{j\omega t} = \hat{\xi}_x(x)e^{j\omega t} \tag{2.63}$$

where A_3 is a real constant representing the amplitude obtained from the vector summation of A_1 and A_2 and α represents the phase between vector A_1 and vector A_3 (see Figure 1.10). Only two boundary condition types are meaningful for longitudinal vibrations. They are fixed or free. At a fixed boundary, $\xi = 0$ and at a free boundary, $\dfrac{\partial \xi}{\partial x} = 0$. Thus, the solution to Equation (2.62) or (2.63) for clamped (fixed) ends is:

$$\xi_x(x,t) = \hat{\xi}_x(x)\mathrm{e}^{\mathrm{j}\omega t} = \left[A \sin \frac{n\pi x}{L} \right] \mathrm{e}^{\mathrm{j}\omega t} \; ; \qquad n = 1, 2, \ \ldots \tag{2.64}$$

where L is the length of the beam and n is the mode number.

For free ends, the solution is:

$$\xi_x(x,t) = \hat{\xi}_x(x)\mathrm{e}^{\mathrm{j}\omega t} = \left[A \cos \frac{n\pi x}{L} \right] \mathrm{e}^{\mathrm{j}\omega t} \; ; \qquad n = 1, 2, \ \ldots \tag{2.65}$$

Equations (2.64) and (2.65) are effectively the beam mode shape functions (if the constant, A, is set $= 1$).

2.3.3.2 Torsional Waves (Transverse Shear Waves)

Solids, unlike fluids, can resist shear deformation and thus are capable of transmitting shear type waves. In a beam, these waves appear as torsional waves, where the motion is characterised by a twisting of the cross section about the longitudinal axis of the beam.

Consider a section of infinitesimal length, δx, of a uniform beam, as shown in Figure 2.11.

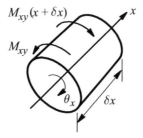

FIGURE 2.11 Beam element showing torques generated by shear waves, with positive angular displacement, θ_x, positive in the anti-clockwise direction when looking towards the origin.

When the beam is twisted, the torque changes with distance along it due to torsional vibration. The difference in torque between the two ends of an element will be equal to the polar moment of inertia multiplied by the angular acceleration. Thus:

$$M_{xy}(x + \delta x) - M_{xy}(x) = \rho_m J \delta x \frac{\partial^2 \theta_x}{\mathrm{d}t^2} \tag{2.66}$$

where M_{xy} is the twisting moment (or torque) acting on the beam, J is the polar second moment of area of the beam cross section, ρ_m is the density of the beam material and θ_x is the angle of rotation about the longitudinal x-axis. It is also clear that:

$$M_{xy}(x + \delta x) - M_{xy}(x) = \frac{\partial M_{xy}}{\partial x} \delta x \tag{2.67}$$

The angular deflection of a shaft is related to the torque (twisting moment M_{xy}) by (see Figure 2.11):

$$M_{xy} = C \frac{\partial \theta_x}{\partial x} \tag{2.68}$$

where C is the torsional stiffness of the beam. Differentiating Equation (2.68) with respect to x gives:

$$\frac{\partial M_{xy}}{\partial x} = C\frac{\partial^2 \theta_x}{\partial x^2} \tag{2.69}$$

Equating (2.66) and (2.67), and substituting Equation (2.69) into the result gives:

$$C\frac{\partial^2 \theta_x}{\partial x^2}\delta x = \rho_m J\frac{\partial^2 \theta_x}{\partial t^2}\delta x \tag{2.70}$$

or:

$$\frac{\partial^2 \theta_x}{\partial x^2} = \frac{\rho_m J}{C}\frac{\partial^2 \theta_x}{\partial t^2} \tag{2.71}$$

which is the wave equation for torsional waves.

Values for the torsional stiffness, C, of rectangular-section bars are given in Table 2.1 for various ratios of the thickness, h, to width, b, where b is larger than h.

TABLE 2.1 Torsional stiffness of solid rectangular bars

$\dfrac{b}{h}$	$\dfrac{C}{Gh^3 b}$
1	0.141
1.2	0.166
1.5	0.196
2	0.230
3	0.263
4	0.281
5	0.291
10	0.312
20	0.333

From Equation (2.71) it can be seen that the phase speed of a torsional wave in the beam is:

$$c_s^2 = \frac{C}{\rho_m J} \tag{2.72}$$

For a circular-section beam, $C = GJ$ and Equation (2.72) becomes:

$$c_s^2 = G/\rho_m \tag{2.73}$$

where G is the shear modulus for the material. The same boundary conditions and solutions apply as for longitudinal waves, with the displacement, $\xi_x(x)$, replaced by the angular displacement, $\theta_x(x)$ and c_L replaced with c_s (see Equations (2.64) and (2.65)). The torsional wavenumber, k_s, at frequency, ω, is:

$$k_s = \omega/c_s = \sqrt{\frac{\rho_m \omega^2}{G}} \tag{2.74}$$

2.3.3.3   Bending Waves

Bending waves are by far the most important in terms of sound radiation, as they result in a displacement, normal to the surface of the beam, which couples well with any adjacent fluid.

Consider a beam subject to bending, as shown in Figure 2.12. The forces acting on a small segment, δx, are also shown in this figure, and the corresponding displacements and rotations are shown in Figure 2.13.

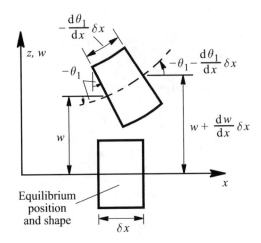

FIGURE 2.12 Forces on an element of a beam.

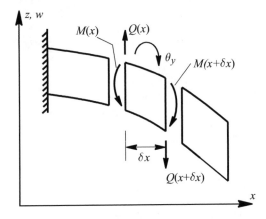

FIGURE 2.13 Displacements of a beam segment of length, δx.

To simplify the notation to follow, the displacement, w_z, in the z-direction will be denoted w. From Figure 2.13, it can be seen that the vertical displacement, w, and the angle of rotation of the element about the y-axis are related by:

$$\theta_y = \theta_1 + \theta_2 = -\frac{\partial w}{\partial x} \tag{2.75}$$

where θ_1 is the slope of the beam due to rotation of the section and θ_2 is the additional slope due to shear.

The resultant shear force acting on the small segment is:

$$-Q_x(x+\delta x) + Q_x(x) = -\frac{\partial Q_x(x)}{\partial x}\delta x \tag{2.76}$$

and the motion of the segment obeys Newton's second law; thus:

$$-\frac{\partial Q_x}{\partial x}\delta x = \rho_m S \delta x \frac{\partial^2 w}{\partial t^2} \tag{2.77}$$

where w is the vertical displacement of the beam at location x, S is the cross-sectional area of the beam and ρ_m is the density of the beam material. The total deflection, w, of the beam element consists of two components; w_1 due to the bending of the beam and w_2 due to the shearing of the beam. The shear motion changes the slope of the beam by an additional angle θ_2, which does not contribute to the rotation of the beam section resulting from the bending slope θ_1.

By examination of Figure 2.13, it can be seen that the resultant bending moment, which tends to rotate the element in a clockwise direction, is:

$$M_x(x + \delta x) - M_x(x) = \frac{\partial M_x}{\partial x} \delta x \tag{2.78}$$

The shear forces, $Q_x(x + \delta x)$ and $Q_x(x)$ exert an additional moment on the element given approximately by $Q_x(x)\delta x$. These two moments may be equated to the rotary inertia of the element using Newton's second law of motion. Thus:

$$\frac{\partial M_x}{\partial x} \delta x + Q_x(x)\delta x = I_m \ddot{\theta}_1 = \rho_m J \delta x \ddot{\theta}_1 \tag{2.79}$$

where I_m is the mass moment of inertia of the element, δx, about the y-axis and J is the second moment of area of the cross section about the transverse axis in the neutral plane. For a circular-section solid rod of radius, r, and mass, m, $I_m = mr^2/2$ and $J = \pi r^4/4$.

Equation (2.79) simplifies to:

$$\frac{\partial M_x}{\partial x} + Q_x = \rho_m J \ddot{\theta}_1 \tag{2.80}$$

Classical beam theory (also known as Euler-Bernoulli theory) would assume that $\ddot{\theta}_1 = 0$. Indeed, for the static case this is also true and the shear force is related to the bending moment by:

$$Q_x = -\frac{\partial M_x}{\partial x} \tag{2.81}$$

The relationship between the bending moment, M, and the deflection, w_1, which is produced as a result, will now be developed. Consider the same element as shown in Figure 2.14. Now consider a segment, δz, of this beam element. The strain in this segment due to bending deformation is:

$$\epsilon = 2z \frac{\partial \theta_1}{\partial x} \frac{\delta x}{2} \frac{1}{\delta x} \tag{2.82}$$

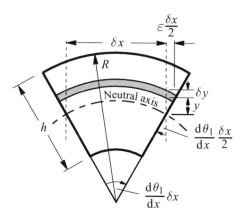

FIGURE 2.14 Beam element undergoing pure bending.

Using Equation (2.75) the following is obtained:

$$\epsilon = -z \frac{\partial^2 w_1}{\partial x^2} \tag{2.83}$$

Assuming that the relationship between stress and strain in the shaded element of Figure 2.14 is the same as that for a longitudinal beam, the following is obtained:

$$\sigma = E\epsilon \tag{2.84}$$

where E is Young's modulus of elasticity.

The bending moment generated by the shaded volume of Figure 2.14 is:

$$M_x = b \int_{-h/2}^{h/2} \sigma z \, \mathrm{d}z \tag{2.85}$$

where b is the width of the beam and h is its thickness.

Substituting Equations (2.83) and (2.84) into Equation (2.85):

$$M_x = -b \int_{-h/2}^{h/2} z E \frac{\partial^2 w_1}{\partial x^2} z \, \mathrm{d}z = -bE \frac{\partial^2 w_1}{\partial x^2} \int_{-h/2}^{h/2} z^2 \, \mathrm{d}z \tag{2.86}$$

Thus:

$$M_x = -EJ \frac{\partial^2 w_1}{\partial x^2} \tag{2.87}$$

and from Equation (2.81), the shear force may be written as:

$$Q_x = -\frac{\partial M_x}{\partial x} = EJ \frac{\partial^3 w_1}{\partial x^3} \tag{2.88}$$

However, the actual deflection is greater than w_1 due to the shearing of the beam. Timoshenko (1921) postulated that these additional shearing forces produced an additional deflection w_2 of the beam, given by:

$$\frac{\partial w_2}{\partial x} = -\frac{Q_x}{\gamma SG} \tag{2.89}$$

where G is the shear modulus or modulus of rigidity of the beam material, S is the beam cross-sectional area and γ is known as the shape factor (or shear coefficient), which is dependent on the shape of the beam cross section. Shape factors for various beam sections are listed in Table 2.2 (Cowper, 1966). More accurate (and more complex) expressions are provided by Hutchinson (2001). Second moments of area of the same beam sections are also provided in Table 2.2. Second moments of area, J_x, about the x-axis for other beam sections can be calculated using the parallel-axis theorem in which the section is broken into its constituent simple cross-sectional shapes, such as rectangles, circles and triangles. Using the parallel-axis theorem, the second moment of area can be calculated for any section such as an I-beam, using:

$$J = \sum_{i=1}^{N} J_i + S_i d_i^2 \tag{2.90}$$

where N is the number of simple shapes (rectangles, circles and triangles) making up the total section (= 3 for an I-beam), S_i is the cross-sectional area of the ith simple shape and d_i is the minimum distance to the neutral axis of the entire section, of the line passing through the centroid of the simple shape and parallel to the neutral axis of the entire section. An example of finding the neutral axis and second moment of area of the I-beam illustrated in Figure 2.15 will now be demonstrated. Referring to Figure 2.15, the I-beam may be divided into three simple rectangular shapes. The given dimensions are h_1, h_2, h_3, b_1, b_2 and b_3.

TABLE 2.2 Shape factors, γ, and second moments of area, J_x, for beam bending vibration

Cross-sectional shape	Shape factor, γ (Cowper, 1966)	Second moment of area, J_x
Solid circle	$\dfrac{6(1+\nu)}{7+6\nu}$	$\dfrac{\pi r^4}{4}$
Hollow circle, $m = r_i/r_o$	$\dfrac{6(1+\nu)(1+m^2)^2}{(7+6\nu)(1+m^2)^2 + (20+12\nu)m^2}$	$\dfrac{\pi}{4}\left(r_o^4 - r_i^4\right)$
Thin-wall circular tube	$\dfrac{2(1+\nu)}{4+3\nu}$	$\pi r^3 t$ $r \approx r_o \approx r_i$ and $t \ll r$
Solid rectangle	$\dfrac{10(1+\nu)}{12+11\nu}$	$\dfrac{bh^3}{12}$
Thin-wall rectangular tube	$\dfrac{20(1+\nu)}{48+39\nu}$	$\dfrac{th^3}{6} + \dfrac{tbh^2}{2}$ $h \approx h_i \approx h_o$ $b \approx b_i \approx b_o$ $t \ll b$ and $t \ll h$
Triangle about centroid axis	—	$\dfrac{bh^3}{36}$
Triangle about base	—	$\dfrac{bh^3}{12}$
I-beam	varies between 0.42 and 0.5	See text for calculation procedure

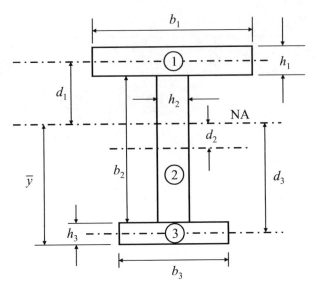

FIGURE 2.15 I-beam section.

Thus:

$$S_1 = b_1 h_1; \qquad S_2 = b_2 h_2; \qquad S_3 = b_3 h_3 \tag{2.91}$$

$$y_1 = h_3 + h_2 + h_1/2; \qquad y_2 = h_3 + h_2/2; \qquad y_3 = h_3/2 \tag{2.92}$$

The height of the centroid of the entire I-beam section above its base is:

$$\bar{y} = \frac{\sum\limits_{i=1}^{3} y_i S_i}{\sum\limits_{i=1}^{3} S_i} \tag{2.93}$$

The distances to be used in the parallel-axis equation (Equation (2.90)) are:

$$d_1 = h_3 + b_2 + h_1/2 - \bar{y}; \qquad d_2 = h_3 + b_2/2 - \bar{y}; \qquad d_3 = \bar{y} - h_3/2 \tag{2.94}$$

$$J_1 = b_1 h_1^3/12; \qquad J_2 = h_2 b_2^3/12; \qquad J_3 = b_3 h_3^3/12 \tag{2.95}$$

The additional deflection of Equation (2.89) increases the slope of the beam by $\theta_2 = -\dfrac{\partial w_2}{\partial x}$, but does not cause a rotation of the element.

Thus, in summary, the following equations describe the motion of the beam:

$$\frac{\partial Q_x}{\partial x} = -\rho_m S \frac{\partial^2 w}{\partial t^2} \tag{2.96}$$

where ρ_m is the density of the beam material and S is the beam cross-sectional area.

$$\frac{\partial M_x}{\partial x} + Q_x = \rho_m J \ddot{\theta}_1 \tag{2.97}$$

$$M_x = -EJ \frac{\partial^2 w_1}{\partial x^2} \tag{2.98}$$

$$\frac{\partial w_2}{\partial x} = -\frac{Q_x}{\gamma S G} \tag{2.99}$$

$$w = w_1 + w_2 \tag{2.100}$$

$$\theta_1 = -\frac{\partial w_1}{\partial x} \tag{2.101}$$

The quantities, w_1, w_2, θ_1, Q_x and M_x must be eliminated to give an equation in w, the total displacement. This is done by first using Equations (2.98) to (2.100) to write M_x in terms of Q_x and w. This result for M_x is substituted into Equation (2.97) and θ_1 is eliminated from that equation using Equations (2.99), (2.100) and (2.101). Q_x is then eliminated from this result using Equation (2.96). When this is done, the following is obtained:

$$EJ\frac{\partial^4 w}{\partial x^4} + \rho_m S\frac{\partial^2 w}{\partial t^2} - \left(\rho_m J + \frac{\rho_m J E}{\gamma G}\right)\frac{\partial^4 w}{\partial x^2 \partial t^2} + \frac{\rho_m^2 J}{\gamma G}\frac{\partial^4 w}{\partial t^4} = 0 \tag{2.102}$$

which is the beam unforced equation of motion for bending waves.

The first two terms in Equation (2.45) represent the classical beam equation for bending waves. In this case (with the remaining terms ignored), the bending wave speed is:

$$c_b^4 = \omega^2 EJ/\rho_m S \tag{2.103}$$

The third term in Equation (2.102) is the correction for transverse shear and the fourth term is the correction for rotary inertia. Equation 2.103, although not exact (as it assumes that the effects of rotary inertia and shear can be treated separately), gives excellent results, even for very thick beams. On the other hand, the classical wave equation (first two terms in Equation (2.102)) results in significant errors if the ratio of the wavelength to the equivalent section radius exceeds 0.1. That is, good results may be expected with classical beam theory provided that the bending wavelength is 10 times greater than the equivalent section radius. The bending wave speed in the beam is related to the bending wavenumber, k_b, by $c_b = \omega/k_b$ (see Figure 2.16).

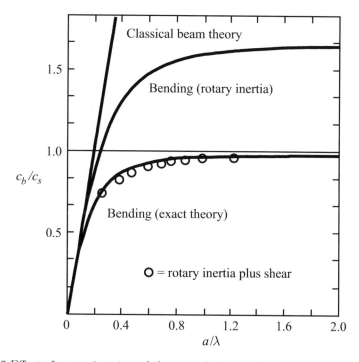

FIGURE 2.16 Effect of rotary inertia and shear on the calculated bending wave speed in a beam.

If an external force, of $q(x)$ force units per unit length, is applied to the beam in the positive z-direction, Equation (2.96) becomes:

$$-\frac{\partial Q_x}{\partial x} + q(x) = \rho_m S \frac{\partial^2 w}{\partial t^2} \qquad (2.104)$$

and Equation (2.102) becomes:

$$EJ\frac{\partial^4 w}{\partial x^4} + \rho_m S \frac{\partial^2 w}{\partial t^2} - \left(\rho_m J + \frac{\rho_m JE}{\gamma G}\right) \frac{\partial^4 w}{\partial x^2 \partial t^2} + \frac{\rho_m^2 J}{\gamma G} \frac{\partial^4 w}{\partial t^4}$$
$$= -\frac{JE}{\gamma SG} \frac{\partial^2 q(x)}{\partial x^2} + q(x) + \frac{\rho_m J}{\gamma SG} \frac{\partial^2 q(x)}{\partial t^2} \qquad (2.105)$$

If classical beam theory were used, then only the first two terms on the left-hand side of the equation and only the second term on the right would remain.

Note that a point force, F, applied to the beam at $x = x_0$ may be represented by:

$$q(x) = F\delta(x - x_0) \qquad (2.106)$$

where $\delta(\ )$ is the Dirac delta function.

If we apply an external moment to the beam (of $M_e(x)$ units of moment per unit length) Equation (2.97) becomes:

$$M_e + \frac{\partial M}{\partial x} + Q_x = \rho_m J \ddot{\theta}_1 \qquad (2.107)$$

and the equation of motion is:

$$EJ\frac{\partial^4 w}{\partial x^4} + \rho_m S \frac{\partial^2 w}{\partial t^2} - \left(\rho_m J + \frac{\rho_m JE}{\gamma G}\right) \frac{\partial^4 w}{\partial x^2 \partial t^2} + \frac{\rho_m^2 J}{\gamma G} \frac{\partial^4 w}{\partial t^4} = \frac{\partial M_e(x)}{\partial x} \qquad (2.108)$$

A line moment, M_L, applied across the beam at $x = x_0$ may be represented by:

$$M_e(x) = M_L \delta(x - x_0) \qquad (2.109)$$

For a beam which is very thin compared to its width, the quantity, E, in the preceding equations must be replaced by $E/(1 - \nu^2)$, to give the one-dimensional plate equation.

It is interesting to note that if the y- and z-axes were interchanged, so that the beam displacement along the y-axis were of interest, the right-hand sign convention would change Equation (2.75) to:

$$\theta_z = \frac{\partial w}{\partial x} \qquad (2.110)$$

Equation (2.97) to:

$$\frac{\partial M_x}{\partial x} + Q_x = -\rho_m J \ddot{\theta}_1 \qquad (2.111)$$

Equation (2.101) to:

$$\theta_1 = \frac{\partial w_1}{\partial x} \qquad (2.112)$$

and Equation (2.107) to:

$$M_e - \frac{\partial M_x}{\partial x} - Q_x = \rho_m J \ddot{\theta}_1 \qquad (2.113)$$

where it is assumed that the conventions for positive internal shear force and bending moment remain the same.

Note that the equations of motion remain unchanged except for Equation (2.108) involving an externally applied moment M_e, where the positive sign on the right-hand side of the equation is now replaced with a negative sign, and in all equations, the second moment of area, J, will be about the z-axis rather than the y-axis.

The bending wave speed, c_b, in a beam is related to the bending wavenumber, k_b, by $c_b = \omega/k_b$. An expression for c_b, which includes the effects of shear and rotary inertia (Timoshenko beam theory), can be determined by substituting the expression for a simple harmonic progressive wave:

$$w(x,t) = A e^{j(\omega t - k_b x)} = \hat{w}(x) e^{j\omega t} \tag{2.114}$$

into Equation (2.102) to give:

$$\alpha^4 k_b^4 = \omega^2 + \beta k_b^2 \omega^2 - \delta \omega^4 \tag{2.115}$$

where:

$$\alpha^4 = \frac{EJ}{\rho_m S}; \qquad \beta = \left(\frac{J}{S} + \frac{JE}{\gamma SG} \right); \qquad \text{and} \qquad \delta = \frac{\rho_m J}{\gamma SG} \tag{2.116}$$

As Equation (2.115) is invalid when the corrections for shear and rotary inertia are large, the quantity, k_b^2 on the right-hand side may be replaced with ω/α^2 (its zero order approximation). Equation (2.115) then can be written as:

$$k_b^4 = \frac{\omega^2 + \beta \omega^3/\alpha^2 - \delta \omega^4}{\alpha^4} \tag{2.117}$$

Hence:

$$k_b = \pm j \Delta \text{ and } \pm \Delta \tag{2.118}$$

where:

$$\Delta = (\omega^2 + \beta \omega^3/\alpha^2 - \delta \omega^4)^{1/4}/\alpha \tag{2.119}$$

Thus, the wave speed, c_b, for bending waves, with corrections for shear and rotary inertia included, is:

$$c_b = \frac{\omega}{k_b} = \frac{\alpha\, \omega^{1/2}}{\left(1 + \beta \dfrac{\omega}{\alpha^2} - \delta \omega^2 \right)^{1/4}} \tag{2.120}$$

The complete solution for the unforced equation of motion for bending waves in a beam (Equation (2.102)), including the effects of shear and rotary inertia, is, for frequencies satisfying $\omega < \omega_c$:

$$w(x,t) = \left[A_1 e^{-jax/L} + A_2 e^{jax/L} + A_3 e^{-bx/L} + A_4\, e^{bx/L} \right] e^{j\omega t} = \hat{w}(x) e^{j\omega t} \tag{2.121}$$

or, in terms of transcendental and hyperbolic functions:

$$w(x,t) = \left[A \cos(ax/L) + B \sin(ax/L) + C \cosh(bx/L) + D \sinh(bx/L) \right] e^{j\omega t} = \hat{w}(x) e^{j\omega t} \tag{2.122}$$

where:

$$\omega_c = \sqrt{\frac{1}{\delta}} = \sqrt{\frac{\gamma GS}{\rho_m J}} \tag{2.123}$$

When $\omega \geq \omega_c$, Equation (2.122) becomes:

$$w(x,t) = \left[\hat{A} \cos(ax/L) + \hat{B} \sin(ax/L) + \hat{C} \cos{(bx/L)} + \hat{D} \sin{(bx/L)} \right] e^{j\omega t} = \hat{w}(x) e^{j\omega t} \tag{2.124}$$

In Equations (2.121), (2.122) and (2.124), the constants, A_1 to A_4, A to D and $\hat{A}$ to $\hat{D}$ may all be complex, depending on the boundary conditions.

When $\omega < \omega_c$, the coefficients a and b may be evaluated using (Han et al., 1999):

$$a = \sqrt{B_1 + B_2 + \sqrt{(B_1 - B_2)^2 + B_3}} \tag{2.125}$$

and:

$$b = \sqrt{-B_1 - B_2 + \sqrt{(B_1 - B_2)^2 + B_3}} \qquad (2.126)$$

When $\omega \geq \omega_c$, the coefficient a is unchanged, but the coefficient b changes to $\hat{b}$, where $\hat{b} = jb$, so that (Han et al., 1999):

$$\hat{b} = \sqrt{B_1 + B_2 - \sqrt{(B_1 - B_2)^2 + B_3}} \qquad (2.127)$$

where:

$$B_1 = \frac{\rho_m L^2 \omega^2}{2E}; \quad B_2 = \frac{2B_1(1+\nu)}{\gamma} = B_1 \varepsilon^2; \quad B_3 = \frac{\rho_m L^4 \omega^2 S}{EJ} \qquad (2.128)$$

where:

$$\varepsilon^2 = 2(1+\nu)/\gamma \qquad (2.129)$$

After considerable algebraic manipulation, the preceding equations may be used to express the coefficient a in terms of the coefficient b, and this relationship can then be used to solve the characteristic frequency equations to obtain resonance frequencies for various boundary conditions at the ends of the beam. Thus, from Equations (2.125) to (2.128), for $\omega < \omega_c$:

$$\frac{(\varepsilon^2 b^2 + a^2)(a^2 \varepsilon^2 + b^2)}{(a^2 - b^2)(1 + \varepsilon^2)} = s_r^2 \qquad (2.130)$$

For, $\omega \geq \omega_c$, b is replaced with $j\hat{b}$, to give:

$$\frac{(-\varepsilon^2 \hat{b}^2 + a^2)(a^2 \varepsilon^2 - \hat{b}^2)}{(a^2 + \hat{b}^2)(1 + \varepsilon^2)} = s_r^2 \qquad (2.131)$$

where the slenderness ratio, s_r, is related to the beam length, L, cross-sectional area, S, and second moment of area, J, as:

$$s_r = L\sqrt{S/J} \qquad (2.132)$$

and values for γ are given in Table 2.2 for various cross-sectional shapes.

With a expressed in terms of b for frequencies satisfying $\omega < \omega_c$ and in terms of $\hat{b}$ for frequencies satisfying $\omega \geq \omega_c$, characteristic frequency equations can be written in terms of one variable, b or $\hat{b}$. These are provided for various boundary conditions in Section 2.3.3.4. It is useful to note that the case of $\omega < \omega_c$ is the same as the case $a < a_c$, where:

$$a_c = s\sqrt{\left(\frac{\varepsilon^2 + 1}{\varepsilon^2}\right)} \qquad (2.133)$$

Boundary conditions to be used to solve Equations (2.121) and (2.122) for various beam end support conditions are summarised below.

$$\begin{cases} \text{Clamped}: & w(x,t) = 0 \text{ at } x = 0,\, L \\[2mm] & \dfrac{\partial w(x,t)}{\partial x} = 0 \text{ at } x = 0,\, L \end{cases} \qquad (2.134)$$

$$\begin{cases} \text{Simply} & w(x,t) = 0 \text{ at } x = 0,\, L \\[2mm] \text{supported}: & \dfrac{\partial^2 w(x,t)}{\partial x^2} = 0 \text{ at } x = 0,\, L \end{cases} \qquad (2.135)$$

$$\begin{cases} \text{Free}: & \dfrac{\partial^2 w(x,t)}{\partial x^2} = 0 \text{ at } x = 0,\, L \\[2mm] & \dfrac{\partial^3 w(x,t)}{\partial x^3} = 0 \text{ at } x = 0,\, L \end{cases} \qquad (2.136)$$

Of course, both ends of the beam do not have to have the same boundary conditions. Any combination of the preceding boundary conditions is possible.

If the effects of rotary inertia are ignored, but the effects of shear still included, the equation of motion is the same as Equation (2.122) and Equations (2.125) and (2.126) for a and b, respectively, are the same except that $B_1 = 0$. In this case, the parameter, b, can be written in terms of a as:

$$b = as_r \sqrt{\frac{1}{a^2 \varepsilon^2 + s_r^2}} \tag{2.137}$$

where s_r is defined in Equation (2.132). Note that unlike for the Timoshenko model (which includes both shear and rotary inertia), there is only one frequency range, as neither $\hat{b}$ nor ω_c exists.

At low frequencies, when the bending wavelength is much smaller than a beam cross-sectional dimension, shear and rotary inertia corrections can both be ignored. In this case, B_1 and B_2 of Equations (2.125) and (2.126) are equal to zero and the classical model (or Euler-Bernoulli model) is valid. The solutions to the equations of motion given in Equations (2.121) and (2.122) are still applicable, except that $a = b = k_b L$, so that:

$$\left(\frac{a}{L}\right)^4 = \left(\frac{b}{L}\right)^4 = k_b^4 = \frac{\omega^2 \rho_m S}{EJ}; \qquad c_b = \frac{\omega}{k_b} = \alpha \sqrt{\omega} \tag{2.138}$$

and the group wave speed for bending waves is:

$$c_g = \left[\frac{\partial \omega}{\partial k_b}\right]^{-1} = 2\omega^{1/2} \left[\frac{\rho_m S}{EJ}\right]^{-1/4} = 2c_b \tag{2.139}$$

Note that for a rectangular-section beam there will be two bending waves with displacements perpendicular to one another. The two equations describing the motion are identical to Equation (2.102), but each type of motion will be characterised by a different second moment of area, J. For motion in the z-direction, J_{yy} is used and vice versa. If damping is included by replacing E with $E(1 + j\eta)$ in the beam equation of motion, the solution may still be expressed in the form of Equation (2.121), but with k_b replaced by a complex bending wavenumber, k_b', defined for small η, as:

$$k_b' \approx k_b(1 - j\eta/4) \tag{2.140}$$

and the solution for a single travelling wave is written in terms of a real amplitude, A, as:

$$w(x,t) = Ae^{j(\omega t - k_b x + \beta)} e^{-k_b \eta x/4} = \hat{w}(x)e^{j\omega t} \tag{2.141}$$

showing that the travelling wave attenuates exponentially with increasing distance, x, from its source and where β is an arbitrary phase angle which may be set equal to 0.

2.3.3.4 Summary of Beam Resonance Frequency Formulae

Classical Beam Theory (Euler-Bernoulli Theory)
Equations for the resonance frequencies, mode shapes and corresponding bending wavenumbers, k_b, for classical beam theory are summarised in Table 2.3 and the characteristic equations and corresponding roots are provided in Table 2.4.

Classical Beam Theory Plus Effects of Shear
Shear effects are much larger than rotary inertia effects so in most cases it is sufficient to include shear effects only, thus avoiding the added complexity of including rotary inertia effects as well. Including the effects of shear is a complex process for all but a simply supported beam. For a

TABLE 2.3 Resonance frequencies, wavenumbers (k_b) and mode shape formulae for isotropic thin beams, where n is an integer, beginning at 1 for the first mode, where The characteristic equations and corresponding roots, a_n, are listed in Table 2.4

Wave type	Boundary conditions	Resonance frequency (Hz)	k_b at resonance	Mode shape
Longit-udinal	Clamped at both ends	$\dfrac{1}{2\pi}\sqrt{\dfrac{E}{\rho_m}}\left(\dfrac{n\pi}{L}\right)$	$\dfrac{n\pi}{L}$	$\sin\left(\dfrac{n\pi x}{L}\right)$
Longit-udinal	Free at both ends	$\dfrac{1}{2\pi}\sqrt{\dfrac{E}{\rho_m}}\left(\dfrac{n\pi}{L}\right)$	$\dfrac{n\pi}{L}$	$\cos\left(\dfrac{n\pi x}{L}\right)$
Torsional	Clamped at both ends	$\dfrac{1}{2\pi}\sqrt{\dfrac{G}{\rho_m}}\left(\dfrac{n\pi}{L}\right)$	$\dfrac{n\pi}{L}$	$\sin\left(\dfrac{n\pi x}{L}\right)$
Torsional	Free at both ends	$\dfrac{1}{2\pi}\sqrt{\dfrac{G}{\rho_m}}\left(\dfrac{n\pi}{L}\right)$	$\dfrac{n\pi}{L}$	$\cos\left(\dfrac{n\pi x}{L}\right)$
Bending	Simply supported at both ends	$\dfrac{1}{2\pi}\sqrt{\dfrac{EJ}{\rho_m S}}\left(\dfrac{n\pi}{L}\right)^2$	$\dfrac{n\pi}{L}$	$\sin\left(\dfrac{n\pi x}{L}\right)$
Bending	Clamped at both ends	$\dfrac{1}{2\pi}\sqrt{\dfrac{EJ}{\rho_m S}}\left(\dfrac{a_n}{L}\right)^2$	$\dfrac{a_n}{L}$	$\cosh(a_n x/L) - \cos(a_n x/L)$ $-b_n\left[\sinh(a_n x/L) - \sin(a_n x/L)\right]$ $b_n = \dfrac{\cosh a_n - \cos a_n}{\sinh a_n - \sin a_n}$
Bending	Free at both ends	$\dfrac{1}{2\pi}\sqrt{\dfrac{EJ}{\rho_m S}}\left(\dfrac{a_n}{L}\right)^2$	$\dfrac{a_n}{L}$	$\cosh(a_n x/L) + \cos(a_n x/L)$ $+b_n\left[\sinh(a_n x/L) + \sin(a_n x/L)\right]$ $b_n = \dfrac{\cosh a_n - \cos a_n}{\sinh a_n - \sin a_n}$
Bending	Cantilever clamped-free	$\dfrac{1}{2\pi}\sqrt{\dfrac{EJ}{\rho_m S}}\left(\dfrac{a_n}{L}\right)^2$	$\dfrac{a_n}{L}$	$\cosh(a_n x/L) - \cos(a_n x/L)$ $-b_n\left[\sinh(a_n x/L) - \sin(a_n x/L)\right]$ $b_n = \dfrac{\cosh a_n + \cos a_n}{\sinh a_n + \sin a_n}$
Bending	clamped-simply supported	$\dfrac{1}{2\pi}\sqrt{\dfrac{EJ}{\rho_m S}}\left(\dfrac{a_n}{L}\right)^2$	$\dfrac{a_n}{L}$	$\cos(a_n x/L) - \cosh(a_n x/L)$ $+b_n\left[\sinh(a_n x/L) - \sin(a_n x/L)\right]$ $b_n = \dfrac{\cos a_n + \cosh a_n}{\sin a_n + \sinh a_n}$

TABLE 2.4 Characteristic equations and corresponding first five roots, for various boundary conditions for bending vibration of isotropic thin beams, where n is the integer mode number

Boundary conditions	Characteristic equation	Characteristic equation roots, a_n				
		$n = 1$	$n = 2$	$n = 3$	$n = 4$	$n = 5$
Simply supported at both ends	$\sin(a_n) = 0$	π	2π	3π	4π	5π
Clamped at both ends	$\cos(a_n)\cosh(a_n) = 1$	4.730041	7.853205	10.995608	14.137165	17.279
Free at both ends	$\cos(a_n)\cosh(a_n) = 1$	4.730041	7.853205	10.995608	14.137165	17.279
Cantilever clamped-free	$\cos(a_n)\cosh(a_n) = -1$	1.875104	4.694091	7.854757	10.995541	14.137
clamped-simply supported	$\tan(a_n) - \tanh(a_n) = 0$	3.926602	7.068583	10.210176	13.351768	16.493

beam simply supported at both ends, the equation for the resonance frequency of the nth mode is (Han et al., 1999):

$$f_n = \frac{1}{2\pi}\left(\frac{n\pi}{L}\right)^2 \sqrt{\frac{EJ}{\rho_m S\left[1 + \left(\frac{n\pi}{L}\right)^2 \frac{EJ}{\gamma SG}\right]}} \tag{2.142}$$

The characteristic equations for various beam end conditions (boundary conditions) are given in Table 2.5. Using Equation (2.137), the characteristic equations can be solved for both a and b. Each pair of solutions, a_n and b_n, corresponds to a resonant vibration mode. The corresponding mode shape is obtained by substituting the solutions, a_n for a and b_n for b into Equation (2.122). The resonance frequency corresponding to a particular solution pair (a_n and b_n) is given by (Han et al., 1999):

$$f_n = \frac{1}{2\pi}\sqrt{a_n^2 - b_n^2}\sqrt{\frac{E\gamma}{2\rho_m L^2(1+\nu)}} \tag{2.143}$$

Classical Beam Theory Plus Effects of Shear and Rotary Inertia (Timoshenko Model)
The characteristic equations for the Timoshenko beam model (classical model plus effects of shear and rotary inertia) for various beam end conditions (boundary conditions) are given in Table 2.6 for the condition, $a < a_c$. The characteristic equations for the condition, $a \geq a_c$, are given in Table 2.7.

Using Equation (2.130), the characteristic equations for the specified boundary conditions for the condition, $a < a_c$, can be solved for both a_n and b_n (where n is an integer representing

TABLE 2.5 Characteristic equations for various boundary conditions for bending vibration of isotropic beams, including the effects of shear (Han et al., 1999)

Boundary conditions	Characteristic equation
Simply supported at both ends	$\sin(a_n)\sinh(b_n) = 0$
Clamped at both ends	$(b_n^6 - a_n^6)\sin(a_n)\sinh(b_n) + 2a_n^3 b_n^3 \cos(a_n)\cosh(b_n) - 2a_n^3 b_n^3 = 0$
Free at both ends	$(b_n^2 - a_n^2)\sin(a_n)\sinh(b_n) - 2a_n b_n \cos(a_n)\cosh(b_n) - 2a_n b_n = 0$
Cantilever clamped-free	$(b_n^2 - a_n^2)a_n b_n \sin(a_n)\sinh(b_n) + (b_n^4 + a_n^4)\cos(a_n)\cosh(b_n) + 2a_n^2 b_n^2 = 0$

TABLE 2.6 Characteristic equations for various boundary conditions for bending vibration of isotropic beams, including the effects of shear and rotary inertia (Timoshenko model), when $a < a_c$ (Han et al., 1999)

Boundary conditions	Characteristic equation
Simply supported at both ends	$\sin(a_n)\sinh(b_n) = 0$
Clamped at both ends	$\left[\dfrac{(a_n^2 - b_n^2)(\varepsilon^2 a_n^2 + \varepsilon^2 b_n^2 + \varepsilon^2 a_n b_n - a_n b_n)(\varepsilon^2 a_n^2 + \varepsilon^2 b_n^2 - \varepsilon^2 a_n b_n + a_n b_n)}{2a_n b_n (b_n^2 + \varepsilon^2 a_n^2)(a_n^2 + \varepsilon^2 b_n^2)}\right] \times$ $\times [\sin(a_n)\sinh(b_n)] - \cos(a_n)\cosh(b_n) + 1 = 0$
Free at both ends	$\left[\dfrac{(a_n^2 - b_n^2)(a_n^2 + b_n^2 + \varepsilon^2 a_n b_n - a_n b_n)(a_n^2 + b_n^2 - \varepsilon^2 a_n b_n + a_n b_n)}{2a_n b_n (b_n^2 + \varepsilon^2 a_n^2)(a_n^2 + \varepsilon^2 b_n^2)}\right] \times$ $\times [\sin(a_n)\sinh(b_n)] - \cos(a_n)\cosh(b_n) + 1 = 0$
Cantilever clamped-free	$(a_n^2 - b_n^2)\sin(a_n)\sinh(b_n) - 2a_n b_n$ $-a_n b_n \dfrac{(a_n^4 + \varepsilon^4 a_n^4 + 4\varepsilon^2 a_n^2 b_n^2 + \varepsilon^4 b_n^4 + b_n^4)}{(b_n^4 + \varepsilon^2 a_n^2)(a_n^2 + \varepsilon^2 b_n^2)}\cos(a_n)\cosh(b_n) = 0$

TABLE 2.7 Characteristic equations for various boundary conditions for bending vibration of isotropic beams, including the effects of shear and rotary inertia (Timoshenko model), when $a \geq a_c$ (Han et al., 1999)

Boundary conditions	Characteristic equation
Simply supported at both ends	$\sin(a_n)\sin(\hat{b}_n) = 0$
Clamped at both ends	$\dfrac{(a_n^2 + \hat{b}_n^2)\left[(\varepsilon^2 a_n^2 - \varepsilon^2 \hat{b}_n^2)^2 + (\varepsilon^2 a_n \hat{b}_n - a_n \hat{b}_n)^2\right]}{2a_n \hat{b}_n(-\hat{b}_n^2 + \varepsilon^2 a_n^2)(a_n^2 - \varepsilon^2 \hat{b}_n^2)} \sin(a_n)\sin(\hat{b}_n)$ $- \cos(a_n)\cos(\hat{b}_n) + 1 = 0$
Free at both ends	$\dfrac{(a_n^2 + \hat{b}_n^2)\left[(a_n^2 - \hat{b}_n^2)^2 + (\varepsilon^2 a_n \hat{b}_n - a_n \hat{b}_n)^2\right]}{2a_n \hat{b}_n(-\hat{b}_n^2 + \varepsilon^2 a_n^2)(a_n^2 - \varepsilon^2 \hat{b}_n^2)} \sin(a_n)\sin(\hat{b}_n)$ $- \cos(a_n)\cos(\hat{b}_n) + 1 = 0$
Cantilever clamped-free	$(a_n^2 + \hat{b}_n^2)\sin(a_n)\sin(\hat{b}_n) - 2a_n \hat{b}_n$ $-a_n \hat{b}_n \dfrac{(a_n^4 + \varepsilon^4 a_n^4 - 4\varepsilon^2 a_n^2 \hat{b}_n^2 + \varepsilon^4 \hat{b}_n^4 + \hat{b}_n^4)}{(-\hat{b}_n^4 + \varepsilon^2 a_n^2)(a_n^2 - \varepsilon^2 \hat{b}_n^2)} \cos(a_n)\cos(\hat{b}_n) = 0$

the mode number). If $a \geq a_c$, then a_n and $\hat{b}_n$ are found using Equation (2.130) and the relevant characteristic equation in Table 2.7. Each pair of solutions, a_n and b_n or a_n and $\hat{b}_n$, corresponds to a resonant vibration mode. The corresponding mode shape is obtained by substituting the solutions, a_n for a and b_n for b, into Equation (2.122) for the condition $a < a_c$ and substituting the solutions, a_n for a and $\hat{b}_n$ for $\hat{b}$, into Equation (2.124) for the condition $a \geq a_c$. The resonance frequency corresponding to a particular solution pair (either a_n and b_n or a_n and $\hat{b}_n$, depending on the frequency) is given for clamped-clamped, free-free and clamped-free boundary conditions by (Han et al., 1999):

$$f_n = \begin{cases} \dfrac{1}{2\pi}\sqrt{\dfrac{a_n^2 - b_n^2}{1 + \varepsilon^2}}\sqrt{\dfrac{E}{\rho_m L^2}} & a_n < a_c \\[4mm] \dfrac{1}{2\pi}\sqrt{\dfrac{a_n^2 + \hat{b}_n^2}{1 + \varepsilon^2}}\sqrt{\dfrac{E}{\rho_m L^2}} & a_n \geq a_c \end{cases} \qquad (2.144)$$

where ε is defined in Equation (2.129).

When the boundary conditions are simply supported an additional imaginary solution occurs over the entire frequency range, which is denoted $\hat{b}_n^{(2)}$. Thus for simply supported ends, the resonance frequencies include those given in Equation (2.144), plus additional resonance frequencies, $f_n^{(2)}$, over the entire frequency range, given by (Han et al., 1999):

$$f_n^{(2)} = \frac{1}{2\pi}\sqrt{\frac{a_n^2 + (\hat{b}_n^{(2)})^2}{1 + \varepsilon^2}}\sqrt{\frac{E}{\rho_m L^2}} \qquad (2.145)$$

2.3.4 Waves in Thin Plates

The coordinate system and notation convention for displacements, stresses, forces and moments for waves in thin plates is shown in Figure 2.17. The sign convention followed is consistent with that of Leissa (1969). Internal forces and stresses are defined in terms of quantities per unit plate width (or unit length) and the externally applied force is defined as force per unit area. The three well-known equilibrium equations (Timoshenko and Woinowsky-Krieger, 1959), $\tau_{xy} = \tau_{yx}$, $\tau_{zx} = \tau_{xz}$ and $\tau_{yz} = \tau_{zy}$, have been included in Figure 2.17(d). Twisting moments are defined such that $M_{xy} = M_{yx}$. They are equal because the shear stresses causing them are equal.

In the following subsections, it will be assumed that the plates are isotropic. Orthotropic plates are also of interest, as stiffened plates can often be modelled as such. These are adequately treated by Leissa (1969) and will not be considered further here.

2.3.4.1 Longitudinal Waves

The derivation of the wave equation for longitudinal wave propagation in one direction in a thin plate is similar to the derivation for a beam. Figure 2.9 and Equation (2.58) also apply to longitudinal propagation in the x-direction in a plate. The only difference is that the relationship between stress and strain is:

$$\sigma_{xx} = \frac{E}{(1 - \nu^2)}(\epsilon_{xx} + \nu\epsilon_{yy}) \tag{2.146}$$

For wave propagation in the x-direction, ϵ_{yy} is assumed to be negligible. Thus, the wave equation for longitudinal wave propagation in the x-direction in a thin plate is:

$$\frac{\partial^2 \xi_x}{\partial x^2} = \frac{\rho_m(1 - \nu^2)}{E}\frac{\partial^2 \xi_x}{\partial t^2} \tag{2.147}$$

and the corresponding phase speed is:

$$c_L = \sqrt{E/\rho_m(1 - \nu^2)} \tag{2.148}$$

The longitudinal wavenumber, k_L, at frequency, ω, is:

$$k_L = \omega/c_L = \sqrt{\frac{\rho_m\omega^2(1 - \nu^2)}{E}} \tag{2.149}$$

The displacement, ξ_x, in the x-direction is:

$$\xi_x(x) = Ae^{j(\omega t - k_L x)} = \hat{\xi}_x(x)e^{j\omega t} \tag{2.150}$$

where $\hat{\xi}_x(x)$ is a complex amplitude and A is real.

For propagation in the y-direction, the subscript, x, in Equation (2.147) is replaced with y. Note that Equation (2.147) contains no correction for shear or rotary inertia and is only valid for thin plates, less than 0.1 wavelengths thick.

Boundary conditions and solutions are identical to those found for longitudinal waves in a beam for classical beam theory (see Equations (2.64) and (2.65), and Table 2.3).

2.3.4.2 Transverse Shear Waves

In-plane shear waves travelling in thin plates are influenced very little by the free surfaces of the plate; thus, the shear wave speed and the wave equation are very similar to those for a large volume of solid. Although in-plane shear waves in thin plates are difficult to generate by applied forces, they can play a significant role in the transportation of vibrational energy through a plate structure such as a ship. This is because some bending wave energy is converted to shear wave

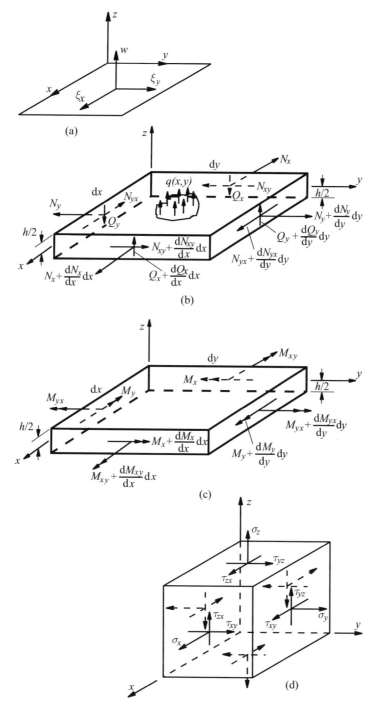

FIGURE 2.17 (a) Coordinate system and displacements for a thin plate. w is the displacement for a bending wave, ζ_y and ζ_x are the in-plane displacements in the y- and x-directions, respectively; (b) forces (intensities) acting on a plate element; (c) moments (intensities) acting on a plate element; (d) notation and positive directions of stress.

energy at structural junctions and discontinuities, some of which in turn is converted back to bending wave energy at other junctions.

For a shear wave propagating in the x-direction, the corresponding displacement is in the y-direction. This will be denoted ξ_y, and the resulting situation for the plate element is shown in Figure 2.18, where (x, y) is in the plane of the plate.

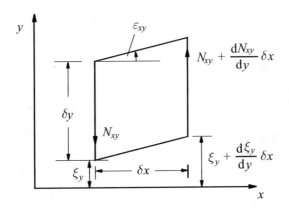

FIGURE 2.18 Shear stresses and displacement associated with motion only in the y-direction.

From Figure 2.18, it can be seen that the net shear force per unit width acting on the element (of thickness, δ_z) in the y-direction is:

$$[N_{xy}(x + \delta x) - N_{xy}] = \frac{\partial N_{xy}}{\partial x} \delta x \qquad (2.151)$$

The equation of motion for an element of thickness, h, can then be written as:

$$\rho_m \delta_x \delta_y h \frac{\partial^2 \xi_y}{\partial t^2} = \frac{\partial N_{xy}}{\partial x} \delta_x \delta_y \qquad (2.152)$$

where N_{xy} is the shear force per unit plate width (or length in the y-direction), ξ_y is the transverse displacement of the element in the y-direction and ρ_m is the density of the plate material.

The shear force, N_{xy}, is obtained by integrating the shear stress over the plate thickness, h. Thus:

$$N_{xy} = \int_{-h/2}^{h/2} \sigma_{xy} \mathrm{d}z \qquad (2.153)$$

For shear wave propagation only, this simplifies to:

$$N_{xy} = h\sigma_{xy} \qquad (2.154)$$

Thus, Equation (2.152) may be written as:

$$\rho_m \frac{\partial^2 \xi_y}{\partial t^2} = \frac{\partial \sigma_{xy}}{\partial x} \qquad (2.155)$$

The shear strain, ϵ_{xy}, is related to the shear stress, σ_{xy}, by the shear modulus, G, as follows:

$$\sigma_{xy} = G\epsilon_{xy} \qquad (2.156)$$

From Equation (2.48) the following is obtained:

$$\epsilon_{xy} = \frac{\partial \xi_y}{\partial x} \qquad (2.157)$$

as the quantity, $\dfrac{\partial \xi_x}{\partial y}$, is zero in this case. Substituting Equations (2.156) and (2.157) into Equation (2.155) gives the following wave equation for a transverse in-plane shear wave propagating in a thin plate in the x-direction:

$$\frac{\partial^2 \xi_y}{\partial x^2} = \frac{\rho_m}{G} \frac{\partial^2 \xi_y}{\partial t^2} \tag{2.158}$$

The displacement, $\xi_y(x)$, in the y-direction at location, x, is:

$$\xi_y(x) = A e^{j(\omega t - k_s x)} = \hat{\xi}_y(x) e^{j\omega t} \tag{2.159}$$

where $\hat{\xi}_y(x)$ is a complex amplitude and A is real.

The frequency independent phase speed is:

$$c_s = \sqrt{\frac{G}{\rho_m}} \tag{2.160}$$

Boundary conditions and solutions are identical with those found for longitudinal waves, except that $E/(1 - \nu^2)$ is replaced with G, and the displacement is normal (but in the plane of the plate) to the direction of wave propagation. The shear wavenumber, k_s, at frequency, ω, is:

$$k_s = \omega/c_s = \sqrt{\frac{\rho_m \omega^2}{G}} \tag{2.161}$$

2.3.4.3 Bending Waves

Referring to Figure 2.18, the following relationships between stress, and forces and bending moments acting on the plate element, can be written:

$$N_x = \int_{-h/2}^{h/2} \sigma_{xx} dz \tag{2.162}$$

$$N_y = \int_{-h/2}^{h/2} \sigma_{yy} dz \tag{2.163}$$

$$N_{xy} = N_{yx} = \int_{-h/2}^{h/2} \sigma_{xy} dz \tag{2.164}$$

$$M_x = \int_{-h/2}^{h/2} \sigma_{xx} z dz \tag{2.165}$$

$$M_y = \int_{-h/2}^{h/2} \sigma_{yy} z dz \tag{2.166}$$

$$M_{xy} = M_{yx} = \int_{-h/2}^{h/2} \sigma_{xy} z dz \tag{2.167}$$

The preceding quantities have the units of force or moment per unit length.

Before proceeding further, it is necessary to relate these quantities to the plate displacements w, ξ_x and ξ_y. Consider the edge view of a portion of plate shown in Figure 2.19.

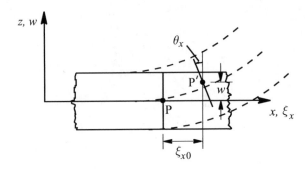

FIGURE 2.19 Shear stresses and displacement associated with motion only in the y-direction.

The centre line of the plate at point, P, is displaced longitudinally by a distance, ξ_{xo}, and laterally by w. From the figure it is clear that the longitudinal displacement in the x-direction of any point in the cross section at vertical location z is:

$$\xi_x = \xi_{xo} - z\frac{\partial w}{\partial x} \tag{2.168}$$

and the longitudinal displacement in the y-direction is:

$$\xi_y = \xi_{yo} - z\frac{\partial w}{\partial y} \tag{2.169}$$

The in-plane and shear strains can be related to the quantities in the preceding equations using Equation (2.48), which gives:

$$\epsilon_{xx} = \frac{\partial \xi_x}{\partial x} \tag{2.170}$$

$$\epsilon_{yy} = \frac{\partial \xi_y}{\partial y} \tag{2.171}$$

$$\epsilon_{xy} = \frac{\partial \xi_y}{\partial x} + \frac{\partial \xi_x}{\partial y} \tag{2.172}$$

The stresses for an isotropic plate are related to the strains as follows:

$$\sigma_{xx} = \frac{E}{1-\nu^2}\left(\epsilon_{xx} + \nu\epsilon_{yy}\right) \tag{2.173}$$

$$\sigma_{yy} = \frac{E}{1-\nu^2}\left(\epsilon_{yy} + \nu\epsilon_{xx}\right) \tag{2.174}$$

$$\sigma_{xy} = G\epsilon_{xy} \tag{2.175}$$

Substituting Equations (2.168) and (2.169) into Equations (2.170) to (2.172) and the results into Equations (2.173) to (2.175) gives expressions for the stresses in terms of the plate mid-plane displacements ξ_{x0}, ξ_{y0} and w. These results can then be substituted into Equations (2.162) to (2.167) to obtain expressions for N_x, N_y, N_{xy}, M_x, M_y and M_{xy} in terms of plate displacements.

For bending wave propagation, $\xi_{x0} = \xi_{y0} = 0$, and the resulting integrations give $N_x = N_y = N_{xy} = 0$. The integration results for the moments are:

$$M_x = -D\left(\frac{\partial^2 w}{\partial x^2} + \nu\frac{\partial^2 w}{\partial y^2}\right) \tag{2.176}$$

$$M_y = -D\left(\frac{\partial^2 w}{\partial y^2} + \nu \frac{\partial^2 w}{\partial x^2}\right) \tag{2.177}$$

$$M_{xy} = M_{yx} = -D(1-\nu)\frac{\partial^2 w}{\partial x \partial y} \tag{2.178}$$

where the bending stiffness, D, is defined as:

$$D = \frac{Eh^3}{12(1-\nu^2)} \tag{2.179}$$

and $\nu =$ Poisson's ratio.

The displacement, $w(x,y,t)$, normal to the surface of the plate at location, (x,y), on the plate at time, t, may be expressed as:

$$w(x,y,t) = A\mathrm{e}^{\mathrm{j}(\omega t - k_{bx}x - k_{by}y)} = \hat{w}(x,y)\mathrm{e}^{\mathrm{j}\omega t} \tag{2.180}$$

where $\hat{w}(x,y)$ is a complex amplitude and A is real.

Referring to Figures 2.17(b) and (c), and summing moments about the x- and y-axes (remembering that for bending waves, $N_x = N_y = N_{xy} = 0$), the following is obtained:

$$Q_x - \frac{\partial M_x}{\partial x} - \frac{\partial M_{xy}}{\partial y} = \frac{\rho_m h^3}{12}\frac{\partial^3 w}{\partial x \partial t^2} \tag{2.181}$$

$$Q_y - \frac{\partial M_{xy}}{\partial x} - \frac{\partial M_y}{\partial y} = \frac{\rho_m h^3}{12}\frac{\partial^3 w}{\partial y \partial t^2} \tag{2.182}$$

where ρ_m is the density of the plate. If it is assumed that the rotary inertia term on the right-hand side of Equations (2.181) and (2.182) is negligible (thin plate), and further, Equations (2.176), (2.177) and (2.178) are substituted for M_x, M_y and M_{xy}, respectively, the following is obtained:

$$Q_x = -D\frac{\partial}{\partial x}\left(\nabla^2 w\right) = -D\frac{\partial}{\partial x}\left(\frac{\partial^2 w}{\partial x^2} + \frac{\partial^2 w}{\partial y^2}\right) \tag{2.183}$$

$$Q_y = -D\frac{\partial}{\partial y}\left(\nabla^2 w\right) = -D\frac{\partial}{\partial y}\left(\frac{\partial^2 w}{\partial x^2} + \frac{\partial^2 w}{\partial y^2}\right) \tag{2.184}$$

where Q_x and Q_y have the units of force per unit length.

Returning to Figure 2.17(b), and summing forces in the z-direction (remembering that for bending waves, $N_x = N_y = N_{xy} = 0$, the following is obtained:

$$\frac{\partial Q_x}{\partial x} + \frac{\partial Q_y}{\partial y} + q(x,y) = \rho_m h\frac{\partial^2 w}{\partial t^2} \tag{2.185}$$

where the external force, $q(x,y)$, has the units of force per unit area. In the following analysis, the $(x,\,y)$ dependence is assumed and the force will be written simply as q.

Substituting Equations (2.183) and (2.184) into Equation (2.185) gives the classical plate equation for an isotropic plate:

$$D\nabla^4 w + \rho_m h\frac{\partial^2 w}{\partial t^2} = q \tag{2.186}$$

which can be rewritten in terms of the plate bending wavenumber, k_b, as:

$$\left[\nabla^4 - k_b^4\right] w = q/D \tag{2.187}$$

where the displacement, w, is defined in Equation (2.180), $\nabla^4 = \nabla^2\nabla^2$ and:

$$k_b^4 = \rho_m h\omega^2/D \tag{2.188}$$

where D is defined in Equation (2.179). In Cartesian coordinates:

$$\nabla^2 w = \frac{\partial^2 w}{\partial x^2} + \frac{\partial^2 w}{\partial y^2} \tag{2.189}$$

In polar coordinates:

$$\nabla^2 w = \frac{\partial^2 w}{\partial r^2} + \frac{1}{r}\frac{\partial w}{\partial r} + \frac{1}{r^2}\frac{\partial^2 w}{\partial \theta^2} \tag{2.190}$$

and the in-plane shearing forces are:

$$Q_r = -D\frac{\partial}{\partial r}\left(\nabla^2 w\right) \tag{2.191}$$

and:

$$Q_\theta = -\frac{D}{r}\frac{\partial}{\partial \theta}\left(\nabla^2 w\right) \tag{2.192}$$

In polar coordinates, the bending and twisting moments corresponding to Equations (2.176) to (2.178) are (Leissa, 1969):

$$M_r = -D\left[\frac{\partial^2 w}{\partial r^2} + \nu\left(\frac{1}{r}\frac{\partial w}{\partial r} + \frac{1}{r^2}\frac{\partial^2 w}{\partial \theta^2}\right)\right] \tag{2.193}$$

$$M_\theta = -D\left[\frac{1}{r}\frac{\partial w}{\partial r} + \frac{1}{r^2}\frac{\partial^2 w}{\partial \theta^2} + \nu\frac{\partial^2 w}{\partial r^2}\right] \tag{2.194}$$

and:

$$M_{r\theta} = -D(1-\nu)\frac{\partial}{\partial r}\left(\frac{1}{r}\frac{\partial w}{\partial \theta}\right) \tag{2.195}$$

Returning to the Cartesian coordinate system, if the external force were replaced with external moments, M_{ex} and M_{ey} (units of moment per unit area), about the x- and y-axes, respectively, then Equations (2.181) and (2.182) become:

$$Q_x - \frac{\partial M_x}{\partial x} - \frac{\partial M_{xy}}{\partial y} + M_{xe} = \frac{\rho_m h^3}{12}\frac{\partial^3 w}{\partial x \partial t^2} \tag{2.196}$$

$$Q_y - \frac{\partial M_{xy}}{\partial x} - \frac{\partial M_y}{\partial y} + M_{ye} = \frac{\rho_m h^3}{12}\frac{\partial^3 w}{\partial y \partial t^2} \tag{2.197}$$

and consequently, the wave equation becomes:

$$D\nabla^4 w + \rho_m h\frac{\partial^2 w}{\partial t^2} = -\frac{\partial M_{ex}}{\partial y} + \frac{\partial M_{ey}}{\partial x} \tag{2.198}$$

or:

$$\left[\nabla^4 - k_b^4\right]w = -\frac{1}{D}\left[\frac{\partial M_{ex}}{\partial y} + \frac{\partial M_{ey}}{\partial x}\right] \tag{2.199}$$

Note that if the external force, q, were acting simultaneously with the external moments, it would be simply added to the right-hand side of Equation (2.198).

When analysing the free vibrations of plates, the right-hand sides of Equations (2.187) and (2.199) are set equal to zero and the resulting equations are factorised to give:

$$\left[\nabla^2 + k_b^2\right]\left[\nabla^2 - k_b^2\right]w = 0 \tag{2.200}$$

The complete solution is then $w = \hat{w}e^{\mathrm{j}\omega t} = w_1 + w_2 = \hat{w}_1 e^{\mathrm{j}\omega t} + \hat{w}_2 e^{\mathrm{j}\omega t}$, where w_1 and w_2 are solutions to:

$$\begin{cases} \left[\nabla^2 + k_b^2\right]w_1 = 0 \\ \left[\nabla^2 - k_b^2\right]w_2 = 0 \end{cases} \tag{2.201}$$

Equations (2.200) and (2.201) are also valid for the polar coordinate system, which is used to analyse circular and annular plates.

Rectangular Plates

For *classical* plate theory, two boundary conditions are needed to solve the equations of motion. They usually take the form of specification of one boundary condition from each of the following groups:

$$(w, V_v) \quad \text{and} \quad \left(\frac{\partial w}{\partial v}, M_v \right) \tag{2.202}$$

where:

$$V_v = Q_v + \frac{\partial M_{vs}}{\partial s} \tag{2.203}$$

where v is the in-plane coordinate normal to the edge of the plate and s is the in-plane coordinate parallel to the edge of the plate to which the boundary condition applies. Thus, for the boundary condition along an edge parallel to the x-axis, $v = x$ and $s = y$, and for a boundary condition along an edge parallel to the y-axis, $v = y$ and $s = x$.

$$\begin{cases} \text{For simply supported edges}: & w = M_v = 0 \\ \text{For clamped edges}: & w = \dfrac{\partial w}{\partial v} = 0 \\ \text{For free edges}: & V_v = M_v = 0 \end{cases} \tag{2.204}$$

The solution to the classical wave equation for rectangular plates, where A_{mn} is a real constant, dependent on the plate vibration amplitude, is:

$$w_{mn}(x, y) = A_{mn} X_m(x) Y_n(y) e^{j\omega t}; \quad m, /, n = 1, 2, \ \dots \tag{2.205}$$

For a plate simply supported at $x = 0$ and $x = L_x$, the mode shape in the x-direction is:

$$X_m(x) = \sin \frac{m\pi x}{L_x}; \quad m = 1, 2, \ \dots . \tag{2.206}$$

where L_x is the plate dimension in the x-direction. The complete solution for the displacement, $w_{mn}(x, y)$, corresponding to mode, (mn), of a simply supported plate is then:

$$w_{mn}(x, y) = w_{1mn}(x, y) + w_{2mn}(x, y) = A_{mn} \sin\left(\frac{m\pi x}{L_1} \right) \sin\left(\frac{n\pi y}{L_2} \right) \tag{2.207}$$

where L_1 and L_2 are the plate dimensions, m and n range from 1 to ∞, the number of nodal lines in the x-direction (excluding the two plate edges) is $m - 1$ and the number of nodal lines in the y-direction (excluding the two plate edges) is $n - 1$.

For a plate clamped at $x = 0$ and $x = L_x$, the mode shape in the x-direction is (Leissa, 1969):

$$X_m(x) = \cos \gamma_{1m} \left(\frac{x}{L_x} - \frac{1}{2} \right) + \frac{\sin(\gamma_{1m}/2)}{\sinh\left(\gamma_{1m}/2\right)} \cosh \gamma_{1m} \left(\frac{x}{L_x} - \frac{1}{2} \right); \quad m = 2, 4, 6, \dots \tag{2.208}$$

$$X_m(x) = \sin \gamma_{2m} \left(\frac{x}{L_x} - \frac{1}{2} \right) - \frac{\sin(\gamma_{2m}/2)}{\sinh(\gamma_{2m}/2)} \sinh \gamma_{2m} \left(\frac{x}{L_x} - \frac{1}{2} \right); \quad m = 1, 3, 5, \dots \tag{2.209}$$

γ_{1m} are solutions of:

$$\tan(\gamma_{1m}/2) + \tanh(\gamma_{1m}/2) = 0; \quad m = 1, 3, 5, \dots \tag{2.210}$$

and γ_{2m} are solutions of:

$$\tan(\gamma_{2m}/2) - \tanh(\gamma_{2m}/2) = 0; \qquad m = 2, 4, 6, \ldots \tag{2.211}$$

The integer, $m-1$, represents the number of nodal lines (excluding plate edges). Similar expressions apply for $Y(y)$, with x replaced by y and m replaced by n.

For a plate free at $x = 0$ and $x = L_x$, the mode shape in the x-direction is (Leissa, 1969):

$$X_m(x) = \cos \gamma_{1m} \left(\frac{x}{L_x} - \frac{1}{2} \right) - \frac{\sin(\gamma_{1m}/2)}{\sinh\left(\gamma_{1m}/2\right)} \cosh \gamma_{1m} \left(\frac{x}{L_x} - \frac{1}{2} \right); \qquad m = 2, 4, 6, \ldots \tag{2.212}$$

$$X_m(x) = \sin \gamma_{2m} \left(\frac{x}{L_x} - \frac{1}{2} \right) + \frac{\sin(\gamma_{2m}/2)}{\sinh(\gamma_{2m}/2)} \sinh \gamma_{2m} \left(\frac{x}{L_x} - \frac{1}{2} \right); \qquad m = 1, 3, 5, \ldots \tag{2.213}$$

where γ_{1m} and γ_{1m} are defined in Equations (2.210) and (2.211), respectively. The integer, $m-1$, represents the number of nodal lines (excluding plate edges). Similar expressions apply for $Y(y)$, with x replaced by y and m replaced by n.

Expressions for plates with different boundary conditions at $x = 0$ and $x = L_x$ are provided by Leissa (1969).

Obtaining resonance frequencies corresponding to a particular boundary condition is a complex process for all except simply supported conditions. For the clamped edge condition, Leissa (1969) provides the following expressions for the resonance frequency of mode, (m, n). For square plates ($L_x = L_y$):

$$f_{mn} = 2 \left(m + \frac{1}{3} \right)^2 \frac{\pi}{L_x^2} \sqrt{\frac{D}{\rho_m}} \tag{2.214}$$

For rectangular plates, the resonance frequency of the fundamental mode (1,1) is:

$$f_{1,1} = 12 \sqrt{3.5 \left(\frac{1}{L_x^4} + \frac{4}{7L_x^2 L_y^2} + \frac{1}{L_y^4} \right)} \sqrt{\frac{D}{\rho_m}} \tag{2.215}$$

The clamped edge condition (and other combinations of edge condition) for rectangular plates are discussed at length by Leissa (1969), but there are no simple equations to solve for resonance frequencies for other than the first mode, as there were for beams. Thus, only solutions are provided for resonance frequencies for simply-supported and clamped-edge plates of varying aspect ratio in Tables 2.8 and 2.9 for the first 4 vibration modes. Unfortunately, different authors using different approaches obtain slightly different numerical results, with the spread increasing with increasing mode order (see various datasets in Leissa (1969)).

Circular Plates

For a circular plate, the solution for the displacement amplitude of the plate at radial location, r (from the plate centre), and angular location, θ (from the horizontal in the anticlockwise direction), is (Leissa, 1969):

$$w_n(r, \theta) = [A_n J_n(k_b r) + C_n I_n(k_b r)] \cos(n\theta) = C_n \left[\frac{A_n}{C_n} J_n(k_b r) + I_n(k_b r) \right] \cos(n\theta) \tag{2.216}$$

where A_n and C_n are arbitrary coefficients, n is the number of diametral nodal lines, ranging from $n = 0$ to $n = \infty$, $J_n(k_b r)$ is the Bessel function (see Abramowitz and Stegun (1965)) of the first kind of order n, $I_n(k_b r)$ is the modified Bessel function of the first kind of order n, $Y_n(k_b r)$ is the Bessel function (see Abramowitz and Stegun (1965)) of the second kind of order n, $K_n(k_b r)$ is the modified Bessel function of the second kind of order n and k_b is the bending wavenumber defined by Equations (2.188) and (2.179).

To determine the mode shape for a given n, the ratio, A_n/C_n is required and this is a function of the boundary conditions (see Table 2.8).

The characteristic equations and resonance frequency equations for clamped and simply supported circular plate boundary conditions are given in Table 2.8. Characteristic equation solutions for use in the resonance frequency equations are provided for the first few modes in Table 2.10 for circular plates. Results for other boundary conditions as well as the treatment of annular plates with various boundary conditions can be found in Leissa (1969).

TABLE 2.8 Characteristic equations for circular plates and resonance frequency equations for various boundary conditions for bending vibration of isotropic circular and rectangular plates, for classical plate theory (Leissa, 1969)

Boundary conditions	Characteristic equation and resonance frequencies, f_{mn}	$\dfrac{A_n}{C_n}$
Circular, simply supported around entire edge	$\dfrac{J_{n+1}(\gamma)}{J_n(\gamma)} + \dfrac{I_{n+1}(\gamma)}{I_n(\gamma)} = \dfrac{2\gamma}{(1-\nu)}$ $f_{mn} = \dfrac{\gamma_{mn}^2}{2\pi a^2}\sqrt{\dfrac{D}{\rho_m}}$	$-\dfrac{I_n(\gamma)}{J_n(\gamma)}$
Circular, clamped around entire edge	$J_n(\gamma)I_{n+1}(\gamma) + I_n(\gamma)J_{n+1}(\gamma) = 0$ $f_{mn} = \dfrac{\gamma_{mn}^2}{2\pi a^2}\sqrt{\dfrac{D}{\rho_m}}$	$-\dfrac{I_n(\gamma)}{J_n(\gamma)}$
Rectangular, simply supported along all edges	$f_{mn} = \dfrac{1}{2\pi}\sqrt{\dfrac{D}{\rho_m}}\left[\left(\dfrac{m\pi}{L_x}\right)^2 + \left(\dfrac{n\pi}{L_y}\right)^2\right]$	—
Rectangular, clamped along all edges	$f_{mn} = \dfrac{\Lambda_{mn}^2}{2\pi L_y^2}\sqrt{\dfrac{D}{\rho_m}}$	—

For circular plates, γ_{mn} is the mth solution for γ in the characteristic equation for a specified value of n, where n is the number of diametral nodes and m is the number of circular nodes, excluding the plate edge.

For rectangular plates, Λ_{mn} is defined in Table 2.9, $m = 1+$(number of nodal lines) between $x = 0$ and $x = L_x$, and $n = 1+$(number of nodal lines) between $y = 0$ and $y = L_y$, excluding the plate edges in both cases.

TABLE 2.9 Frequency coefficients, $\Lambda_{mn}^2 = 2\pi f_{mn} L_y^2 \sqrt{\rho_m/D}$, for clamped edge boundary conditions for bending vibration of isotropic rectangular thin plates using classical plate theory, for various aspect ratios, where n, m is the integer mode number. $(m-1)$ is the number of nodal lines in the x-direction and $(n-1)$ is the number of nodal lines in the y-direction (excluding the plate edges). (Data from Leissa (1969))

Mode	Frequency coefficient, Λ_{mn}^2 for $L_x/L_y =$						
	1.0	1.2	1.5	2.0	2.5	3.0	∞
$m=1, n=1$	35.98	30.65	27.07	24.56	23.76	23.19	22.37
$m=1, n=2$	73.41	70.50	67.58	65.41	64.49	64.02	61.7
$m=2, n=1$	73.41	59.20	41.95	31.83	27.81	25.78	—
$m=2, n=2$	108.24	95.24	81.57	72.66	68.89	66.96	61.7

TABLE 2.10 First few roots $(\gamma_{mn}^2 = 2\pi f_{mn} a^2 \sqrt{\rho_m/D})$ of the characteristic equation, for clamped edge boundary conditions for bending vibration of isotropic circular thin plates using classical plate theory, where n, m is the integer mode number, $(n-1)$ is the number of diametral nodal lines and $(m-1)$ is the number of circular nodal lines (excluding the plate edges). Numbers in brackets are for a simply supported edge for $\nu = 0.3$. (Data from Leissa (1969))

m	$n=0$	$n=1$	$n=2$	$n=3$	$n=4$
0	10.22(4.977)	21.26(13.94)	34.88(25.65)	51.04	69.67
1	39.77(29.76)	60.82(48.51)	84.58(70.14)	111.01	140.18
2	89.10(74.20)	120.08(102.80)	153.81(134.33)	190.30	229.52
3	158.18(138.34)	199.06(176.84)	242.71(218.24)	289.17	338.41
4	247.00	297.77	351.38	407.72	—

Effects of Shear Deformation and Rotary Inertia

In the same way as was found for beams, the effects of shear deformation and rotary inertia become important for thick plates and/or high frequencies (short wavelengths).

The three fundamental Equations (2.181), (2.182) and (2.185), which were derived using classical theory, can still be made use of. This time, however, the rotary inertia terms on the right-hand side of Equations (2.181) and (2.182) cannot be neglected. The bending displacement, w, is now made up of a bending contribution and a shear deformation contribution. Thus, the angles of rotation of lines normal to the mid-plane before deformation cannot be directly related to the displacement, w, as was done in Equations (2.168) and (2.169), in which shear deformation was neglected. Thus, the in-plane displacements must now be expressed in terms of the rotation angles, ψ_x, of a plane normal to the x-axis and ψ_y, of a plane normal to the y-axis. Thus, Equations (2.168) and (2.169) become:

$$\xi_x = -z\psi_x \tag{2.217}$$

$$\xi_y = -z\psi_y \tag{2.218}$$

where only displacements associated with bending wave propagation have been included.

Using the same procedure as for classical theory, the following expressions for the bending moments are obtained:

$$M_x = -D\left(\frac{\partial \psi_x}{\partial x} + \nu \frac{\partial \psi_y}{\partial y}\right) \tag{2.219}$$

$$M_y = -D\left(\frac{\partial \psi_y}{\partial y} + \nu \frac{\partial \psi_x}{\partial x}\right) \tag{2.220}$$

$$M_{xy} = M_{yx} = -\frac{D(1-\nu)}{2}\left(\frac{\partial \psi_y}{\partial x} + \frac{\partial \psi_x}{\partial y}\right) \tag{2.221}$$

The transverse shear forces, Q_x and Q_y, are now obtained by integrating the transverse shearing stresses over the plate thickness:

$$Q_x = \int_{-h/2}^{h/2} \sigma_{xz}\mathrm{d}z \tag{2.222}$$

$$Q_y = \int_{-h/2}^{h/2} \sigma_{yz}\mathrm{d}z \tag{2.223}$$

Substituting Equation (2.48) into Equation (2.47), then into Equations (2.222) and (2.223), and then integrating results in the following:

$$Q_x = -\kappa^2 Gh\left(\psi_x - \frac{\partial w}{\partial x}\right) \tag{2.224}$$

$$Q_y = -\kappa^2 Gh\left(\psi_y - \frac{\partial w}{\partial y}\right) \tag{2.225}$$

where κ^2 is a constant introduced to account for the shear stresses not being constant over the thickness of the plate. Mindlin (1951) chose κ to make the dynamic theory predictions consistent with the known exact theory of elasticity prediction of the frequency of the fundamental 'thickness-shear' mode of vibration. Thus:

$$\kappa^2 \approx 0.76 + 0.3\nu \tag{2.226}$$

See Mindlin (1951) for a more detailed discussion of this constant.

By using Equations (2.219) to (2.221), (2.224) and (2.225), Equations (2.181), (2.182) and (2.185) can be expressed in terms of ψ_x, ψ_y and w, as follows:

$$\frac{D}{2}\left[(1-\nu)\nabla^2\psi_x + (1+\nu)\frac{\partial \Phi}{\partial x}\right] - \kappa^2 Gh\left(\psi_x + \frac{\partial w}{\partial x}\right) = \frac{\rho_m h^3}{12}\frac{\partial^2 \psi_x}{\partial t^2} \tag{2.227}$$

where:

$$\frac{D}{2}\left[(1-\nu)\nabla^2\psi_y + (1+\nu)\frac{\partial\Phi}{\partial y}\right] - \kappa^2 Gh\left(\psi_y + \frac{\partial w}{\partial y}\right) = \frac{\rho_m h^3}{12}\frac{\partial^2\psi_y}{\partial t^2} \qquad (2.228)$$

$$\kappa^2 Gh(\nabla^2 w + \Phi) + q = \rho_m h\frac{\partial^2 w}{\partial t^2} \qquad (2.229)$$

$$\Phi = \frac{\partial\psi_x}{\partial x} + \frac{\partial\psi_y}{\partial y} \qquad (2.230)$$

A single differential equation in w may be obtained by eliminating ψ_x and ψ_y from the preceding equations. Equations (2.227) and (2.228) are first differentiated with respect to x and y, respectively, and then added to give:

$$\left(D\nabla^2 - G'h - \frac{\rho_m h^3}{12}\frac{\partial^2}{\partial t^2}\right)\Phi = G'h\nabla^2 w \qquad (2.231)$$

where $G' = \kappa^2 G$.

The quantity Φ is then eliminated between Equations (2.231) and (2.229) to give:

$$\left(\nabla^2 - \frac{\rho_m}{G'}\frac{\partial^2}{\partial t^2}\right)\left(D\nabla^2 - \frac{\rho_m h^3}{12}\frac{\partial^2}{\partial t^2}\right)w + \rho_m h\frac{\partial^2 w}{\partial t^2} = \left(1 - \frac{D\nabla^2}{G'h} + \frac{\rho_m h^2}{12G'}\frac{\partial^2}{\partial t^2}\right)q \qquad (2.232)$$

The effect of ignoring shear and rotary inertia is shown in Figure 2.20, where c_b is the bending wave speed and c_s is defined as $c_s = \sqrt{G/\rho_m}$.

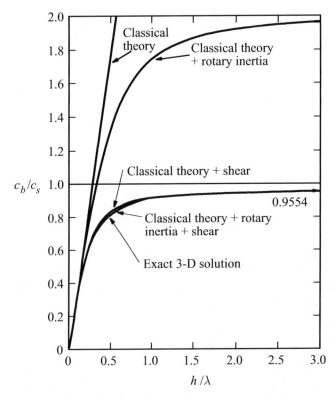

FIGURE 2.20 Effect of rotary inertia and transverse shear on bending wave speed predictions for a plate. The two curves, 'classical theory + rotary inertia + shear' and 'exact 3-D solution' lie on top of one another.

Note that shear deformation by itself accounts for almost all of the discrepancy between classical plate theory and three-dimensional elasticity theory. If rotary inertia terms are omitted from Equation (2.228), the following is obtained:

$$D\left(\nabla^2 - \frac{\rho_m}{G'}\frac{\partial^2}{\partial t^2}\right)\nabla^2 w + \rho_m h\frac{\partial^2 w}{\partial t^2} = \left(1 - \frac{D\nabla^2}{G'h}\right)q \tag{2.233}$$

If transverse shear deformation only is neglected, then the following is obtained:

$$\left(D\nabla^2 - \frac{\rho_m h^3}{12}\frac{\partial^2}{\partial t^2}\right)\nabla^2 w + \rho_m h\frac{\partial^2 w}{\partial t^2} = q \tag{2.234}$$

If both rotary inertia and transverse shear are neglected, then Equation (2.186) is obtained.

In all of the preceding equations q is an externally applied force per unit area. If this were replaced by a point force F, located at (x_0, y_0), then q could be written as:

$$q = F\delta(x - x_0)\delta(y - y_0) \tag{2.235}$$

If the plate were subjected to externally applied moments, Equations (2.181) and (2.182) would be replaced with Equations (2.196) and (2.197), and Equation (2.232) would become:

$$\left(\nabla^2 - \frac{\rho_m}{G'}\frac{\partial^2}{\partial t^2}\right)\left(D\nabla^2 - \frac{\rho_m h^3}{12}\frac{\partial^2}{\partial t^2}\right)w + \rho_m h\frac{\partial^2 w}{\partial t^2} = -\frac{\partial M_{ex}}{\partial y} + \frac{\partial M_{ey}}{\partial x} \tag{2.236}$$

In Equations (2.234) and (2.236), the quantity $\rho_m h$ is often written as m_s, where m_s is the mass per unit area of the plate.

In a similar way as was done for a beam, an external point moment M_{xp} acting about the x-axis at $(x = x_0, y = y_0)$ may be included in Equation (2.236) using:

$$M_{ex}(x, y) = M_{xp}\delta(x - x_0)\delta(y - y_0) \tag{2.237}$$

Similarly, a line moment, M_{xL}, of length b extending parallel to the x-axis at $x = x_0$ may be written as:

$$M_{ex}(x, y) = \frac{M_{xL}}{b}\delta(x - x_0)[u(y - y_1) - u(y - y_2)] \tag{2.238}$$

where y_1 and y_2 represent the beginning and end of the line and $b = y_1 - y_2$. $u()$ is the unit step function (equal to 0 for a negative argument and equal to 1 for a positive argument).

The equations of motion just derived can be used to accurately calculate plate resonance frequencies and mode shapes for a variety of boundary conditions and plate shapes by setting $q = M_{xe} = M_{ye} = 0$. Here only rectangular shaped plates will be considered. Other plate shapes have been considered by a number of researchers (including Mindlin and Deresiewicz (1954)), who considered circular plates), and much of this work has been summarised by Leissa (1969).

Mindlin (1951) showed that in the absence of external surface loading, the plate equation of motion can be simplified. Using the notation of Skudrzyk (1968):

$$(\nabla^2 + k_1^2)w_1 = 0 \tag{2.239}$$

$$(\nabla^2 + k_2^2)w_2 = 0 \tag{2.240}$$

$$(\nabla^2 + k_3^2)H = 0 \tag{2.241}$$

where the normal plate displacement is:

$$w = w_1 + w_2 \tag{2.242}$$

The in-plane plate rotations may be written as:

$$\psi_y = (\alpha_1 - 1)\frac{\partial w_1}{\partial x} + (\alpha_2 - 1)\frac{\partial w_2}{\partial x} + \frac{\partial H}{\partial y} \tag{2.243}$$

$$\psi_x = (\alpha_1 - 1)\frac{\partial w_1}{\partial y} + (\alpha_2 - 1)\frac{\partial w_2}{\partial y} - \frac{\partial H}{\partial x} \tag{2.244}$$

where:

$$\alpha_1, \alpha_2 = \frac{2(k_1^2), (k_2^2)}{(1 - v)k_3^2} \tag{2.245}$$

The wavenumbers k_1, k_2 and k_3 are related to the classical plate bending wavenumber k_b found by substituting the solution:

$$w = A\mathrm{e}^{\mathrm{j}(\omega t - k_b x - k_b y)} = \hat{w}(x,y)\mathrm{e}^{\mathrm{j}\omega t} \tag{2.246}$$

into the classical equation of motion for the unloaded plate. The wavenumbers are defined in the following equations:

$$k_1^2, k_2^2 = \frac{k_b^4}{2}\left\{F + J' \pm \left[(F - J')^2 + 4/k_b^4\right]^{1/2}\right\} \tag{2.247}$$

$$k_3^2 = \frac{2}{(1 - \nu)}\left(J'k_b^4/h - 1/F\right) \tag{2.248}$$

where:

$$k_b^4 = \left(\frac{\omega}{c_b}\right)^4 = \frac{\rho_m h\omega^2}{D} = \frac{12(1 - \nu^2)\rho_m h\omega^2}{Eh^3} \tag{2.249}$$

$$J' = \frac{h^3}{12} \tag{2.250}$$

$$F = 2h^2/(1 - \nu)\pi^2 \tag{2.251}$$

and where h is the plate thickness. Note that A may be a complex quantity, depending upon the plate boundary conditions.

Equations (2.232), (2.233) and (2.234) (with the right-hand sides set equal to zero) can be solved for the plate resonance frequencies and mode shapes for any defined set of plate boundary conditions. These boundary conditions are in the form of specification of one of each of the pairs (M_v, ψ_v) (M_{vs}, ψ_s) and (Q_v, w) where v is the normal and s is the tangent to the boundary edge.

For simply supported edges : $M_v = \psi_s = w = 0$ $\tag{2.252}$

For clamped edges : $\psi_v = \psi_s = w = 0$ $\tag{2.253}$

For free edges : $M_v = M_{vs} = Q_v = 0$ $\tag{2.254}$

where v is the in-plane coordinate normal to the edge of the plate and s is the in-plane coordinate parallel to the edge of the plate to which the boundary condition applies. Thus, for the boundary condition along an edge parallel to the x-axis, $v = x$ and $s = y$, and for a boundary condition along an edge parallel to the y-axis, $v = y$ and $s = x$.

Solutions to the equations of motion for rectangular plates are of the form:

$$w_1 = A_1 \sin(\alpha_1 x)\sin(\beta_1 y)\mathrm{e}^{\mathrm{j}\omega t} \tag{2.255}$$

$$w_2 = A_2 \sin(\alpha_2 x)\sin(\beta_2 y)\mathrm{e}^{\mathrm{j}\omega t} \tag{2.256}$$

$$H = A_3 \cos(\alpha_3 x)\sin(\beta_3 y)\mathrm{e}^{\mathrm{j}\omega t} \tag{2.257}$$

where:

$$\begin{cases} \alpha_1^2 + \beta_1^2 = k_1^2 \\ \alpha_2^2 + \beta_2^2 = k_2^2 \\ \alpha_3^2 + \beta_3^2 = k_3^2 \end{cases} \tag{2.258}$$

The coefficients, α and β, are determined by substitution of the appropriate boundary conditions into the preceding four equations.

For *classical* plate theory discussed earlier, only two boundary conditions are needed to solve the equations of motion; thus, Q_ν and $M_{\nu s}$ combine together into a single boundary condition (see Equation (2.203)), as do ψ_v and ψ_s (see Equation (2.204)).

2.3.5   Waves in Thin Circular Cylinders

The derivation of the wave equations or equations of motion for a thin circular cylindrical shell is a complex procedure and will only be outlined briefly here. For a more detailed treatment the reader is advised to consult the excellent publication by Leissa (1973).

Various researchers have derived these equations from first principles, making various simplifying assumptions along the way. The simplest theory is that due to Donnell and Mushtari (Leissa, 1973). However, the more complicated theory of Goldenveizer–Novozhilov has been shown (Pope, 1971) to give more accurate results for resonance frequencies and mode shapes of thin, simply supported (or shear diaphragm supported) circular cylinders which is an advantage for the analysis of sound transmission into cylindrical enclosures. Both of these will be discussed here, although the derivations will not be included for the Donnell-Mushtari theory. On the other hand, some authors prefer to use Flügge's theory and for this reason the results derived using this approach will also be given, but the derivations will be omitted.

There are three different methods that have been used by various authors to derive the equations of motion for a curved shell (of which the circular cylinder is a special case). The method most widely used applies Newton's laws by summing the forces and moments that act on a shell element in much the same way as was done for a beam element in Section 2.3.3.3. The second method begins with the equations of motion, for an infinitesimal element, from the three dimensional theory of elasticity, and integrates these equations over the shell wall thickness to obtain the equations for a shell element. The third method is a class of variational methods, one of which involves the use of Hamilton's principle, which was discussed earlier. This latter method will be outlined briefly here.

Although the various wave types could be considered independently for beams and plates, for shells it is necessary to consider the displacements of the surface in all three directions (radially, tangentially and axially) simultaneously. This is mainly because a radial displacement of the wall of a shell or cylinder produces tensile or compressive tangential and axial membrane stresses, causing bending waves to couple with longitudinal and shear waves.

The assumptions made in the following analysis are as follows:

1. The cylinder material is isotropic.
2. The cylinder wall thickness is uniform and small compared with the cylinder radius and length.
3. Strains and displacements are sufficiently small so that quantities of second and higher order in the strain displacement relations may be neglected in comparison with first order terms.
4. The transverse normal stress, σ_{zz}, is small and may be neglected in comparison with the other normal stress components.
5. Normals to the undeformed middle surface of the wall of the cylinder remain straight and normal to the deformed middle surface.

The stresses and their sign conventions for a cylindrical shell element are illustrated in Figure 2.21.

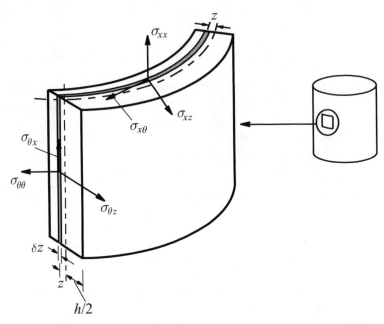

FIGURE 2.21 Cylinder element showing normal and shear stresses.

For the element shown in Figure 2.21, a is its radius of curvature, σ_{xz}, $\sigma_{\theta z}$, $\sigma_{x\theta}$ and $\sigma_{\theta x}$ are shear stresses, σ_{xx} and $\sigma_{\theta\theta}$ are normal stresses in the axial and tangential directions, respectively, and z is the distance of the infinitesimal segment, δz, from the central axis of the shell element. Note that z rather than r is used as the radial coordinate, as it has its origin at the centre of the shell element rather than the centre of curvature of the element.

The third assumption outlined above implies that the normal stress, $\sigma_{zz} = 0$. The fourth assumption is known as Kirchoff's hypothesis and implies that:

$$\sigma_{xz} = \sigma_{\theta z} = e_{xy} = e_{\theta z} = 0 \qquad (2.259)$$

where e_{ik} represents the shear strain at element, δz. In practice, σ_{xz} and $\sigma_{\theta z}$ are small but non-zero, as their integrals must supply the shearing forces needed for equilibrium.

The strain displacement equations for a circular cylinder may be derived from the general three-dimensional equations of motion (Leissa, 1973), and for the segment, δz, they may be written for Donnell–Mushtari, Goldenveizer–Novozhilov and Flügge shell theories as:

$$e_{xx} = \epsilon_{xx} + z\kappa_x \qquad (2.260)$$

and for Goldenveizer–Novozhilov and Flügge shell theories only, as:

$$e_{\theta\theta} = \frac{1}{(1 + z/a)}(\epsilon_{\theta\theta} + z\kappa_\theta) \qquad (2.261)$$

and for Donnell–Mushtari shell theory only, as:

$$e_{\theta\theta} = \epsilon_{\theta\theta} + z\kappa_\theta \qquad (2.262)$$

and for Goldenveiser–Novozhilov and Flügge shell theories only, as:

$$e_{x\theta} = \frac{1}{(1 + z/a)}\left[\epsilon_{x\theta} + z\left(1 + \frac{z}{2a}\right)\tau\right] \qquad (2.263)$$

and for Donnell–Mushtari shell theory only, as:

$$e_{x\theta} = \epsilon_{x\theta} + z\tau \tag{2.264}$$

where e_{xx}, $e_{\theta\theta}$ and $e_{x\theta}$ are the normal and shear strains of the arbitrary segment, δz, and ϵ_{xx}, $\epsilon_{\theta\theta}$ and $\epsilon_{x\theta}$ are the normal and shear strains of the surface in the middle of the wall thickness (mid-surface, $z = 0$). τ is the angular twist of this mid-surface and κ_x and κ_θ are the changes in curvature of the same surface. These six latter quantities are (for both Goldenveiser and Flügge shell theories):

$$\epsilon_{xx} = \frac{\partial \xi_x}{\partial x} \tag{2.265}$$

$$\epsilon_{\theta\theta} = \frac{1}{a}\frac{\partial \xi_\theta}{\partial \theta} + \frac{w}{a} \tag{2.266}$$

$$\epsilon_{x\theta} = \frac{1}{a}\frac{\partial \xi_x}{\partial \theta} + \frac{\partial \xi_\theta}{\partial x} \tag{2.267}$$

$$\tau = -\frac{2}{a}\frac{\partial^2 w}{\partial x \partial \theta} + \frac{2}{a}\frac{\partial \xi_\theta}{\partial x} \tag{2.268}$$

$$\kappa_x = -\frac{\partial^2 w}{\partial x^2} \tag{2.269}$$

$$\kappa_\theta = \frac{1}{a^2}\left[\frac{\partial \xi_\theta}{\partial \theta} - \frac{\partial^2 w}{\partial \theta^2}\right] \tag{2.270}$$

Note the presence of the additional strain caused by the radial displacement, which is not taken into account in Equation (2.47). In the preceding equations, ξ_x, ξ_θ and w are the axial, tangential and radial displacements, respectively, of the cylinder (see Figure 2.22). Note that w is positive outwards. For Donnell–Mushtari theory, the same previous six equations apply except that the term involving ξ_θ in Equations (2.268) and (2.270) is omitted.

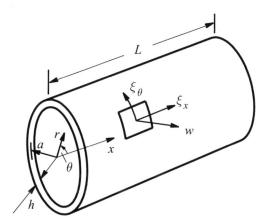

FIGURE 2.22 Circular cylinder coordinates and displacements.

The equations of motion of a thin cylindrical shell will now be derived using the Goldenveiser–Novozhilov method. Applying the Kirchoff hypothesis to the expression for strain energy derived from the theory of elasticity for a circular cylinder, the following is obtained (Leissa, 1973):

$$E_p = \frac{1}{2}\iiint\limits_V (\sigma_{xx}e_{xx} + \sigma_{\theta\theta}e_{\theta\theta} + \sigma_{x\theta}e_{x\theta})\mathrm{d}V \tag{2.271}$$

where dV is an elemental volume, which, when expressed in cylindrical shell coordinates, is:

$$dV = (1 + z/a)a\,dx\,d\theta\,dz \tag{2.272}$$

Before Equation (2.271) can be further expanded, expressions for the stresses in terms of the strains are needed. Using the assumptions outlined previously, the well known three-dimensional form of Hooke's law can be written for the segment, δz, as:

$$e_{xx} = \frac{1}{E}(\sigma_{xx} - \nu\sigma_{\theta\theta}) \tag{2.273}$$

$$e_{\theta\theta} = \frac{1}{E}(\sigma_{\theta\theta} - \nu\sigma_{xx}) \tag{2.274}$$

$$e_{x\theta} = \frac{2(1+\nu)}{E}\sigma_{x\theta} \tag{2.275}$$

where ν is Poisson's ratio. Inverting the preceding three equations gives:

$$\sigma_{xx} = \frac{E}{(1-\nu^2)}(e_{xx} + \nu e_{\theta\theta}) \tag{2.276}$$

$$\sigma_{\theta\theta} = \frac{E}{(1-\nu^2)}(e_{\theta\theta} + \nu e_{xx}) \tag{2.277}$$

$$\sigma_{x\theta} = \frac{E}{2(1+\nu)}e_{x\theta} \tag{2.278}$$

Substituting Equations (2.276) to (2.278) into Equation (2.271) gives:

$$E_p = \frac{E}{2(1-\nu^2)}\iiint_V \left[e_{xx}^2 + e_{\theta\theta}^2 + 2\nu e_{xx}e_{\theta\theta} + \frac{(1-\nu)}{2}e_{x\theta}^2\right]dV \tag{2.279}$$

Substituting further the Equations (2.260), (2.261), (2.263) and (2.272) for ϵ_{xx}, $\epsilon_{\theta\theta}$, $\epsilon_{x\theta}$ and dV, respectively, into Equation (2.279) gives:

$$E_p = \frac{E}{2(1-\nu^2)}\int_{-h/2}^{h/2}\int_0^{2\pi}\int_0^L (Q_0 + zQ_1 + z^2Q_2)a\,dx\,d\theta\,dz \tag{2.280}$$

where:

$$Q_0 = (\epsilon_{xx} + \epsilon_{\theta\theta})^2 - 2(1-\nu)(\epsilon_{xx}\epsilon_{\theta\theta} - \epsilon_{x\theta}^2/4) \tag{2.281}$$

$$Q_1 = 2(\epsilon_{xx}\kappa_x + \epsilon_{\theta\theta}\kappa_\theta) + 2\nu(\epsilon_{xx}\kappa_\theta + \epsilon_{\theta\theta}\kappa_x) + (1-\nu)\tau\epsilon_{x\theta} + \frac{1}{a}(\epsilon_{xx}^2 - \epsilon_{\theta\theta}^2) \tag{2.282}$$

$$Q_2 = (\kappa_x + \kappa_\theta)^2 - 2(1-\nu)(\kappa_x\kappa_\theta - \tau^2/4) + \frac{2}{a}(\epsilon_{xx}\kappa_x - \epsilon_{\theta\theta}\kappa_\theta)$$
$$- \frac{(1-\nu)\tau\epsilon_{x\theta}}{2a} + \frac{\epsilon_{\theta\theta}^2}{a^2} + \frac{(1-\nu)\epsilon_{x\theta}^2}{2a^2} \tag{2.283}$$

Integrating Equation (2.280) with respect to z gives:

$$E_p = \frac{Eh}{2(1-\nu^2)}\int_0^{2\pi}\int_0^L \left(Q_0 + \frac{h^2}{12}Q_2\right)a\,dx\,d\theta \tag{2.284}$$

The equations of motion for the cylinder may now be derived by invoking Hamilton's variational principle. That is:

$$\delta \int_{t_1}^{t_2} (E_k - E_p)\mathrm{d}t = 0 \tag{2.285}$$

The kinetic energy E_k of the cylinder is:

$$E_k = \frac{1}{2}\rho_m h \int_0^{2\pi} \int_0^L \left[\left(\frac{\partial \xi_x}{\partial t}\right)^2 + \left(\frac{\partial \xi_\theta}{\partial t}\right)^2 + \left(\frac{\partial w}{\partial t}\right)^2\right] a\, \mathrm{d}x\, \mathrm{d}\theta \tag{2.286}$$

Substituting Equations (2.265) to (2.270) into Equations (2.281) and (2.283), then substituting the result into Equation (2.284) gives the potential energy in terms of the displacements, ξ_x, ξ_θ and w and their partial derivatives. Substituting this result and Equation (2.286) into Equation (2.285) gives the following:

$$\delta \int_{t_1}^{t_2} \int_0^{2\pi} \int_0^L \mathrm{F}\left(\xi_x,\ \xi_\theta,\ w,\ \frac{\partial \xi_x}{\partial x},\ \frac{\partial \xi_x}{\partial \theta},\ \frac{\partial \xi_x}{\partial t},\ \frac{\partial \xi_\theta}{\partial x},\ \frac{\partial \xi_\theta}{\partial \theta},\ \frac{\partial \xi_\theta}{\partial t},\right.$$
$$\left.\frac{\partial w}{\partial x},\ \frac{\partial w}{\partial \theta},\ \frac{\partial w}{\partial t},\ \frac{\partial^2 w}{\partial x^2},\ \frac{\partial^2 w}{\partial \theta^2},\ \frac{\partial^2 w}{\partial x \partial \theta}\right) \mathrm{d}x\, \mathrm{d}\theta\, \mathrm{d}t = 0 \tag{2.287}$$

where the parameters, ξ_x, $\dots$, $\dfrac{\partial^2 w}{\partial \theta^2}$ are functions of x, θ and t and the function, F, is equal to $(E_k - E_p)/(\mathrm{d}x\, \mathrm{d}\theta)$. From the calculus of variations discussed in Section 2.2.2, the conditions that allow Equation (2.287) to be satisfied are the Lagrange equations:

$$\frac{\partial \mathrm{F}}{\partial \xi_x} - \frac{\partial}{\partial x}\left(\frac{\partial \mathrm{F}}{\partial \xi_x^x}\right) - \frac{\partial}{\partial \theta}\left(\frac{\partial \mathrm{F}}{\partial \xi_x^\theta}\right) - \frac{\partial}{\partial t}\left(\frac{\partial \mathrm{F}}{\partial \xi_x^t}\right) = 0 \tag{2.288}$$

$$\frac{\partial \mathrm{F}}{\partial \xi_\theta} - \frac{\partial}{\partial x}\left(\frac{\partial \mathrm{F}}{\partial \xi_\theta^x}\right) - \frac{\partial}{\partial \theta}\left(\frac{\partial \mathrm{F}}{\partial \xi_\theta^\theta}\right) - \frac{\partial}{\partial t}\left(\frac{\partial \mathrm{F}}{\partial \xi_\theta^t}\right) = 0 \tag{2.289}$$

$$\frac{\partial \mathrm{F}}{\partial w} - \frac{\partial}{\partial x}\left(\frac{\partial \mathrm{F}}{\partial w^x}\right) - \frac{\partial}{\partial \theta}\left(\frac{\partial \mathrm{F}}{\partial w^\theta}\right) - \frac{\partial}{\partial t}\left(\frac{\partial \mathrm{F}}{\partial w^t}\right) + \frac{\partial^2}{\partial x^2}\left(\frac{\partial \mathrm{F}}{\partial w^{xx}}\right)$$
$$+ \frac{\partial^2}{\partial x\, \partial \theta}\left(\frac{\partial \mathrm{F}}{\partial w^{x\theta}}\right) + \frac{\partial^2}{\partial \theta^2}\left(\frac{\partial \mathrm{F}}{\partial w^{\theta\theta}}\right) = 0 \tag{2.290}$$

where, for example, $\dfrac{\partial \mathrm{F}}{\partial \xi_x^x}$ denotes the partial derivative of the function, F, with respect to the function, $\dfrac{\partial \xi_x}{\partial x}$, and $w^{x\theta} = \dfrac{\partial^2 w}{\partial x \partial \theta}$.

Replacing F by $(E_k - E_p)/(\mathrm{d}x\, \mathrm{d}\theta)$ and substituting Equations (2.284) and (2.286) for E_p and E_k, respectively, into Equations (2.288) to (2.290) gives the required cylinder equations of motion (or the wave equations). These are then:

$$\nu\frac{\partial \xi_x}{\partial x} + \frac{1}{a}\frac{\partial \xi_\theta}{\partial \theta} + \frac{w}{a} + \frac{h^2}{12a^2}\left[-a(2-\nu)\frac{\partial^3 \xi_\theta}{\partial x^2 \partial \theta} - \frac{1}{a}\frac{\partial^3 \xi_\theta}{\partial \theta^3} + a^3\frac{\partial^4 w}{\partial x^4}\right.$$
$$\left. + 2a\frac{\partial^4 w}{\partial x^2 \partial \theta^2} + \frac{1}{a}\frac{\partial^4 w}{\partial \theta^4}\right] = \frac{-\rho_m a(1-\nu^2)}{E}\frac{\partial^2 w}{\partial t^2} \tag{2.291}$$

$$a\frac{\partial^2 \xi_x}{\partial x^2} + \frac{(1-\nu)}{2a}\frac{\partial^2 \xi_x}{\partial \theta^2} + \frac{(1+\nu)}{2}\frac{\partial^2 \xi_\theta}{\partial x\,\partial\theta} + \nu\frac{\partial w}{\partial x} = \frac{\rho_m a(1-\nu^2)}{E}\frac{\partial^2 \xi_x}{\partial t^2} \qquad (2.292)$$

$$\frac{(1+\nu)}{2}\frac{\partial^2 \xi_x}{\partial x\,\partial\theta} + \frac{a(1-\nu)}{2}\frac{\partial^2 \xi_\theta}{\partial x^2} + \frac{1}{a}\frac{\partial^2 \xi_\theta}{\partial \theta^2} + \frac{1}{a}\frac{\partial w}{\partial\theta}$$
$$+ \frac{h^2}{12a^2}\left[2(1-\nu)a\frac{\partial^2 \xi_\theta}{\partial x^2} + \frac{1}{a}\frac{\partial^2 \xi_\theta}{\partial \theta^2} - a(2-\nu)\frac{\partial^3 w}{\partial x^2\,\partial\theta} - \frac{1}{a}\frac{\partial^3 w}{\partial\theta^3}\right] = \frac{\rho_m a(1-\nu^2)}{E}\frac{\partial^2 \xi_\theta}{\partial t^2} \quad (2.293)$$

In matrix form, the equations of motion are:

$$\begin{bmatrix} a_{11} & a_{12} & a_{13} \\ a_{21} & a_{22} & a_{23} \\ a_{31} & a_{32} & a_{33} \end{bmatrix}\begin{bmatrix} \xi_x \\ \xi_\theta \\ w \end{bmatrix} + \frac{h^2}{12a^2}\begin{bmatrix} b_{11} & b_{12} & b_{13} \\ b_{21} & b_{22} & b_{23} \\ b_{31} & b_{32} & b_{33} \end{bmatrix}\begin{bmatrix} \xi_x \\ \xi_\theta \\ w \end{bmatrix} = 0 \qquad (2.294)$$

where:

$$a_{11} = a\frac{\partial^2}{\partial x^2} + \frac{(1-\nu)}{2a}\frac{\partial^2}{\partial\theta^2} - \frac{\rho_m a(1-\nu^2)}{E}\frac{\partial^2}{\partial t^2} \qquad (2.295)$$

$$a_{12} = \frac{(1+\nu)}{2}\frac{\partial^2}{\partial x\,\partial\theta} \qquad (2.296)$$

$$a_{13} = \nu\frac{\partial}{\partial x} \qquad (2.297)$$

$$a_{21} = \frac{(1+\nu)}{2}\frac{\partial^2}{\partial x\,\partial\theta} \qquad (2.298)$$

$$a_{22} = \frac{a(1-\nu)}{2}\frac{\partial^2}{\partial x^2} + \frac{1}{a}\frac{\partial^2}{\partial\theta^2} - \frac{\rho_m a(1-\nu^2)}{E}\frac{\partial^2}{\partial t^2} \qquad (2.299)$$

$$a_{23} = \frac{1}{a}\frac{\partial}{\partial\theta} \qquad (2.300)$$

$$a_{31} = \nu\frac{\partial}{\partial x} \qquad (2.301)$$

$$a_{32} = \frac{1}{a}\frac{\partial}{\partial\theta} \qquad (2.302)$$

$$a_{33} = \frac{1}{a} + \frac{h^2}{12a^2}\left[a^3\frac{\partial^4}{\partial x^4} + 2a\frac{\partial^4}{\partial x^2\partial\theta^2} + \frac{1}{a}\frac{\partial^4}{\partial\theta^4}\right] + \frac{\rho_m a(1-\nu^2)}{E}\frac{\partial}{\partial t^2} \qquad (2.303)$$

$$b_{11} = b_{12} = b_{13} = b_{21} = b_{31} = b_{33} = 0 \qquad (2.304)$$

$$b_{22} = 2a(1-\nu)\frac{\partial^2}{\partial x^2} + \frac{1}{a}\frac{\partial^2}{\partial\theta^2} \qquad (2.305)$$

$$b_{32} = b_{23} = -a(2-\nu)\frac{\partial^3}{\partial x^2\partial\theta} - \frac{1}{a}\frac{\partial^3}{\partial\theta^3} \qquad (2.306)$$

If the Donnell–Mushtari theory had been used to derive the equations of motion, the coefficients, a_{ij}, would remain the same but all b_{ij} would be zero. If Flügge's theory had been used, the coefficients, a_{ij}, would remain the same but the coefficients, b_{ij}, would be replaced with the following:

$$b_{12} = b_{21} = 0 \qquad (2.307)$$

$$b_{11} = \frac{(1-\nu)}{2a}\frac{\partial^2}{\partial\theta^2} \qquad (2.308)$$

$$b_{13} = -a^2\frac{\partial^3}{\partial x^3} + \frac{(1-\nu)}{2}\frac{\partial^3}{\partial x\,\partial\theta^2} \qquad (2.309)$$

$$b_{22} = \frac{3a(1-\nu)}{2} \frac{\partial^2}{\partial x^2} \tag{2.310}$$

$$b_{23} = b_{32} = -\frac{a(3-\nu)}{2} \frac{\partial^3}{\partial x^2 \partial \theta} \tag{2.311}$$

$$b_{31} = a^2 \frac{\partial^3}{\partial x^3} + \frac{(1-\nu)}{2} \frac{\partial^3}{\partial x \partial \theta^2} \tag{2.312}$$

$$b_{33} = a + 2a \frac{\partial^2}{\partial \theta^2} \tag{2.313}$$

Equation (2.294) is used together with appropriate boundary conditions to determine the resonance frequencies and mode shapes for a particular cylinder. Generally only modes involving the radial displacement, w, are of interest and analyses are usually restricted accordingly.

The solution for harmonic vibration involves assuming a form of solution for ξ_x, ξ_θ and w and then substituting this back into Equation (2.294). The determinant of the a_{ij} coefficient matrix is then set equal to zero which allows the eigen frequencies to be determined. These frequencies are then used together with the assumed solutions for ξ_x, ξ_θ and w in Equation (2.294) (see Equations (2.377) to (2.379) and (2.380) to (2.382)) to determine the corresponding mode shapes.

2.3.5.1 Boundary Conditions

These are specified in terms of cylinder displacements at each end and the forces and moments acting on the cylinder at each end. Four boundary conditions must be specified for each end, one from each of the pairs listed below:

$$\xi_x = 0 \text{ or } N_x = 0 \tag{2.314}$$

$$\xi_\theta = 0 \text{ or } N_{x\theta} + \frac{M_{x\theta}}{a} = 0 \tag{2.315}$$

$$w = 0 \text{ or } Q_x + \frac{1}{a} \frac{\partial M_{x\theta}}{\partial \theta} = 0 \tag{2.316}$$

$$\frac{\partial w}{\partial x} = 0 \text{ or } M_x = 0 \tag{2.317}$$

The quantities, N_x, $N_{x\theta}$, Q_x, $M_{x\theta}$ and M_x, have not yet been defined. The first two are in-plane forces, the third is a force normal to the cylinder surface, the fourth is a twisting moment and the fifth term is a bending moment. They are defined in Figure 2.23 for an element of a cylinder. The sign convention follows that of Leissa (1973).

Consider the faces of the element shown in Figure 2.21. The resultant forces per unit length acting on each face can be calculated by integrating the stresses over the face thickness. For the vertical face:

$$N_\theta = \int_{-h/2}^{h/2} \sigma_{\theta\theta} \, \mathrm{d}z \tag{2.318}$$

$$N_{\theta x} = \int_{-h/2}^{h/2} \sigma_{\theta x} \, \mathrm{d}z \tag{2.319}$$

$$Q_\theta = \int_{-h/2}^{h/2} \sigma_{\theta z} \, \mathrm{d}z \tag{2.320}$$

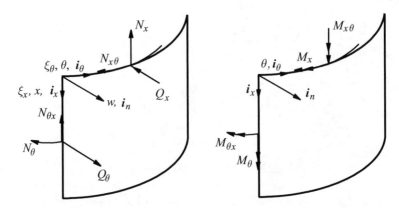

FIGURE 2.23 Forces and moments acting on a cylindrical element.

For the top surface, the situation is a little different, as the surface is curved. The arc length of the middle surface, representing an incremental angle, $d\theta$, is:

$$ds_c = a\,d\theta \tag{2.321}$$

where a is the cylinder radius. The arc length for other surfaces parallel to the middle surface and spaced a distance, z, from it is:

$$ds_z = a(1 + z/a)\,d\theta \tag{2.322}$$

The difference between the two is $(1 + z/a)$. Thus, the net forces per unit length on the top horizontal face of Figure 2.21 are:

$$N_x = \int_{-h/2}^{h/2} \sigma_{xx}(1 + z/a)\,dz \tag{2.323}$$

$$N_{x\theta} = \int_{-h/2}^{h/2} \sigma_{x\theta}(1 + z/a)\,dz \tag{2.324}$$

$$Q_x = \int_{-h/2}^{h/2} \sigma_{xz}(1 + z/a)\,dz \tag{2.325}$$

Similarly, the moment of the infinitesimal force, $\sigma_{xx}ds_c\,dz$, about the θ line through the centre of the section is simply $z\sigma_{xx}ds_c\,dz$. The moment resultant, M_x, is obtained by integrating the moment over the thickness and dividing by $a\,d\theta$. Thus:

$$M_x = \int_{-h/2}^{h/2} \sigma_{xx}(1 + z/a)z\,dz \tag{2.326}$$

$$M_{\theta x} = \int_{-h/2}^{h/2} \sigma_{\theta x}z\,dz \tag{2.327}$$

$$M_\theta = \int_{-h/2}^{h/2} \sigma_{\theta\theta} z \, \mathrm{d}z \tag{2.328}$$

$$M_{x\theta} = \int_{-h/2}^{h/2} \sigma_{x\theta}(1 + z/a) z \, \mathrm{d}z \tag{2.329}$$

To express the net forces and moments in terms of in-plane and normal cylinder displacements, they must first be expressed in terms of the strain components. To do this it is necessary to return to the functional of Equation (2.271) and take its variation:

$$\delta E_p = \iiint_{V'} \left(\sigma_{xx}\delta e_{xx} + \sigma_{\theta\theta}\delta e_{\theta\theta} + \sigma_{x\theta}\delta e_{x\theta} \right) \mathrm{d}V \tag{2.330}$$

Substituting Equations (2.260), (2.261) and (2.263) for the total strains into Equation (2.330) gives (for Goldenveiser–Novozhilov theory):

$$\delta E_p = \int_0^L \int_0^{2\pi} \int_{-h/2}^{h/2} \left[\sigma_{xx}(1 + z/a)(\delta\epsilon_{xx} + z\delta\kappa_x) + \sigma_{\theta\theta}(\delta\epsilon_{\theta\theta} + z\delta\kappa_\theta) \right.$$

$$\left. + \sigma_{x\theta}(\delta\epsilon_{x\theta} + z(1 + z/2a)\tau) \right] a \, \mathrm{d}z \, \mathrm{d}\theta \, \mathrm{d}x \tag{2.331}$$

Making use of the definitions for moments and forces of Equations (2.318) to (2.329), Equation (2.331) can be written as:

$$\delta E_p = \int_0^L \int_0^{2\pi} \left[N_x\delta\epsilon_{xx} + N_\theta\delta\epsilon_{\theta\theta} + N_{\theta x}\delta\epsilon_{x\theta} + M_x\delta\kappa_x + M_\theta\delta\kappa_\theta \right.$$

$$\left. + \frac{1}{2}\left(M_{x\theta} + M_{\theta x} \right)\delta\tau \right] a \, \mathrm{d}\theta \, \mathrm{d}x \tag{2.332}$$

Returning to Equation (2.284), Leissa (1973) shows that the last four terms in Q_2 (see Equation (2.283)) can be neglected and the following is obtained:

$$E_p = \frac{Eh}{2(1 - \nu^2)} \int_0^L \int_0^{2\pi} \left\{ (\epsilon_{xx} + \epsilon_{\theta\theta})^2 - 2(1 - \nu)\left(\epsilon_{xx}\epsilon_{\theta\theta} - \epsilon_{x\theta}^2/4 \right) \right.$$

$$\left. + \frac{h^2}{12}\left[(\kappa_x + \kappa_\theta)^2 - 2(1 - \nu)(\kappa_x\kappa_\theta - \tau^2/4) \right] \right\} a \, \mathrm{d}\theta \, \mathrm{d}x \tag{2.333}$$

Taking the variation of Equation (2.333) yields:

$$\delta E_p = \frac{Eh}{1 - \nu^2} \int_0^L \int_0^{2\pi} \left\{ (\epsilon_{xx} + \nu\epsilon_{\theta\theta})\,\delta\epsilon_{xx} + (\epsilon_{\theta\theta} + \nu\epsilon_{xx})\,\delta\epsilon_{\theta\theta} + \frac{(1 - \nu)}{2}\epsilon_{x\theta}\,\delta\epsilon_{x\theta} \right.$$

$$\left. + \frac{h^2}{12}\left[(\kappa_x + \nu\kappa_\theta)\,\delta\kappa_x + (\kappa_\theta + \nu\kappa_x)\delta\kappa_\theta + \frac{(1 - u)}{2}\tau\,\delta\tau \right] \right\} a \, \mathrm{d}\theta \, \mathrm{d}x \tag{2.334}$$

Comparing Equations (2.332) and (2.334) gives expressions (corresponding to the Goldenveiser–Novozhilov analysis method) for the moments and forces in terms of cylinder strains as:

$$N_x = \frac{Eh}{(1 - \nu^2)}\left(\epsilon_{xx} + \nu\epsilon_{\theta\theta} \right) \tag{2.335}$$

$$N_\theta = \frac{Eh}{(1-\nu^2)}(\epsilon_{\theta\theta} + \nu\epsilon_{xx}) \tag{2.336}$$

$$N_{\theta x} = \frac{Eh}{2(1+\nu)}\epsilon_{x\theta} \tag{2.337}$$

$$M_x = \frac{Eh^3}{12(1-\nu^2)}(\kappa_x + \nu\kappa_\theta) \tag{2.338}$$

$$M_\theta = \frac{Eh^3}{12(1-\nu^2)}(\kappa_\theta + \nu\kappa_x) \tag{2.339}$$

$$\frac{1}{2}(M_{x\theta} + M_{\theta x}) = \frac{Eh^3\tau}{24(1+\nu)} \tag{2.340}$$

From symmetry of the stress tensor in Figure 2.23, $\sigma_{x\theta} = \sigma_{\theta x}$. Thus, using Equations (2.319), (2.324) and (2.329) it can be shown that:

$$N_{x\theta} = N_{\theta x} + \frac{M_{\theta x}}{a} \tag{2.341}$$

Also, from Equations (2.327) and (2.329) the following is obtained:

$$M_{\theta x} = M_{x\theta} + \frac{1}{a}\int_{-h/2}^{h/2} \sigma_{x\theta}\, z^2 \, \mathrm{d}z \tag{2.342}$$

It can be easily shown (Leissa, 1973) that the second term on the right-hand side of Equation (2.342) is very small compared with the first. Thus, with the help of Equation (2.340), the following is obtained:

$$M_{x\theta} = M_{\theta x} = \frac{Eh^3\tau}{24(1+\nu)} \tag{2.343}$$

Substituting Equations (2.337) and (2.343) into Equation (2.341) gives:

$$N_{x\theta} = \frac{Eh}{2(1+\nu)}\left(\epsilon_{x\theta} + \frac{h^2}{12a}\tau\right) \tag{2.344}$$

Equations (2.335) to (2.340), (2.343) and (2.344) also apply to the analysis using Donnell–Mushtari theory, except for Equation (2.344), where the term containing τ is omitted. If the Flügge analysis method is used, Equations (2.335) to (2.339), (2.343) and (2.344) become:

$$N_x = \frac{Eh}{(1-\nu^2)}\left[\epsilon_{xx} + \nu\epsilon_{\theta\theta} + \frac{h^2\kappa_x}{12a}\right] \tag{2.345}$$

$$N_\theta = \frac{Eh}{(1-\nu^2)}\left[\epsilon_{\theta\theta} + \nu\epsilon_{xx} - \frac{h^2}{12a}\left(\kappa_\theta - \frac{\epsilon_{\theta\theta}}{a}\right)\right] \tag{2.346}$$

$$N_{x\theta} = \frac{Eh}{2(1+\nu)}\left[\epsilon_{x\theta} + \frac{h^2\tau}{24a}\right] \tag{2.347}$$

$$N_{\theta x} = \frac{Eh}{2(1+\nu)}\left[\epsilon_{x\theta} - \frac{h^2}{12a}\left(\frac{\tau}{2} - \frac{\epsilon_{x\theta}}{a}\right)\right] \tag{2.348}$$

$$M_x = \frac{Eh^3}{12(1-\nu^2)}\left[\kappa_x + \nu\kappa_\theta + \frac{\epsilon_{xx}}{a}\right] \tag{2.349}$$

$$M_\theta = \frac{Eh^3}{12(1-\nu^2)}\left[\kappa_\theta + \nu\kappa_x - \frac{\epsilon_{\theta\theta}}{a}\right] \tag{2.350}$$

$$M_{x\theta} = \frac{Eh^3\tau}{24(1+\nu)} \tag{2.351}$$

$$M_{\theta x} = \frac{Eh^3}{24(1+\nu)}\left(\tau - \frac{\epsilon_{x\theta}}{a}\right) \tag{2.352}$$

Expressions for Q_x and Q_θ, in terms of the other forces and moments, will be derived later by consideration of the equations of motion of a shell element (see Equations (2.374) and (2.372)). However, for convenience, the results will be written down here. Thus:

$$Q_x = \frac{\partial M_x}{\partial x} + \frac{1}{a}\frac{\partial M_{\theta x}}{\partial \theta} \tag{2.353}$$

$$Q_\theta = \frac{1}{a}\frac{\partial M_\theta}{\partial \theta} + \frac{\partial M_{x\theta}}{\partial x} \tag{2.354}$$

Note that Equations (2.345) and (2.346) only apply if there is no external loading acting on the cylinder, and they are valid for both Flügge and Goldenveiser–Novozhilov shell theories. However, the Donnell–Mushtari theory assumes that $Q_x = Q_\theta = 0$.

The quantities on the left-hand side of Equations (2.335) to (2.339), (2.343) and (2.344) can be expressed in terms of the displacements, ξ_x, ξ_θ and w, by making use of the relationships expressed in Equations (2.265) to (2.270) to give the following:

$$N_x = \frac{Eh}{a(1-\nu^2)}\left(\frac{\partial \xi_x}{\partial x} + \nu\frac{\partial \xi_\theta}{\partial \theta} + \nu w\right) \tag{2.355}$$

$$N_\theta = \frac{Eh}{a(1-\nu^2)}\left(\frac{\partial \xi_\theta}{\partial \theta} + w + \nu\frac{\partial \xi_x}{\partial x}\right) \tag{2.356}$$

$$N_{x\theta} = \frac{Eh}{2a(1+\nu)}\left[\frac{\partial \xi_x}{\partial \theta} + \frac{\partial \xi_\theta}{\partial x} + \frac{h^2}{6a^2}\left(-\frac{\partial^2 w}{\partial x\,\partial \theta} + \frac{\partial \xi_\theta}{\partial x}\right)\right] \tag{2.357}$$

$$N_{\theta x} = \frac{Eh}{2a(1+\nu)}\left(\frac{\partial \xi_x}{\partial \theta} + \frac{\partial \xi_\theta}{\partial x}\right) \tag{2.358}$$

$$M_{x\theta} = M_{\theta x} = \frac{Eh^3}{(1+\nu)}\left(-\frac{\partial^2 w}{\partial x\,\partial \theta} + \frac{\partial \xi_\theta}{\partial x}\right) \tag{2.359}$$

$$Q_x = -\frac{Eh}{a(1-\nu^2)}\left(\frac{\partial^2 \xi_x}{\partial x^2} + \frac{\nu\partial^2 \xi_\theta}{\partial x\,\partial \theta} + \nu\frac{\partial w}{\partial x}\right) - \frac{Eh}{2a(1+\nu)}\left(\frac{\partial^2 \xi_x}{\partial \theta^2} + \frac{\partial^2 \xi_\theta}{\partial x\,\partial \theta}\right) \tag{2.360}$$

$$Q_\theta = -\frac{Eh}{a(1-\nu^2)}\left(\frac{\partial^2 \xi_\theta}{\partial \theta^2} + \frac{\partial w}{\partial \theta} + \nu\frac{\partial^2 \xi_x}{\partial x\,\partial \theta}\right)$$
$$- \frac{Eh}{2a(1+\nu)}\left[\frac{\partial^2 \xi_x}{\partial x\,\partial \theta} + \frac{\partial^2 \xi_\theta}{\partial x^2} + \frac{h^2}{6a^2}\left(-\frac{\partial^3 w}{\partial x^2\,\partial \theta} + \frac{\partial^2 \xi_\theta}{\partial x^2}\right)\right] \tag{2.361}$$

The boundary condition which is closest to the equivalent of a simply supported plate boundary condition is referred to as the shear diaphragm (or SD) condition where:

$$w = M_x = N_x = \xi_\theta = 0 \tag{2.362}$$

That is, the cylinder is closed at the end with a thin flat circular cover plate. The plate has considerable stiffness in its own plane, thus restraining the ν and w components of cylinder displacement. As the end plate is not very stiff in its transverse plane, it would generate very little bending moment, M_x, and very little longitudinal membrane force, N_x, which explains the choice of boundary conditions for this case.

For a cylinder that has free ends, all of the boundary conditions on the left-hand side of Equations (2.314) to (2.317) would be satisfied.

2.3.5.2   Cylinder Equations of Motion: Alternative Derivation

Referring to the cylinder element of thickness, h, shown in Figure 2.21, and considering its equilibrium under the influence of internal force and moment resultants and external applied forces and moments, two equations of motion can be written, one involving forces and the other involving moments, as follows (Leissa, 1973):

$$\frac{\partial \boldsymbol{F}_x}{\partial x} dx + \frac{\partial \boldsymbol{F}_\theta}{\partial \theta} d\theta + \boldsymbol{q}\, a\, dx\, d\theta = 0 \qquad (2.363)$$

$$\frac{\partial \boldsymbol{M}_{xT}}{\partial x} dx + \frac{\partial \boldsymbol{M}_{\theta T}}{\partial \theta} d\theta - (\boldsymbol{F}_x \times \boldsymbol{i}_\theta) \frac{a\, d\theta}{2} - (\boldsymbol{F}_\theta \times \boldsymbol{i}_x) \frac{dx}{2}$$
$$+ \left(\boldsymbol{F}_x + \frac{\partial \boldsymbol{F}_x}{\partial x} dx \right) \times \left(dx \boldsymbol{i}_x + \frac{a\, d\theta}{2} \boldsymbol{i}_\theta \right) + \left(\boldsymbol{F}_\theta + \frac{\partial \boldsymbol{F}_\theta}{\partial \theta} d\theta \right) \left(a d\theta \boldsymbol{i}_\theta + \frac{dx}{2} \boldsymbol{i}_x \right) + \boldsymbol{M}_e\, a\, dx\, d\theta = 0$$
$$(2.364)$$

The external forces, $\boldsymbol{q}$, per unit area and moments, $\boldsymbol{M}_e$, per unit area are defined as (see Figure 2.23):

$$\boldsymbol{q} = q_x \boldsymbol{i}_x + q_\theta \boldsymbol{i}_\theta + q_n \boldsymbol{i}_n \qquad (2.365)$$

$$\boldsymbol{M}_e = M_{ex} \boldsymbol{i}_x + M_{e\theta} \boldsymbol{i}_\theta + M_{en} \boldsymbol{i}_n \qquad (2.366)$$

where $\boldsymbol{i}_x$, $\boldsymbol{i}_\theta$ and $\boldsymbol{i}_n$ are unit vectors defined in Figure 2.23. The total internal forces $\boldsymbol{F}_x$ and $\boldsymbol{F}_\theta$ are defined as:

$$\boldsymbol{F}_x = (N_x \boldsymbol{i}_x + N_{x\theta} \boldsymbol{i}_\theta + Q_x \boldsymbol{i}_n) a\, d\theta \qquad (2.367)$$

$$\boldsymbol{F}_\theta = (N_{\theta x} \boldsymbol{i}_x + + N_\theta \boldsymbol{i}_\theta + Q_\theta \boldsymbol{i}_n)\, dx \qquad (2.368)$$

The total internal moments $\boldsymbol{M}_{xT}$ and $\boldsymbol{M}_{\theta T}$ are defined as:

$$\boldsymbol{M}_{xT} = (-M_{x\theta} \boldsymbol{i}_x + M_x \boldsymbol{i}_\theta) a\, d\theta \qquad (2.369)$$

$$\boldsymbol{M}_{\theta T} = (-M_\theta \boldsymbol{i}_x + M_{\theta x} \boldsymbol{i}_\theta)\, dx \qquad (2.370)$$

Using Equations (2.365), (2.367) and (2.368), Equation (2.363) can be expanded into its three scalar components as follows (Leissa, 1973):

$$\frac{\partial N_x}{\partial x} + \frac{1}{a} \frac{\partial N_{\theta x}}{\partial \theta} + q_x = 0 \qquad (2.371)$$

$$\frac{1}{a} \frac{\partial N_\theta}{\partial \theta} + \frac{\partial N_{x\theta}}{\partial x} + \frac{Q_\theta}{a} + q_\theta = 0 \qquad (2.372)$$

Using Equations (2.365) to (2.370), Equation (2.364) can be expanded into its three scalar components (Leissa, 1973):

$$-\frac{N_\theta}{a} + \frac{\partial Q_x}{\partial x} + \frac{1}{a} \frac{\partial Q_\theta}{\partial \theta} + q_n = 0 \qquad (2.373)$$

$$\frac{\partial M_x}{\partial x} + \frac{1}{a} \frac{\partial M_{\theta x}}{\partial \theta} - Q_x + M_{e\theta} = 0 \qquad (2.374)$$

$$\frac{1}{a} \frac{\partial M_\theta}{\partial \theta} + \frac{\partial M_{x\theta}}{\partial x} - Q_\theta + M_{ex} = 0 \qquad (2.375)$$

$$N_{x\theta} - N_{\theta x} - \frac{M_{\theta x}}{a} = 0 \qquad (2.376)$$

Using Equations (2.374) and (2.375) with $M_{e\theta} = M_{ex} = 0$, Equations (2.345) and (2.346) are obtained.

Eliminating Q_θ and Q_x from Equations (2.372) and (2.373) by using Equations (2.374) and (2.375), the number of equations of motion can be reduced to three. (Note that Equation (2.376) is satisfied identically and is not a useful equation of motion.) Substitution of expressions for N_x, N_θ, $N_{x\theta}$, M_x, M_θ and $M_{x\theta}$ in terms of displacements, ξ_x, ξ_θ and w, into the three equations so obtained will result in the equations of motion previously derived using Hamilton's principle (Equations (2.291) to (2.293)).

2.3.5.3 Solution of the Equations of Motion

Functions of the following forms are found to describe the motion of the cylinder for all types of boundary condition:

$$\xi_x = U_n e^{\lambda s} \cos n\theta e^{j\omega t} \tag{2.377}$$

$$\xi_\theta = V_n e^{\lambda s} \sin n\theta e^{j\omega t} \tag{2.378}$$

$$w = W_n e^{\lambda s} \cos n\theta e^{j\omega t} \tag{2.379}$$

where U_n, V_n, W_n, λ and s are undetermined coefficients and n is an integer describing the circumferential displacement distribution, where $n =$ half the number of vibration nodes around the cylinder circumference (see Figure 2.24).

Note for shear diaphragm boundary conditions, the solutions become:

$$\xi_x = U_n \cos(\lambda s) \cos n\theta e^{j\omega t} \tag{2.380}$$

$$\xi_\theta = V_n \sin(\lambda s) \sin n\theta e^{j\omega t} \tag{2.381}$$

$$w = W_n \sin(\lambda s) \cos n\theta e^{j\omega t} \tag{2.382}$$

where $\lambda = m\pi a/L$ and $s = x/a$, so that $\lambda s = m\pi x/L$.

To find the resonance frequencies and mode shapes, the following steps are implemented (Warburton, 1965):

1. Substitute Equations (2.377) to (2.379) into the equations of motion (2.294). This will produce a quartic equation in w^2 with coefficients that are a function of ρ_m, a, ν, ω, E, h and n.

2. As there will be eight roots for λ (for each value of n) from the quartic equation derived in item 1 above, the quantity, $W_n e^{\lambda s}$ may be expressed in terms of eight real constants as:

$$W_n e^{\lambda s} = \sum_{r=1}^{8} B_r e^{\lambda_r s} \tag{2.383}$$

where λ_r is rth root of the quartic in λ^2 and B_r is the rth constant.

3. Use the equations of motion to find the ratios, ξ_x/w and $\xi_{\theta\theta}/w$, and then the quantities, $U_n e^{\lambda s}$ and $V_n e^{\lambda s}$, can also be written in terms of the constants, B_r.

4. Substitute the solutions for u, v and w into the eight boundary condition equations (four for each end of the cylinder) in eight unknown coefficients. The characteristic frequency equation is then found by setting the determinant of the unknown coefficients equal to zero. The eigen frequencies corresponding to a specific value of n are then found by solving this characteristic frequency equation. There will be three roots for each value of n.

For shear diaphragm end conditions, substituting Equations (2.380) to (2.382) into the equations of motion gives for the Donnell-Mushtari theory (Leissa, 1969):

$$
\begin{bmatrix}
\left[-\lambda^2 - \dfrac{(1-\nu)}{2}n^2 + \Omega^2\right] & \dfrac{(1+\nu)}{2}\lambda n & \nu\lambda \\[2mm]
\dfrac{(1+\nu)}{2}\lambda n & \left[-n^2 - \dfrac{(1-\nu)}{2}\lambda^2 + \Omega^2\right] & -n \\[2mm]
-\nu\lambda & n & \zeta
\end{bmatrix}
\begin{bmatrix} U_n \\ V_n \\ w \end{bmatrix}
=
\begin{bmatrix} 0 \\ 0 \\ 0 \end{bmatrix}
\tag{2.384}
$$

where $\zeta = \left[1 + \dfrac{h^2}{12a^2}(\lambda^2 + n^2)^2 - \Omega^2\right]$, $\Omega^2 = \rho(1-\nu^2)a^2\omega_m n^2/E$ and E is Young's modulus.

Setting the determinant of the coefficient matrix in Equation (2.384) equal to 0 gives the following solutions for the frequency parameter, Ω (Donnell-Mushtari theory):

$$
\Omega^2 = \begin{cases}
0, \ \left(1 + \dfrac{h^2}{12a^2}\right); & \text{if } n = 0 \\[4mm]
\dfrac{1}{2}\left[1 + n^2 + \dfrac{h^2}{12a^2}(n^2-1)^2 \right. \\[3mm]
\left. \mp\sqrt{\left[1 + n^2 + \dfrac{h^2}{12a^2}(n^2-1)^2\right]^2 - \dfrac{h^2}{3a^2}n^2(n-1)^2}\right]; & \text{if } n \neq 0
\end{cases}
\tag{2.385}
$$

where the root, $\Omega^2 = 0$ for $n = 0$ corresponds to a rigid body torsional rotation of the cylinder and the frequency parameter, Ω^2, are solutions of:

$$
\Omega^6 - \left(K_2 + \dfrac{h^2}{12a^2}\Delta K_2\right)\Omega^4 + \left(K_1 + \dfrac{h^2}{12a^2}\Delta K_1\right)\Omega^2 - \left(K_0 + \dfrac{h^2}{12a^2}\Delta K_0\right) = 0
\tag{2.386}
$$

where K_0, K_1 and K_2 are the same for all shell theories, while the modifying constants, ΔK_0, ΔK_1 and ΔK_2, differ between theories. The constants, K_0, K_1 and K_2, are given by (Leissa, 1973):

$$
\begin{cases}
K_0 = \dfrac{1}{2}\left(1-\nu\right)\left[(1-\nu^2)\lambda^4 + \dfrac{h^2}{12a^2}\left(n^2 + \lambda^2\right)^4\right] \\[3mm]
K_1 = \dfrac{1}{2}\left(1-\nu\right)\left[(3+2\nu)\lambda^2 + n^2 + (n^2 + \lambda^2)^2 + \dfrac{(3-\nu)}{(1-\nu)}\dfrac{h^2}{12a^2}\left(n^2 + \lambda^2\right)^3\right] \\[3mm]
K_2 = 1 + \dfrac{1}{2}\left(3-\nu\right)\left(n^2 + \lambda^2\right) + \dfrac{h^2}{12a^2}\left(n^2 + \lambda^2\right)^2
\end{cases}
\tag{2.387}
$$

where $\lambda = m\pi a/L$, $(m-1)$ is the number of nodal circles in the axial, x-direction (excluding the ends) and n represents half of the number of nodes around the circumference (see Figure 2.24).

For Donnell-Mushtari theory, $\Delta K_2 = \Delta K_1 = \Delta K_0 = 0$, and for Flügge theory, $\Delta K_2 = \Delta K_1 = 0$ and ΔK_0 is given by (Leissa, 1973):

$$
\Delta K_0 = \dfrac{1}{2}(1-\nu)\left[(4-2\nu)\lambda^2 n^2 + n^4 - 2\nu\lambda^6 - 6\lambda^4 n^2 - (8-2\nu)\lambda^2 n^4 - 2n^6\right]
\tag{2.388}
$$

For Goldenveiser–Novozhilov theory, the constants, ΔK_0, ΔK_1 and ΔK_2, are given by (Leissa, 1973):

$$
\begin{cases}
\Delta K_0 = \dfrac{1}{2}\left(1-\nu\right)\left[4(1-\nu^2)\lambda^4 + 4\lambda^2 n^2 + n^4 - (4-2\nu)(2+\nu)\lambda^4 n^2 - 8\lambda^2 n^4 - 2n^6\right] \\[3mm]
\Delta K_1 = 2(1-\nu)\lambda^2 + n^2 + 2(1-\nu)\lambda^4 - (2-\nu)\lambda^2 n^2 - \dfrac{1}{2}\left(3+\nu\right)n^4 \\[3mm]
\Delta K_2 = 2(1-\nu)\lambda^2 + n^2
\end{cases}
\tag{2.389}
$$

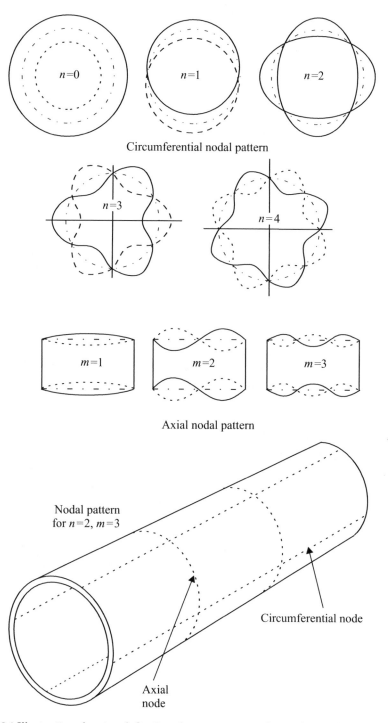

FIGURE 2.24 Illustration showing deflection shapes corresponding to low-order bending vibration modes of a circular-section cylinder with shear diaphragm ends. Any given point on the cylinder vibrates between the solid line and the dashed line. The rest position is shown by the dash-dot line.

Mode shapes may be found by returning to Equation (2.384) (Donnell-Mushtari theory), choosing two of the equations, and solving for the amplitude ratios, U_n/w and V_n/w:

$$\left[\begin{array}{c} U_n/w \\ V_n/w \end{array} \right] = \left[\begin{array}{cc} \left[-\lambda^2 - \dfrac{(1-\nu)}{2}n^2 + \Omega^2 \right] & \dfrac{(1+\nu)}{2}\lambda n \\[3mm] \dfrac{(1+\nu)}{2}\lambda n & \left[-n^2 - \dfrac{(1-\nu)}{2}\lambda^2 + \Omega^2 \right] \end{array} \right]^{-1} \left[\begin{array}{c} -\nu\lambda \\ n \end{array} \right] \qquad (2.390)$$

Three frequency parameters, Λ, exist for each combination of n and λ and the lowest of these will usually result in U_n/w and V_n/w ratios less than 1, indicating that the motion is primarily radial in the w-direction. The lowest frequency parameters, corresponding to motion primarily in the w-direction, are provided in Table 2.11. Amplitude ratios, U/w and V/w, according to Flügge theory are provided in Table 2.12, where it can be seen that there are some modes corresponding to the lowest frequency parameter for which motion is not dominated by motion in the w-direction.

TABLE 2.11 Lowest frequency parameters (corresponding to primarily radial motion in the w-direction) for shear diaphragm supports at each end of the cylinder and $\nu = 0.3$, according to 3-D theory (after Leissa (1973))

a/h	n	$L/(ma)$					
		0.1	0.25	1.0	4.0	20.0	100.0
	0	10.459	2.2851	0.95808	0.46465	0.092930	0.018586
	1	10.467	2.2938	0.85641	0.25701	0.016106	0.00066503
20	2	10.491	2.3204	0.67549	0.12125	0.039233	0.034771
	3	10.533	2.3660	0.53929	0.12988	0.10948	0.10919
	4	10.590	2.4323	0.49234	0.21910	0.20901	0.20871
	0	1.1113	0.95799	0.94920	0.46465	0.092930	0.018586
	1	1.1105	0.95199	0.84495	0.25688	0.016101	0.00266482
500	2	1.1089	0.93446	0.65215	0.11269	0.0054524	0.0015623
	3	1.1063	0.90673	0.48103	0.058009	0.0050372	0.0043863
	4	1.1028	0.87076	0.35412	0.035393	0.0085341	0.0084030

TABLE 2.12 Amplitude ratios for the lowest frequencies for shear diaphragm supports at each end of the cylinder and $\nu = 0.3$, according to Flügge theory (after Leissa (1973))

a/h	n	$L/(ma)$							
		0.25		1.0		4.0		20.0	
		U_n/w	V_n/w	U_n/w	V_n/w	U_n/w	V_n/w	U_n/w	V_n/w
	0	0.0274	0	0.1051	0	0.5607	0	2.802	0
	1	0.0286	0.0235	0.1759	0.3583	0.9937	1.304	0.4073	1.033
20	2	0.0320	0.0455	0.2481	0.4273	0.4108	0.5755	0.1054	0.5044
	3	0.0372	0.0648	0.2533	0.3727	0.2010	0.3609	0.0473	0.3350
	4	0.0433	0.0807	0.2241	0.3025	0.1250	0.2630	0.0268	0.2511
	0	0.0240	0	0.1041	0	0.5605	0	2.802	0
	1	0.0251	0.0217	0.1746	0.3565	0.9941	1.304	0.4075	1.033
500	2	0.0281	0.0421	0.2462	0.4253	0.4105	0.5754	0.1054	0.5043
	3	0.0327	0.0598	0.2508	0.3705	0.2093	0.3604	0.0471	0.3346
	4	0.0381	0.0743	0.2211	0.3002	0.1241	0.2623	0.0266	0.2505

2.3.5.4 Effect of Longitudinal and Circumferential Stiffeners

Provided that the stiffeners are relatively closely spaced and only low-order long-wavelength modes are considered, the resulting increased stiffness may be smeared out along and over the shell to produce an orthotropic shell constructed from isotropic materials (Mikulas and McElman, 1965). In this case, the equations of motion may be written as in Equation (2.294), with different coefficients, a_{ij}, defined below, with a the cylinder radius (see Figure 2.22):

$$a_{11} = a\frac{\partial^2}{\partial x^2} + \frac{C_{66}}{aC_{11}}\frac{\partial^2}{\partial \theta^2} - \frac{\rho_m ha}{C_{11}}\frac{\partial^2}{\partial t^2} \qquad (2.391)$$

$$a_{12} = \frac{(C_{12}+C_{22})}{C_{11}}\frac{\partial^2}{\partial x\partial \theta} \qquad (2.392)$$

$$a_{13} = \frac{C_{12}}{C_{11}}\frac{\partial}{\partial x} \qquad (2.393)$$

$$a_{21} = \frac{(C_{12}+C_{22})}{C_{11}}\frac{\partial^2}{\partial x\partial \theta} \qquad (2.394)$$

$$a_{22} = \frac{a(C_{66}+D_{66})}{C_{11}}\frac{\partial^2}{\partial x^2} + \frac{(C_{22}+D_{22})}{aC_{11}}\frac{\partial^2}{\partial \theta^2} - \frac{\rho_m ha}{C_{11}}\frac{\partial}{\partial t^2} \qquad (2.395)$$

$$a_{23} = \frac{C_{22}}{aC_{11}}\frac{\partial}{\partial \theta} - \frac{D_{22}}{aC_{11}}\frac{\partial^3}{\partial \theta^3} - \frac{(D_{12}+D_{66})}{C_{11}}\frac{\partial^3}{\partial x\partial \theta^2} \qquad (2.396)$$

$$a_{31} = \frac{C_{12}}{C_{11}}\frac{\partial}{\partial x} \qquad (2.397)$$

$$a_{32} = \frac{C_{22}}{aC_{11}}\frac{\partial}{\partial \theta} - \frac{D_{22}}{aC_{11}}\frac{\partial^3}{\partial \theta^3} - \frac{a(D_{12}+D_{66})}{C_{11}}\frac{\partial^3}{\partial x^2\partial \theta} \qquad (2.398)$$

$$a_{33} = \frac{C_{22}}{aC_{11}} + a^3\frac{D_{11}}{C_{11}}\frac{\partial^4}{\partial x^4} + \frac{D_{22}}{a}\frac{\partial^4}{\partial \theta^4} + \frac{a(2D_{12}+D_{66})}{C_{11}}\frac{\partial^4}{\partial x^2\partial \theta^2} + \frac{\rho_m ha}{C_{11}}\frac{\partial^2}{\partial t^2} \qquad (2.399)$$

where C_{11}, C_{12}, C_{22} and C_{66} are the extensional stiffness constants and D_{11}, D_{12}, D_{22} and D_{66} are the bending stiffness constants defined by:

$$C_{11} = \frac{E_{LS}}{L_{R\theta}}\left[A_L + hL_{R\theta}/(1-\nu^2)\right] \qquad (2.400)$$

$$C_{12} = \nu C_{11} \qquad (2.401)$$

$$C_{22} = \frac{1}{L_{Rx}}\left[E_F A_F + E_{LS}hL_{Rx}/(1-\nu^2)\right] \qquad (2.402)$$

$$C_{66} = (1-\nu)C_{11}/2 \qquad (2.403)$$

$$D_{11} = \frac{E_{LS}}{L_{R\theta}}\left[J_{Lx} + J_{Sx}(1-\nu^2)\right] \qquad (2.404)$$

$$D_{12} = \nu D_{11} \qquad (2.405)$$

$$D_{22} = \frac{1}{L_{Rx}}\left[E_F J_{F\theta} + E_{LS}J_{SS}/(1-\nu^2)\right] \qquad (2.406)$$

$$D_{66} = 2(1-\nu)D_{11} \qquad (2.407)$$

$$J_{F\theta} = J_F + A_F(y_F + h - r_\theta)^2 \qquad (2.408)$$

$$J_{Lx} = J_L + A_L(y_L + h - r_x)^2 \qquad (2.409)$$

and where

$$J_{SS} = 3L_{Rx}h^3/48 + 3\beta L_{Rx}h(r_\theta - h/2)^2/4 \qquad (2.410)$$

$$J_{Sx} = L_{R\theta}h^3/12 + L_{R\theta}h(r_x - h/2)^2 \qquad (2.411)$$

$$r_\theta = \frac{A_F(y_F + h) + 3\beta L_{Rx}h^2/8}{A_F + 3\beta L_{Rx}h/4} \qquad (2.412)$$

$$r_x = \frac{A_L(y_L + h) + L_{R\theta}h^2/2}{A_L + L_{R\theta}h} \qquad (2.413)$$

where:

A_F and A_L	$=$	cross-sectional areas of rings (frames) and stringers (longerons), respectively.
J_F and J_L	$=$	second moments of area of frames and stringers about their own centroidal axes, respectively.
J_{Lx} and J_{Sx}	$=$	second moments of area of the stringers and skins, respectively, about the centroidal axis of the skin/stringer cross-section.
J_{FS} and J_{SS}	$=$	second moments of area of the frames and skins, respectively, about the centroidal axis of the frame/skin cross-section.
h	$=$	skin (wall) thickness.
y_F and y_L	$=$	distances from the centroidal axes of the frames and stringers, respectively, to the underside of the skin.
E_F and E_{LS}	$=$	moduli of elasticity of the frames and stringers (and skins), respectively.
L_{Rx} and $L_{R\theta}$	$=$	lengths of the repeating sections in the axial and circumferential directions, respectively.
β	$=$	0 if the skin is attached to the stringers but not to the frames.
	$=$	1 if the skin is attached to the stringers and frames.

2.3.5.5   Other Complicating Effects

Various complicating effects such as uniform preloading and inclusion of the effects of rotary inertia and shear deformation in the analysis are beyond the scope of this text but are considered elsewhere (Leissa, 1973).

Inclusion of the effects of rotary inertia and shear deformation results in a displacement vector containing five instead of three components and the equations of motion (2.209) are now represented by a (5×5) instead of a (3×3) matrix, which adds considerable algebraic complexity to the solution. The effect of including rotary inertia and shear deformation in the analysis is only significant for relatively thick $(a/h < 10)$ cylinders or high order modes.

3

Sound Radiation and Propagation Fundamentals

3.1 Introduction

In practical terms, we are interested in being able to predict vibration levels in structures and associated sound radiation and propagation from a knowledge of the excitation force acting on a structure or acoustic medium. These types of predictions have myriad applications, ranging from predicting the expected sound levels in a community as a result of an industrial noise source to predicting the acoustic environment in the payload bay of a rocket. The analyses discussed in this chapter are based on the Green's function approach and underpin many software tools that are used by practitioners to make such predictions. The remainder of this chapter introduces acoustical Green's functions for sound propagation in an acoustic medium, structural Green's functions for propagation of waves in structures and radiation Green's functions for sound radiation from vibrating structures. This is followed by a general discussion of the application of these functions.

3.2 Green's Functions

The Green's function technique is a convenient approach to analysing sound radiation and propagation problems, whether sound radiation by a structure into free space is considered or whether sound transmission through a structure into an enclosed space is of interest. Physically, a Green's function is simply a transfer function that relates the response at one point in an acoustic medium or a structure to an excitation by a unit point source at another point. The value of the Green's function for a particular physical system is dependent on the location of the source and observation points and the frequency of excitation. Note that here, neither the type of excitation source nor the type of response has been defined. This will be done when specific examples are considered.

Although Green's functions are not unique to acoustics and vibrations problems (Morse and Feshbach, 1953), the discussion here will be restricted to these types of problem in the interests of clarity and relevance. In particular, problems involving sound radiation from vibrating structures, sound transmission through structures into enclosed spaces, vibration transmission through connected structures, and sound propagation in an acoustic medium will be considered.

A vibrating structure in contact with a compressible fluid such as air or water will generate pressure fluctuations in the fluid which, in turn, will react back on the structure and modify its vibration behaviour. This loading by the pressure waves in the fluid is known as radiation loading and generally for structures radiating into air it can be ignored, except for some cases in the high-frequency range. Consequently, in most cases, the dynamic response of a structure can be evaluated as though it were vibrating in a vacuum and the pressure field generated by the vibrating structure can be evaluated independently by equating the velocity of the fluid to that of the structure at the structure/fluid interface (see Section 5.3.1).

However, for structures radiating into relatively dense fluids such as water or oil, the forces acting on the structure are significantly modified by the radiation loading, and since the acoustic pressure is dependent upon the structural response, a feedback coupling between the fluid and structure exists. Thus, in these cases, the structural vibration and acoustic pressure responses must be evaluated simultaneously. For complex structures this is usually done using finite element analysis (see Section 5.3).

Another type of problem involves sound radiation into an enclosed space, where the response of the enclosed space is coupled to the response of the structure through which the sound is transmitted. In this instance, the response of the coupled system is derived from the mode shapes of the structure vibrating in a vacuum and the mode shapes of the enclosed space are calculated with the assumption that the boundaries enclosing it are perfectly rigid. The two responses are then coupled together at the boundaries of the enclosure where the external to internal pressure difference across the boundary is related to the normal velocity of the structure by using the structural Green's function. Of course, this method is not mathematically rigorous (as the assumption of rigid enclosure boundaries for the acoustic mode shape calculations results in small errors, as does the assumption of a surrounding vacuum for the calculation of the structural mode shapes) but the results obtained are generally sufficiently accurate to justify its use (see Section 5.3.1). For sound propagation in an acoustic medium, the response at a particular location in the medium due to sources acting at other locations can be calculated using the appropriate Green's function. Sound propagation in ducts, both plane wave and higher order mode propagation, and the implementation of active acoustic sources can be analysed by use of these Green's functions.

The solutions to the types of problems just mentioned are conveniently expressed in terms of acoustic Green's functions and structural Green's functions. Note that some authors prefer to refer to structural Green's functions as influence coefficients in an attempt to avoid confusion with acoustic Green's functions but this is done at the expense of additional complication of the terminology and will not be done here.

The underlying assumptions in the development of the solutions to acoustic propagation, transmission and radiation problems discussed in this text are listed below:

1. Linearity: for a structure, each component of stress is a linear function of the corresponding strain component, and for a fluid, the acoustic pressure fluctuations about the mean are a linear fraction of the corresponding density fluctuations about the mean.

2. Dissipation: frictional dissipation of energy is assumed to take place in solid structures, as this is a necessary requirement if meaningful solutions are to be obtained for the structural response. The dissipation mechanism will be simulated here by using a small structural loss factor associated with the stress component proportional to the strain rate. That is, a complex modulus of elasticity, $E(1+j\eta)$, will be assumed, where E is Young's modulus and η is the structural loss factor.

3. Homogeneity: the structure and fluid are both regarded as homogeneous. This assumption is not valid if sound propagation over large distances is considered. However, it does provide good results for short distance propagation, which can then be used with an appropriate model of the atmosphere to calculate long-range propagation without further consideration of the source.

4. Inviscid fluid: it is assumed that the acoustic fluid has no viscosity and therefore cannot support shear forces. Thus, the only component of structural displacement

that contributes to the radiated sound field is that which is normal to the surface of the structure. Similarly the acoustic medium can only apply normal loads to a structure.

It is possible to calculate the Green's function from classical analysis only for physically simple systems, and examples of some of these are discussed in the following sections. For more complex systems, finite element (Howard and Cazzolato, 2015) and boundary element methods can be used to numerically evaluate what constitutes an equivalent Green's function (see also Section 5.3).

Once the Green's function has been determined, it can be used to calculate the total system response due to a finite size source by integrating over the boundary of the source. Similarly, if n point sources are considered, the total sound field at any point can be calculated by summing the product of the Green's function corresponding to each source with the source strength of each source.

At this point it will be valuable to consider a rigorous mathematical definition of the acoustic Green's function, which may be defined as the solution to the inhomogeneous scalar Helmholtz equation (wave equation for a periodic disturbance with simple harmonic time dependence) for an acoustic medium containing a periodic driving source of unit strength. In other words, it is the solution of the inhomogeneous wave equation (or inhomogeneous scalar Helmholtz equation) with a singularity at the source point.

3.3 Acoustic Green's Functions

3.3.1 Acoustic Green's Function: Unbounded Medium

The acoustic Green's function for an unbounded medium is defined as the solution of:

$$\nabla^2 G(\boldsymbol{r}, \boldsymbol{r}_0, \omega) + k_a^2 G(\boldsymbol{r}, \boldsymbol{r}_0, \omega) = -\delta(\boldsymbol{r} - \boldsymbol{r}_0) \tag{3.1}$$

This wave equation is discussed in detail in Section 2.2.1 and by Morse and Feshbach (1953); Pierce (1981).

The function on the right-hand side of Equation (3.1) is the three-dimensional Dirac delta function, representing a unit point source at vector location, $\boldsymbol{r}_0$. The Dirac delta function allows a discontinuous point source to be described mathematically in terms of source strength per unit volume. In other words, it concentrates a uniformly distributed source onto a single point. The reason for doing this is to allow a uniformly distributed source to be more easily handled mathematically. Thus, if $\hat{q}'(\boldsymbol{r})$ is the source strength amplitude per unit volume at any location, $\boldsymbol{r}$, it may be expressed in terms of a point source amplitude, $\hat{q}(\boldsymbol{r}_0)$ at location $\boldsymbol{r}_0$, using the Dirac delta function as:

$$\hat{q}'(\boldsymbol{r}) = \hat{q}(\boldsymbol{r})\delta(\boldsymbol{r} - \boldsymbol{r}_0) \tag{3.2}$$

Integrating $\hat{q}'$ over any enclosed volume gives:

$$\iiint\limits_V \hat{q}'(\boldsymbol{r})\mathrm{d}\boldsymbol{r} = \iiint\limits_V \hat{q}(\boldsymbol{r})\delta(\boldsymbol{r} - \boldsymbol{r}_0)\mathrm{d}\boldsymbol{r} = \hat{q}(\boldsymbol{r}_0); \quad \boldsymbol{r}_0 \text{ in } V$$

$$= \frac{1}{2}\hat{q}(\boldsymbol{r}_0); \quad \boldsymbol{r}_0 \text{ on boundary of } V \tag{3.3}$$

$$= 0; \quad \boldsymbol{r}_0 \text{ outside } V$$

Note that $\hat{q}'$ has the units of T^{-1} and q the units of $\mathrm{L}^3\mathrm{T}^{-1}$.

It is clear from Equation (3.3) that the function, $\delta(\boldsymbol{r} - \boldsymbol{r}_0)$, has the dimensions L^{-3}. Thus, the dimensions of the Green's function in Equation (3.1) must be L^{-1}. The Dirac delta function:

$$\delta(\boldsymbol{r} - \boldsymbol{r}_0) = \delta(x - x_0)\delta(y - y_0)\delta(z - z_0) \tag{3.4}$$

is thus defined as a very high, very large and very narrow step function of source strength centred at $\boldsymbol{r}_0$ and with an area of unity under each of the curves of force versus $(x - x_0)$, $(y - y_0)$ and $(z - z_0)$.

A solution of Equation (3.1) is found by application of Gauss's integral theorem (Pierce, 1981), and is:

$$G(\boldsymbol{r}, \boldsymbol{r}_0, \omega) = \frac{\mathrm{e}^{-\mathrm{j}k_a R}}{4\pi R} \tag{3.5}$$

in which case the unit volume flow of the source is defined as:

$$q(\boldsymbol{r}_0, t) = \lim_{R \to 0} \left(-4\pi R^2 \frac{\partial G}{\partial R} \right) \mathrm{e}^{\mathrm{j}\omega t} = \hat{q}(\boldsymbol{r}_0)\mathrm{e}^{\mathrm{j}\omega t} \tag{3.6}$$

where:

$$R = |\boldsymbol{r} - \boldsymbol{r}_0| \tag{3.7}$$

Note that in some textbooks, the Green's function is defined without the 4π term in the denominator. In this case, the quantity 4π is included in the right-hand side of Equation (3.1). Also, those texts that use negative ($\mathrm{e}^{-\mathrm{i}\omega t}$) rather than positive ($\mathrm{e}^{\mathrm{j}\omega t}$) time dependence show the Green's function as $\mathrm{e}^{\mathrm{i}k_a R}/(4\pi R)$.

Equation (3.5) is known as the free field Green's function for an acoustic medium; that is, for any three-dimensional gas, liquid or solid supporting longitudinal wave propagation.

The Green's function (as well as being a solution of Equation (3.1)) must also satisfy the Sommerfeld radiation condition, to ensure that only outward travelling waves are represented. That is:

$$\lim_{R \to 0} R \left(\frac{\partial G}{\partial R} + \mathrm{j}k_a G \right) = 0 \tag{3.8}$$

The solution to Equations (3.1) and (3.8) (given by Equation (3.5)) is not subject to any boundary condition at finite range and thus is referred to as the free space Green's function. Remember that the Green's function represents the effect of a unit point source at any point in the system, on the response at any other point in the system.

As the units of the Green's function are L^{-1}, the pressure response at any location, $\boldsymbol{r}$, in the acoustic medium due to a point source of volume velocity amplitude, $\hat{q}(\boldsymbol{r}_0, \omega)$ ($\mathrm{m}^3\ \mathrm{s}^{-1}$) as a function of frequency, ω, is obtained by multiplying the product of the Green's function and the source strength by $\rho\omega$. Thus:

$$p(\boldsymbol{r}, t) = \hat{p}(\boldsymbol{r})\mathrm{e}^{\mathrm{j}\omega t} = \mathrm{j}\rho\omega G(\boldsymbol{r}, \boldsymbol{r}_0, \omega)\hat{q}(\boldsymbol{r}_0, \omega)\mathrm{e}^{\mathrm{j}\omega t} \tag{3.9}$$

It may seem that a point source is a fairly idealised case to be considering. However, the pressure response at any point in an acoustic medium due to a distributed source can be found by integrating the product of the source distribution, $\hat{q}(\boldsymbol{r}_0, \omega)$, with the Green's function over the space of the distributed source. Thus (with the time dependence ($\mathrm{e}^{\mathrm{j}\omega t}$) omitted):

$$\hat{p}(\boldsymbol{r}, \omega) = \mathrm{j}\rho\omega \iiint\limits_V G(\boldsymbol{r}, \boldsymbol{r}_0, \omega)\hat{q}'(\boldsymbol{r}_0, \omega)\mathrm{d}\boldsymbol{r}_0 \tag{3.10}$$

which is the solution to the inhomogeneous wave equation with homogeneous boundary conditions, where $\hat{q}'(\boldsymbol{r}_0, \omega)$ is the volume velocity amplitude per unit volume at location, $\boldsymbol{r}_0$.

In most cases, the source distribution is over a defined surface, so the volume integral in Equation (3.10) can usually be replaced with a surface integral.

3.3.2 Reciprocity of Green's Functions

Before reciprocity can be discussed, it is necessary to introduce Green's theorem which is a special case of Gauss's theorem and relates an area integral to a volume integral. Note that Gauss's theorem states that for an incompressible fluid, the fluid generated per unit time by all sources in a given volume is equal to the fluid that leaves the volume per unit time through its boundary. That is, for any two scalar functions, $A(\boldsymbol{r})$ and $B(\boldsymbol{r})$, of position, $\boldsymbol{r}$:

$$\iint_S \left[A\nabla B - B\nabla A \right] \cdot \mathrm{d}S = \iiint_V \left[A\nabla^2 B - B\nabla^2 A \right] \mathrm{d}V \tag{3.11}$$

As stated in Section 3.3.1 the Green's function, $G(\boldsymbol{r}, \boldsymbol{r}_0, \omega)$, satisfies the equation:

$$\nabla^2 G(\boldsymbol{r}, \boldsymbol{r}_0, \omega) + k_a^2 G(\boldsymbol{r}, \boldsymbol{r}_0, \omega) = -\delta(\boldsymbol{r} - \boldsymbol{r}_0) \tag{3.12}$$

However, the Green's function, $G(\boldsymbol{r}, \boldsymbol{r}_1, \omega)$, satisfies the equation:

$$\nabla^2 G(\boldsymbol{r}, \boldsymbol{r}_1, \omega) + k_a^2 G(\boldsymbol{r}, \boldsymbol{r}_1, \omega) = -\delta(\boldsymbol{r} - \boldsymbol{r}_1) \tag{3.13}$$

If the first equation is multiplied by $G(\boldsymbol{r}, \boldsymbol{r}_1, \omega)$ and the second equation by $G(\boldsymbol{r}, \boldsymbol{r}_0, \omega)$ and the difference (Equation (3.13) minus Equation (3.12)) is integrated over the volume, V, enclosed by an arbitrary boundary surface of area, S, the following is obtained:

$$-\iiint_V \left[G(\boldsymbol{r}, \boldsymbol{r}_0, \omega)\nabla^2 G(\boldsymbol{r}, \boldsymbol{r}_1, \omega) - G(\boldsymbol{r}, \boldsymbol{r}_1, \omega)\nabla^2 G(\boldsymbol{r}, \boldsymbol{r}_0, \omega) \right] \mathrm{d}V$$

$$= \iiint_V G(\boldsymbol{r}, \boldsymbol{r}_0, \omega)\delta(\boldsymbol{r} - \boldsymbol{r}_1)\mathrm{d}V - \iiint_V G(\boldsymbol{r}, \boldsymbol{r}_1, \omega)\delta(\boldsymbol{r} - \boldsymbol{r}_0)\mathrm{d}V \tag{3.14}$$

Using Green's theorem and the definition of the delta function, the preceding equation can be written as:

$$-\iint_S [G(\boldsymbol{r}, \boldsymbol{r}_0, \omega)\nabla G(\boldsymbol{r}, \boldsymbol{r}_1, \omega) - G(\boldsymbol{r}, \boldsymbol{r}_1, \omega)\nabla G(\boldsymbol{r}, \boldsymbol{r}_0, \omega)] \, \mathrm{d}S = G(\boldsymbol{r}_1, \boldsymbol{r}_0, \omega) - G(\boldsymbol{r}_0, \boldsymbol{r}_1, \omega)$$

$$\tag{3.15}$$

However, as will be shown in Section 3.3.3, the functions, G, by definition, must satisfy one of the following types of boundary condition on the surface, S:

$$G = 0 \qquad \frac{\partial G}{\partial \boldsymbol{n}} = 0 \quad \text{or} \quad \frac{\partial G}{\partial \boldsymbol{n}} \bigg/ G = \text{const} \tag{3.16}$$

where $\dfrac{\partial G}{\partial \boldsymbol{n}}$ represents the normal gradient of G at the boundary surface S. Therefore, the integrand of Equation (3.15) vanishes and the following expression remains:

$$G(\boldsymbol{r}_1, \boldsymbol{r}_0, \omega) = G(\boldsymbol{r}_0, \boldsymbol{r}_1, \omega) \tag{3.17}$$

The physical interpretation of this relationship is that if a source at $\boldsymbol{r}_0$ produces a certain response at $\boldsymbol{r}_1$, it would produce the same response at $\boldsymbol{r}_0$ if it were moved to $\boldsymbol{r}_1$. This is known as reciprocity and it is fundamental to many acoustical analyses.

3.3.3 Acoustic Green's Function for a Three-Dimensional Bounded Fluid

Equation (3.1), of which the Green's function is a solution, describes a medium that is homogeneous everywhere except at one point, the source point. When the point is on the boundary of a medium, the Green's function may be used to satisfy boundary conditions (for the homogeneous wave equation) that require neither the acoustic response nor the gradient of the acoustic response to be zero on the boundary. These are referred to as inhomogeneous boundary conditions. Conversely, when the point is a source point within the medium and not on the boundary, the Green's function is used to satisfy the inhomogeneous wave equation with homogeneous boundary conditions. It is implicit in the use of Green's functions solutions to the inhomogeneous wave equation with homogeneous boundary conditions or the homogeneous wave equation with inhomogeneous boundary conditions that the two conditions of inhomogeneity do not coexist. If they do, then solutions must be obtained for only one inhomogeneity condition at once and the two solutions added together to give the solution corresponding to the coexistence of an inhomogeneous equation (where a source point is contained within the medium) and an inhomogeneous boundary (where a source point is on the boundary). Thus, it is implicit in the Green's function solution of the inhomogeneous wave equation representation of a point excitation source in an acoustic medium that, on the boundary of the medium, at least one of the following conditions must be satisfied:

$$G = 0 \text{ and } \frac{\partial G}{\partial \boldsymbol{n}} = 0 \ \text{ or } \ \frac{\partial G}{\partial \boldsymbol{n}} \Big/ G = \text{const} \tag{3.18}$$

where $\boldsymbol{n}$ is the unit vector normal to the bounding surface. These conditions are referred to as homogeneous boundary conditions.

For the homogeneous wave equation with inhomogeneous boundary conditions, it is implicit that on the boundary surface of the medium, the function, $G(\boldsymbol{r}_0, \boldsymbol{r}, \omega)$, has specified values (not everywhere zero) or that $\dfrac{\partial G}{\partial \boldsymbol{n}}$ $(= \boldsymbol{n} \cdot \nabla G)$ has specified values (not everywhere zero) or that:

$$aG + b(\partial G / \partial \boldsymbol{n}) = F \tag{3.19}$$

where a and b are constants and F is the forcing function (inhomogeneous term) incorporating the forces acting on the boundary.

The solution, in terms of acoustic pressure, to the inhomogeneous wave equation with homogeneous boundary conditions, is represented by Equation (3.10). For a volume of fluid enclosed within a bounding surface, containing volume velocity sources, the total solution for the pressure amplitude response at frequency, ω, is Equation (3.10) plus the solution for the harmonic form of Equation (1.54), which is the homogeneous wave equation with inhomogeneous boundary conditions, written as:

$$\nabla^2 \hat{p}(\boldsymbol{r}, \omega) + k_a^2 \hat{p}(\boldsymbol{r}, \omega) = 0 \tag{3.20}$$

To solve Equation (3.20) with boundary conditions at finite surfaces, Equation (3.1) is multiplied by $\hat{p}(\boldsymbol{r}, \omega)$ and Equation (3.20) by $G(\boldsymbol{r}, \boldsymbol{r}_0, \omega)$, and the first result subtracted from the second to obtain:

$$G(\boldsymbol{r}, \boldsymbol{r}_0, \omega) \nabla^2 \hat{p}(\boldsymbol{r}, \omega) - \hat{p}(\boldsymbol{r}, \omega) \nabla^2 G(\boldsymbol{r}, \boldsymbol{r}_0, \omega) = \hat{p}(\boldsymbol{r}, \omega) \delta(\boldsymbol{r} - \boldsymbol{r}_0) \tag{3.21}$$

where $\boldsymbol{r}_0 = \boldsymbol{r}_s$ is now a point on the boundary surface. If $\boldsymbol{r}$ and $\boldsymbol{r}_0$ are now interchanged ($\boldsymbol{r}$ and $\boldsymbol{r}_0$ are any points in the volume enclosed by the boundary), and reciprocity is used ($G(\boldsymbol{r}, \boldsymbol{r}_0, \omega) = G(\boldsymbol{r}_0, \boldsymbol{r}, \omega)$ and $\delta(\boldsymbol{r} - \boldsymbol{r}_0) = \delta(\boldsymbol{r}_0 - r)$) and if an integration is performed over the volume defined by x_0, y_0 and z_0, the following is obtained:

$$\iiint\limits_{V} \left[G(\boldsymbol{r}, \boldsymbol{r}_0, \omega) \nabla^2 \hat{p}(\boldsymbol{r}_0, \omega) - \hat{p}(\boldsymbol{r}_0, \omega) \nabla^2 G(\boldsymbol{r}, \boldsymbol{r}_0, \omega) \right] \mathrm{d}\boldsymbol{r} = \iiint\limits_{V} \hat{p}(\boldsymbol{r}_0, \omega) \delta(\boldsymbol{r} - \boldsymbol{r}_0) \mathrm{d}\boldsymbol{r}_0 \tag{3.22}$$

Using Gauss's integral theorem and Green's theorem (sometimes referred to as Green's identity) (Junger and Feit, 1986), the first integral in Equation (3.22) can be written as a surface integral over the bounding surface and the second integral is (from Equation (3.3)) equal to $\hat{p}(\boldsymbol{r}, \omega)$. Thus, Equation (3.22) can be written as:

$$\hat{p}(\boldsymbol{r}, \omega) = -\iint_{S} \left[G(\boldsymbol{r}, \boldsymbol{x}, \omega) \frac{\partial}{\partial \boldsymbol{n}_s} \hat{p}(\boldsymbol{x}, \omega) - \hat{p}(\boldsymbol{x}, \omega) \frac{\partial}{\partial \boldsymbol{n}_s} G(\boldsymbol{r}, \boldsymbol{x}, \omega) \right] \mathrm{d}\boldsymbol{x} \qquad (3.23)$$

which is the solution to the homogeneous wave equation with inhomogeneous boundary conditions, otherwise known as the Helmholtz integral equation and which forms the basis of the numerical Boundary Element Method (or BEM). BEM is used to calculate sound radiation from a vibrating structure into either free space or an enclosed space. It is usually implemented by discretising the radiating structure and surrounding medium into small elements. The Helmholtz equation for each element is solved numerically by matching pressures and velocities at points where individual elements are connected. Commercially available software is available for implementing this process. More details of the process involved, together with an implementation example, are provided in Chapter 11 of Bies et al. (2017), in Howard and Cazzolato (2015) and in Section 5.3.

In Equation (3.23), S is the area of the boundary surface, $\boldsymbol{x}$ is a vector location on the boundary surface, V is the volume enclosed by the boundary and $\boldsymbol{n}_s$ is the unit vector normal to the local boundary surface, directed into the fluid.

Adding to Equation (3.23), the solution for the inhomogeneous wave equation (due to acoustic sourcs within the bounded volume) with homogeneous boundary conditions (Equation (3.10)), the Kirkhoff–Helmholtz integral equation is obtained:

$$p(\boldsymbol{r}, \omega) = \iint_{S} \left[j\omega\rho\hat{u}_n(\boldsymbol{x}, \omega) G(\boldsymbol{r}, \boldsymbol{x}, \omega) + \hat{p}(\boldsymbol{x}, \omega) \frac{\partial}{\partial \boldsymbol{n}_s} G(\boldsymbol{r}, \boldsymbol{x}, \omega) \right] \mathrm{d}\boldsymbol{x}$$

$$+ \iiint_{V} j\omega\rho\hat{q}'(\boldsymbol{r}_0, \omega) G(\boldsymbol{r}, \boldsymbol{r}_0, \omega) \mathrm{d}\boldsymbol{r}_0 \quad (3.24)$$

where the pressure gradient has been replaced with the component, $\hat{u}_n$, of particle velocity amplitude normal to the local boundary surface multiplied by $-j\omega\rho$. The first integral is evaluated over all of the bounding surfaces and the second is evaluated over the bounded volume. Equation (3.24) is a special harmonic case of a more general integral equation, in which the time dependence is arbitrary and where the phase, $k_a|\boldsymbol{r}|$, is replaced by a time difference, $(t - |\boldsymbol{r}|/c)$.

If the Green's function, G, defined by Equation (3.5), is chosen to satisfy one of the boundary conditions, $G = 0$ or $\partial G / \partial \boldsymbol{n}_s = 0$ over the entire boundary (as well as the wave equation and the Sommerfeld radiation condition), then one of the surface integral terms in Equation (3.24) will disappear.

For the special case of an infinitely baffled, plane surface radiating into free space, choosing the boundary condition, $\partial G / \partial \boldsymbol{n}_s = 0$, and choosing a Green's function consisting of Equation (3.5) multiplied by two to account for an image source (reflection of the source in the plane surface), and ignoring the last term in Equation (3.24) (as all sources are on the surface) allows Equation (3.24) to be written in the form of Rayleigh's well-known integral equation as follows:

$$\hat{p}(\boldsymbol{r}, \omega) = \frac{j\omega\rho}{2\pi} \iint_{S} \frac{\hat{u}_n(\boldsymbol{x}, \omega) \mathrm{e}^{-\mathrm{j}k_a R}}{R} \mathrm{d}S \qquad (3.25)$$

where the incremental vector location, $\mathrm{d}\boldsymbol{x}$, has been replaced by the incremental area, $\mathrm{d}S$, and where R is the distance to the observer location from an incremental area, $\mathrm{d}S$, on the plane surface.

A solution to Equation (3.20) subject to a rigid wall boundary condition, $(\partial p / \partial \boldsymbol{n} = 0)$, can be written as:

$$\hat{p}(\boldsymbol{r})\mathrm{e}^{\mathrm{j}\omega_n t} = A_n \psi_n(\boldsymbol{r})\mathrm{e}^{\mathrm{j}\omega_n t} \tag{3.26}$$

where A_n is a complex constant and ψ_n is the pressure mode shape function for the nth acoustic mode in the rigid walled volume, with a resonance frequency of ω_n (radians/s). Substituting Equation (3.26) into (3.20) gives:

$$\nabla^2 \psi_n(\boldsymbol{r}) + k_n^2 \psi_n(\boldsymbol{r}) = 0 \tag{3.27}$$

where $k_n = \omega_n / c$. For a discrete set of values, k_n, the mode shape functions, ψ_n, satisfy the condition, $\dfrac{\partial \psi_n}{\partial \boldsymbol{n}} = 0$, on the enclosure walls; thus, they can be incorporated into an acoustic Green's function that satisfies the same condition and which can be expressed as:

$$G(\boldsymbol{r}, \boldsymbol{r}_0, \omega) = \sum_{n=0}^{\infty} B_n \psi_n(\boldsymbol{r}) \tag{3.28}$$

where B is a complex constant.

Equation (3.1) can now be written as:

$$-\sum_{n=0}^{\infty} k_n^2 B_n \psi_n(\boldsymbol{r}) + k_a^2 \sum_{n=0}^{\infty} B_n \psi_n(\boldsymbol{r}) = -\delta(\boldsymbol{r} - \boldsymbol{r}_0) \tag{3.29}$$

where the following relations have been used (from Equations (3.28) and (3.27)):

$$\nabla^2 G(\boldsymbol{r}, \boldsymbol{r}_0, \omega) = \sum_{n=0}^{\infty} B_n \nabla^2 \psi_n(\boldsymbol{r}) = -\sum_{n=0}^{\infty} k_n^2 B_n \psi_n(\boldsymbol{r}) \tag{3.30}$$

Using the condition that the natural modes of closed elastic systems are mutually orthogonal, the following is obtained (assuming a uniform mean fluid density):

$$\iiint_V \rho(\boldsymbol{r}) \psi_m(\boldsymbol{r}) \psi_n(\boldsymbol{r}) dV = \begin{cases} 0; & \text{if } m \neq n \\ V\Lambda_n; & \text{if } m = n \end{cases} \tag{3.31}$$

where:

$$\Lambda_n = \frac{1}{V} \iiint_V \rho(\boldsymbol{r}) \psi_n^2(\boldsymbol{r}) \mathrm{d}V \tag{3.32}$$

If the medium has a uniform mean density then $\rho(\boldsymbol{r}) = \rho$. Multiplying Equation (3.29) by $\rho(\boldsymbol{r}) \psi_m(\boldsymbol{r})$, integrating over the fluid volume, V, then setting $n = m$, the following is obtained:

$$B_n \Lambda_n (k_a^2 - k_n^2) = -\psi_n(\boldsymbol{r}_0)\rho \tag{3.33}$$

or:

$$B_n = \frac{\rho \psi_n(\boldsymbol{r}_0)}{V\Lambda_n(k_n^2 - k_a^2)} \tag{3.34}$$

Substituting Equation (3.34) into (3.28) gives the Green's function for an enclosed acoustic space with rigid boundaries as:

$$G(\boldsymbol{r}, \boldsymbol{r}_0, \omega) = \sum_{n=0}^{\infty} \frac{\rho \psi_n(\boldsymbol{r})\psi_n(\boldsymbol{r}_0)}{V\Lambda_n(k_n^2 - k_a^2)} \tag{3.35}$$

The sound pressure at any point in the acoustic medium due to a point source of frequency ω and strength amplitude, $\hat{q}$, located at $\boldsymbol{r}_0$ can be calculated by substituting Equation (3.35) into Equation (3.9) to give:

$$\hat{p}(\boldsymbol{r}, \omega) = \hat{q}(\boldsymbol{r}_0, \omega)\rho^2\omega \sum_{n=0}^{\infty} \frac{\psi_n(\boldsymbol{r})\psi_n(\boldsymbol{r}_0)}{V\Lambda_n(k_n^2 - k_a^2)} = \sum_{n=0}^{\infty} A_n(\omega)\psi_n(\boldsymbol{r}) \tag{3.36}$$

For more complicated distributed sources, Equation (3.35) is substituted into Equation (3.10).

If damping is included and expressed in terms of a loss factor, η, which is modelled as viscous and referred to as a loss factor (twice the critical damping ratio, ζ — see Chapter 5), the Green's function, expressed in terms of frequency, becomes (Fahy and Gardonio, 2007):

$$G(\boldsymbol{r}, \boldsymbol{r}_0, \omega) = \sum_{n=0}^{\infty} \frac{\psi_n(\boldsymbol{r})\psi_n(\boldsymbol{r}_0)}{V\Lambda_n(\omega_n^2 - \omega^2 + \mathrm{j}\eta\omega\omega_n)} \tag{3.37}$$

However, if the damping is modelled as hysteretic (ideal for structures but less representative than viscous for acoustic spaces), the Green's function is written as (Fahy and Gardonio, 2007):

$$G(\boldsymbol{r}, \boldsymbol{r}_0, \omega) = \sum_{n=0}^{\infty} \frac{\psi_n(\boldsymbol{r})\psi_n(\boldsymbol{r}_0)}{V\Lambda_n(\omega_n^2 - \omega^2 + \mathrm{j}\eta\omega_n^2)} \tag{3.38}$$

Equation (3.38) is slightly different to what is found in some other books where the definition, $k_n' = k_n\sqrt{1 + \mathrm{j}\eta}$ is used, rather than beginning with a complex bulk modulus. This changes the sign of the damping term in Equation (3.38). The convention used here was chosen to be consistent with the treatment of structural damping, where the complex Young's modulus is $E' = E(1 + \mathrm{j}\eta)$.

For a particular enclosed volume, the loss factor, η, is related to the enclosure reverberation time (time for a sound field to decay by 60 dB after the source is shut down) by:

$$\eta = \frac{2.2}{T_{60}f} \tag{3.39}$$

where f is the frequency of excitation in Hz, and T_{60} is the 60 dB decay time (seconds).

For a rectangular-shaped enclosure, the modal index, n, can be replaced with a triple index (l, m, n) and the mode shape function, ψ_{lmn}, is:

$$\psi_{lmn}(x, y, z) = \cos\left(\frac{l\pi x}{b}\right)\cos\left(\frac{m\pi y}{b}\right)\cos\left(\frac{n\pi z}{d}\right) \tag{3.40}$$

3.3.4 Acoustic Green's Function for a Source in a Two-Dimensional Duct of Infinite Length

The formulation for the Greens function in a 2-D duct of infinite length is similar to that for a three-dimensional enclosed space, except that the mode shape functions are only defined in two dimensions, as no reflections occur in one of the coordinate directions.

To begin, consider a duct, infinitely long, with a unit point source of sound placed half way along it. To be consistent with the notation used by other authors, the plane of the duct cross section will be denoted the x, y-plane and the duct axis, the z-axis. Expressing the Dirac delta function of Equation (3.1) in Cartesian coordinates gives:

$$\nabla^2 G(\boldsymbol{r}, \boldsymbol{r}_0, \omega) + k_a^2 G(\boldsymbol{r}, \boldsymbol{r}_0, \omega) = -\delta(x - x_0)\delta(y - y_0)\delta(z - z_0) \tag{3.41}$$

where $\boldsymbol{r} = (x, y, z)$ and $\boldsymbol{r}_0 = (x_0, y_0, z_0)$.

As shown by Morse and Ingard (1968), a solution to Equation (3.20) for a duct is:

$$\hat{p}(\boldsymbol{r})\mathrm{e}^{\mathrm{j}\omega_n t} = A_n \psi_n(x, y)\mathrm{e}^{-\mathrm{j}k_{zn}z}\mathrm{e}^{\mathrm{j}\omega_n t} \tag{3.42}$$

Substituting this expression into Equation (3.20) and separating out the transverse mode shape function, $\psi_n(x, y)$, gives:

$$\left(\frac{\partial^2}{\partial x^2} + \frac{\partial^2}{\partial y^2}\right)\psi_n + \kappa_n^2 \psi_n = 0 \tag{3.43}$$

and:

$$k_{zn}^2 + \kappa_n^2 = k_a^2 \tag{3.44}$$

where $\psi_n \equiv \psi_n(x, y)$, κ_n is the wavenumber of the nth mode shape (see Equation (3.70) for a rectangular-section duct) in the plane of the duct cross-section of area, S, and k_{zn} is the wavenumber of the nth mode shape along the duct axis.

Solutions that fit an appropriate boundary condition at the duct walls occur only for a discrete set of values for the wavenumber, κ_n, these values being called characteristic (or eigen) values and the corresponding solutions, ψ_n, being called eigenvectors. Note that in the case of rigid walls, the eigenvectors, ψ_n, satisfy the relation:

$$\iint\limits_{S} \psi_n(x, y)\psi_m(x, y)\mathrm{d}x\,\mathrm{d}y = \begin{cases} 0; & m \neq n \\ S\Lambda_n; & m = n \end{cases} \tag{3.45}$$

The functions, ψ_n, satisfy the condition:

$$\frac{\partial \psi_n}{\partial \boldsymbol{n}} = 0 \tag{3.46}$$

around the duct perimeter for rigid duct walls. Thus, these functions can be incorporated into a Green's function, satisfying the same condition, which may be expressed as:

$$G(\boldsymbol{r}, \boldsymbol{r}_0, \omega) = \sum_{n=0}^{\infty} B_n(z)\psi_n(x, y) \tag{3.47}$$

Thus:

$$\nabla^2 G(\boldsymbol{r}, \boldsymbol{r}_0, \omega) = \sum_{n=0}^{\infty} \left\{ B_n \left[\frac{\partial^2 \psi}{\partial x^2} + \frac{\partial^2 \psi}{\partial y^2}\right] + \psi_n \frac{\partial^2 B_n}{\partial z^2} \right\} \tag{3.48}$$

Substituting Equation (3.43) into Equation (3.48) gives:

$$\nabla^2 G(\boldsymbol{r}, \boldsymbol{r}_0, \omega) = -\sum_{n=0}^{\infty} \left[B_n \psi_n \kappa_n^2 - \psi_n \frac{\partial^2 B_n}{\partial z^2} \right] \tag{3.49}$$

where $\psi_n = \psi_n(x, y)$ and $B_n = B_n(z)$. Substituting Equation (3.49) into Equation (3.41) gives:

$$-\sum_{n=0}^{\infty} \left[B_n \psi_n \kappa_n^2 - \psi_n \frac{\partial^2 B_n}{\partial z^2} \right] + k_a^2 \sum_{n=0}^{\infty} B_n \psi_n = -\delta(\boldsymbol{r} - \boldsymbol{r}_0) \tag{3.50}$$

Multiplying Equation (3.50) by ψ_m, integrating over the cross-sectional area of the duct, and making use of Equation (3.45) gives:

$$-\Lambda_n \left[B_n \kappa_n^2 - \frac{\partial B_n}{\partial z^2} \right] + k_a^2 B_n S\Lambda_n = -\psi_n(x_0, y_0)\delta(z - z_0) \tag{3.51}$$

That is:

$$\left[\frac{\partial}{\partial z^2} + k_{zn}^2\right] B_n = \frac{-\psi_n(x_0, y_0)\delta(z - z_0)}{S\Lambda_n} \tag{3.52}$$

where:

$$k_{nz}^2 = k_a^2 - \kappa_n^2 \tag{3.53}$$

The function, $B_n(z)$, can be found by integrating Equation (3.52) over z from $z_0 - \alpha$ to $z_0 + \alpha$ and then letting α go to zero. Thus:

$$\int_{z_0-\alpha}^{z_0+\alpha} \frac{\partial^2 B_n}{\partial z^2}\,\mathrm{d}z + k_{zn}^2 \int_{z_0-\alpha}^{z_0+\alpha} B_n\mathrm{d}z = -\int_{z_0-\alpha}^{z_0+\alpha} \frac{\psi_n(x_0, y_0)\delta(z - z_0)}{S\Lambda_n}\mathrm{d}z \tag{3.54}$$

or:

$$\left[\frac{\partial B_n}{\partial z}\right]_{z_0-\alpha}^{z_0+\alpha} + k_{zn}^2 \int_{z_0-\alpha}^{z_0+\alpha} B_n dz = -\frac{\psi_n(x_0, y_0)}{S\Lambda_n} \tag{3.55}$$

Before continuing, it is necessary to make an assumption regarding the form of the function, $B_n(z)$, in particular the z dependence. For a duct extending infinitely in both directions from the source point, z_0, the wave travelling on the positive size side of z_0 must be represented by a constant multiplied by $\mathrm{e}^{\mathrm{j}(\omega t - k_{zn}z)}$ and the wave travelling to the left of z_0 must be represented by a constant multiplied by $\mathrm{e}^{\mathrm{j}(\omega t + k_{zn}z)}$. However, as the value of $B_n(z)$ must be continuous across z_0, the constants must be adjusted so that:

$$B_n(z) = D_n\mathrm{e}^{-\mathrm{j}k_{zn}|z-z_0|} \tag{3.56}$$

$$|z - z_0| = \begin{cases} z - z_0; & \text{if } z > z_0 \\ z_0 - z; & \text{if } z < z_0 \end{cases} \tag{3.57}$$

$$\lim_{\alpha \to 0}\left[\frac{\partial B_n}{\partial z}\right]_{z_0-\alpha}^{z_0+\alpha} = \lim_{\alpha \to 0}\left[\left(\frac{\partial B_n}{\partial z}\right)_{z_0+\alpha} - \left(\frac{\partial B_n}{\partial z}\right)_{z_0-\alpha}\right]$$

$$= \lim_{\alpha \to 0}\left[-\mathrm{j}k_{zn}D_n\mathrm{e}^{\mathrm{j}k_{zn}\alpha} - \mathrm{j}k_{zn}D_n\mathrm{e}^{-\mathrm{j}k_{zn}\alpha}\right] = -2\mathrm{j}k_{zn}D_n \tag{3.58}$$

As α is a very small quantity, $B_n(z)$ may be considered a constant over the interval, $z_0 - \alpha$ to $z_0 + \alpha$ (Morse and Ingard, 1968). Thus:

$$k_{zn}^2 \int_{z_0-\alpha}^{z_0+\alpha} B_n\mathrm{d}z = 2k_{zn}^2\alpha B_n \tag{3.59}$$

As $\alpha \to 0$, Equation (3.59) $\to 0$. Thus, in the limit as $\alpha \to 0$, Equation (3.55) becomes:

$$-2\mathrm{j}k_{zn}D_n = \frac{-\psi_n(x_0, y_0)}{S\Lambda_n} \tag{3.60}$$

Combining Equations (3.56) and (3.60) gives:

$$B_n(z) = \frac{-\mathrm{j}\psi_n(x_0, y_0)}{2S\Lambda_n k_{zn}}\mathrm{e}^{-\mathrm{j}k_{zn}|z-z_0|} \tag{3.61}$$

and the acoustic Green's function is:

$$G(\boldsymbol{r}, \boldsymbol{r}_0, \omega) = -\sum_{n=0}^{\infty} \frac{\mathrm{j}\psi_n(x_0, y_0)\psi_n(x, y)\mathrm{e}^{-\mathrm{j}k_{zn}|z-z_0|}}{2S\Lambda_n k_{zn}} \tag{3.62}$$

If the duct is excited by a harmonic sound source of frequency ω, located at $z = 0$, the solution for the pressure at any location (x, y, z) in the duct can be written as:

$$p(x, y, z, \omega, t) = \sum_{n=0}^{\infty} A_n(\omega)\psi_n(x, y)e^{j(\omega t - k_{zn}z)} \qquad (3.63)$$

where the coefficient $A_n(\omega)$ can be evaluated for any source type by substituting Equation (3.62) into Equation (3.9) or (3.10) and setting the result equal to Equation (3.63). Alternatively, the coefficients could be found experimentally as described in Section 3.3.4.1.

For a rectangular section duct, the modal index, n, can be replaced with a double index (m, n), where m is the number of horizontal nodal lines and n is the number of vertical nodal lines in a duct cross section. The quantities in Equation (3.63) are then defined as follows (Morse and Ingard, 1968):

$$\Psi_{mn}(x, y) = \cos\left(\frac{m\pi x}{b}\right)\cos\left(\frac{n\pi y}{d}\right) \qquad (3.64)$$

$$\Lambda_{mn} = \frac{1}{S}\iint_S \Psi_{mn}^2 \, dS \qquad (3.65)$$

$$k_{mn} = \left(k_a^2 - \kappa_{mn}^2\right)^{1/2} = \sqrt{\left(\frac{\omega}{c}\right)^2 - \left(\frac{\pi m}{b}\right)^2 - \left(\frac{\pi n}{d}\right)^2} \qquad (3.66)$$

where b and d are the duct cross-sectional dimensions, and S is the duct cross-sectional area.

3.3.4.1 Experimental Determination of the Sound Pressure for Waves Propagating in One Direction

For tonal noise and only plane waves, the sound pressure associated with the wave propagating towards the duct exit can be determined by measuring the maximum and minimum sound pressure associated with the standing wave in the duct. The mean square sound pressure associated with the wave propagating towards the duct exit is then:

$$p^2 = p_{\min} \times p_{\max} \qquad (3.67)$$

where $p_{\min}$ and $p_{\max}$ are RMS quantities at the frequency of interest.

For broadband random noise, the sound pressure associated with the wave propagating towards the duct exit can be calculated by measuring the cross correlation function, $R_{12}(\tau)$ (see Section 7.3.16), between two microphones well separated axially in the duct (Shepherd et al., 1986). Two peaks will appear in the function, $R_{12}(\tau)$, at delays τ equal to the sound propagation times between points $\pm z/c$ where z is the microphone separation distance. All parts of the function $R_{12}(\tau)$, except the peak of interest, are then edited out and the result is Fourier transformed to give the power spectral density of the noise propagating away from the source.

When higher order modes are propagating in the frequency range of interest, it is desirable to be able to determine the contributions of each mode to the propagating wave.

One method of determining the contribution from each higher order mode involves measurement of the cross spectrum (see Section 7.3.11) between two microphones located in the same duct cross section at various different positions (Bolleter and Crocker, 1972; Shepherd et al., 1986).

Alternatively, transfer function measurements (see Section 7.3.14) may be made between a single reference microphone and another microphone (or number of microphones) that is moved from point to point over a particular duct cross section (Åbom, 1989). The effect of turbulent pressure fluctuations can be limited by restricting both the reference and scanning microphone locations to a single duct cross section. However, if reflected waves are present, measurements over two cross sections are necessary to resolve the amplitudes of the direct and reflected waves. For cases where it is unnecessary to resolve the direct and reflected amplitudes, and only a single amplitude is needed for each mode, measurements may be restricted to a single duct cross section (with only half the total number of sensors as used over two cross sections), and the result will be the sum of the incident and reflected amplitudes for each mode. Note that to simplify the process and avoid the influence of evanescent waves, measurements should be made in the far field of any noise sources or duct discontinuities. Evanescent waves are those that decay as they propagate away from the source, and their significance is usually limited to the source near field.

The sound pressure at any location (x, y, z) in the duct may be written as:

$$p(x,y,z,t) = \sum_m \sum_n \left[A_{mn}^+ e^{-jk_{mn}^+ z} + A_{mn}^- e^{jk_{mn}^- z} \right] \psi_{mn}(x,y) e^{j\omega t} \tag{3.68}$$

where, if no flow is present, k_{mn} is defined for mode (m, n) in a rectangular-section duct of cross-sectional dimensions, $b \times d$, by:

$$k_{mn}^2 = k_a^2 - \kappa_{mn}^2 \tag{3.69}$$

where:

$$\kappa_{mn} = \sqrt{\left(\frac{\pi m}{b}\right)^2 + \left(\frac{\pi n}{d}\right)^2} \tag{3.70}$$

In the presence of a mean flow of Mach number M, k_{mn} is:

$$k_{mn}^2 = \frac{[(k_a^2 - \kappa_{mn}^2)(1 - M^2)]^{1/2} - k_a M}{1 - M^2} \tag{3.71}$$

where $k_a = \omega/c$. The quantity, M, is defined as positive in the direction of k_a^+ wave propagation and negative in the direction of k_a^- wave propagation. A_{mn}^+ and A_{mn}^- are the complex modal amplitudes characterising the waves propagating to the right and left, respectively, and ψ_{mn} is the mode shape function.

In the frequency domain (taking the Fourier transform of Equation (3.68)) we can write for the acoustic pressure amplitude at frequency, ω, and location, $\boldsymbol{x}$:

$$p(\boldsymbol{x}, \omega) = \sum_m \sum_n \left[A_{mn}^+(\omega) e^{-jk_{mn}^+ z} + A_{mn}^-(\omega) e^{jk_{mn}^- z} \right] \psi_{mn}(x,y) \tag{3.72}$$

where $\boldsymbol{x} = (x, y, z)$. In Equation (3.72), all modes above their cut on frequency, as well as any modes below their cut on frequency that have significant levels at the measurement positions should be included. If N modes are included, then a minimum of $2N$ transfer function measurements will be needed to determine the modal amplitudes of waves propagating in both directions along the duct. Note that the measurements should be independent. Using a matrix formulation, we can write (for a single frequency ω):

$$\hat{\boldsymbol{p}} = \boldsymbol{\Omega A} \tag{3.73}$$

where $\hat{\boldsymbol{p}}$ is a $2N \times 1$ vector containing the complex acoustic pressure amplitudes derived from the transfer function measurements using:

$$\hat{\boldsymbol{p}} = \hat{p}_1 \boldsymbol{H} \tag{3.74}$$

where $\hat{p}_1$ is the acoustic pressure amplitude at the reference location and H is the $2N \times 1$ transfer function vector representing the transfer function between $\hat{p}_1$ and the sound pressure at the measurement locations, represented by the vector, $\hat{\boldsymbol{p}}$. For convenience, the minimum of two cross-sectional planes in the duct can be used with the first N measurements taken in the first plane and the measurements, $N + 1$ to $2N$, taken in the second plane.

In Equation (3.73), $\mathbf{\Omega}$ is a $2N \times 2N$ matrix that represents the transfer function between two measurement planes. Each row corresponds to the modal contributions at the ith microphone location, with each element in the row representing the contribution from one modal component. Thus, row i of the matrix has the form:

$$\left[\psi_{00}(x_i, y_i) \mathrm{e}^{-\mathrm{j} k_{00} z_i} \ \ \psi_{00}(x_i, y_i) \mathrm{e}^{\mathrm{j} k_{00} z_i} \ \ldots\ldots\ \psi_{mn}(x_i, y_i) \mathrm{e}^{-\mathrm{j} k_{mn} z_i} \ \ \psi_{mn}(x_i, y_i) \mathrm{e}^{\mathrm{j} k_{mn} z_i} \right] \quad (3.75)$$

where the oo subscript corresponds to the plane wave and where z_i is the axial distance between the first measurement plane and the measurement point.

The quantity, $\mathbf{A}$, is a $2N \times 1$ vector containing the modal amplitudes of the incident, A_{mni}, and reflected, A_{mnr}, waves at the first measurement plane. Thus:

$$\mathbf{A} = [A_{00i} A_{00r}, \ldots, A_{mni} A_{mnr}]^{\mathrm{T}} \quad (3.76)$$

Rearranging Equation (3.73) provides a solution for the modal amplitudes as follows:

$$\mathbf{A} = \mathbf{\Omega}^{-1} \hat{\boldsymbol{p}} \quad (3.77)$$

If all of the measurements are not independent at the frequency of interest, then the matrix, $\mathbf{\Omega}$, will be singular and will not be invertible to give $\mathbf{A}$. One way around this is to take more measurements, M, than modes to be resolved. In this case, $\hat{\boldsymbol{p}}$ will be an $M \times 1$ vector and $\mathbf{\Omega}$ will be an $M \times N$ matrix. Equation (3.77) can then be written as:

$$\mathbf{A} = [\mathbf{\Omega}^{\mathrm{T}} \mathbf{\Omega}]^{-1} \mathbf{\Omega}^{\mathrm{T}} \hat{\boldsymbol{p}} \quad (3.78)$$

The preceding analysis applies equally well to circular or rectangular-section ducts, provided the correct mode shape functions, ψ_{mn}, are used. For rectangular-section ducts this function is:

$$\Psi_{mn}(x, y) = \cos\left(\frac{m\pi x}{b}\right) \cos\left(\frac{n\pi y}{d}\right) \quad (3.79)$$

where (m, n) are the modal indices, and b and d, are the duct cross-sectional dimensions. Note that if plane waves only are considered, $m = n = 0$.

If N pressure measurements are taken on only one plane (as discussed earlier), so that a single amplitude is obtained for each mode (representing the sum of amplitudes corresponding to both directions of propagation), then $\mathbf{A}$ becomes an $N \times 1$ matrix, given by:

$$\mathbf{A} = [A_{00}, \ \cdots, \ A_{mn}]^{\mathrm{T}} \quad (3.80)$$

and the ith row of the $N \times N$ matrix, $\mathbf{\Omega}$ (representing the ith measurement location), becomes

$$\mathbf{\Omega} = [\psi_0(x_i, y_i), \ \cdots, \ \psi_{mn}(x_i, y_i)]^{\mathrm{T}} \quad (3.81)$$

Equation (3.77) is still used to find the N modal amplitudes represented by the matrix, $\mathbf{A}$, but the matrices are of order N, rather than $2N$.

3.4 Green's Function for a Vibrating Surface

The general two-dimensional wave equation for a surface, vibrating at frequency, ω, can be written as (Cremer et al., 1973):

$$L[\hat{w}(\boldsymbol{x})] - m_s(\boldsymbol{x}) \omega^2 \hat{w}(\boldsymbol{x}) = 0 \quad (3.82)$$

where $L[\ ]$ represents a differential operator ($EJ'\nabla^4$ for a homogeneous thin plate, where J' is the second moment of area of the plate cross section per unit width), $m_s(\boldsymbol{x})$ is the mass per unit area at a point on the vibrating surface at vector location, $\boldsymbol{x} = (x,\ y)$, and $\hat{w}(\boldsymbol{x})$ is the normal displacement amplitude of the surface at vector location, $\boldsymbol{x}$.

As for the acoustic case discussed previously, a solution to Equation (3.82) can be written as:

$$\hat{w}(\boldsymbol{x})e^{j\omega_n t} = A_n \psi_n(\boldsymbol{x})e^{j\omega_n t} \tag{3.83}$$

where A_n is a complex constant and ψ_n is the displacement mode shape function for the nth structural mode, assuming no fluid loading by the surrounding medium. Substituting Equation (3.83) into Equation (3.82) gives for each structural mode:

$$L[\psi_n(\boldsymbol{x})] - m_s(\boldsymbol{x})\omega_n^2 \psi_n(\boldsymbol{x}) = 0 \tag{3.84}$$

If the boundary conditions are such that no energy can be conducted across the boundaries, then the mode shape functions, $\psi_n(\boldsymbol{x})$, are orthogonal; that is:

$$\iint\limits_S m_s(\boldsymbol{x})\psi_n(\boldsymbol{x})\psi_m(\boldsymbol{x})d\boldsymbol{x} = \begin{cases} 0; & \text{if } m \neq n \\ m_n; & \text{if } m = n \end{cases} \tag{3.85}$$

where m_n is known as the modal mass of the nth mode. The functions, ψ_n, satisfy the boundary condition expressed in Equation (3.46) around the boundaries of the surface. Thus, they can be incorporated into a structural Green's function that satisfies the same condition and that can be expressed as:

$$G_s(\boldsymbol{x}, \boldsymbol{x}_0, \omega) = \sum_{n=0}^{\infty} B_n \psi_n(\boldsymbol{x}) \tag{3.86}$$

where the structural Green's function is a solution of:

$$L[G_s(\boldsymbol{x}, \boldsymbol{x}_0, \omega)] - m_s(\boldsymbol{x})\omega^2 G_s(\boldsymbol{x}, \boldsymbol{x}_0, \omega) = \delta(\boldsymbol{x} - \boldsymbol{x}_0) \tag{3.87}$$

and where B_n is a complex constant. Using Equations (3.84) and (3.86), Equation (3.87) can be written as:

$$\sum_{n=0}^{\infty} B_n m_s(\boldsymbol{x})\omega_n^2 \psi_n(\boldsymbol{x}) - m_s(\boldsymbol{x})\omega^2 \sum_{n=0}^{\infty} B_n \psi_n(\boldsymbol{x}) = \delta(\boldsymbol{x} - \boldsymbol{x}_0) \tag{3.88}$$

Multiplying Equation (3.88) by $\psi_m(\boldsymbol{x})$ and integrating over the surface of the vibrating structure gives:

$$(B_n \omega_n^2 m_n - B_n \omega^2 m_n) = \psi(\boldsymbol{x}_0) \tag{3.89}$$

Thus:

$$B_n = \frac{\psi_n(\boldsymbol{x}_0)}{m_n(\omega_n^2 - \omega^2)} \tag{3.90}$$

and the structural Green's function (which is also known as the structural receptance function — see Equation (5.50)) is given by:

$$G_s(\boldsymbol{x}, \boldsymbol{x}_0, \omega) = \sum_{n=0}^{\infty} \frac{\psi_n(\boldsymbol{x})\psi_n(\boldsymbol{x}_0)}{m_n(\omega_n^2 - \omega^2)} \tag{3.91}$$

where the modal mass, m_n, is defined by Equation (3.85), and G_s has the units, M$^-$1T2. The displacement at any point, $\boldsymbol{x} = (x,\ y)$, on the surface can be written as:

$$w(\boldsymbol{x}, \omega, t) = \sum_{n=0}^{\infty} A_n(\omega)\psi_n(\boldsymbol{x})e^{j\omega t} \tag{3.92}$$

where the coefficient $A_n(\omega)$ can be evaluated for any source type by setting the RHS of Equation (3.92) equal to Equation (3.95). Alternatively, the coefficients could be found experimentally using modal analysis as described in Chapter 5.

If structural damping, characterised by a hysteretic loss factor, η_n, is included, then the resonance frequency, ω_n, of mode, n, becomes complex and equal to $\omega_n\sqrt{(1+j\eta)}$. Thus, for a damped structure, the Green's function is (Fahy and Gardonio, 2007):

$$G_s(\boldsymbol{x},\boldsymbol{x}_0,\omega) = \sum_{n=0}^{\infty} \frac{\psi_n(\boldsymbol{x})\psi_n(\boldsymbol{x}_0)}{m_n(\omega_n^2 - \omega^2 + j\omega_n^2\eta)} \tag{3.93}$$

which is the same as the receptance function for a damped system (see Equations (5.67) and (5.100)).

3.5 General Application of Green's Functions

Now that the acoustic and structural Green's functions of interest have been derived, it is useful to show how they can be used to find the response of a structure or acoustic medium to point or distributed excitation forces.

As noted earlier, the structural Green's function is denoted with a subscript, s, and written as $G_s(\boldsymbol{x}, \boldsymbol{x}_0, \omega)$, whereas the acoustic Green's function has no subscript.

3.5.1 Excitation of a Structure by Point Forces

The structural displacement response amplitude, $\hat{w}(\boldsymbol{x})$, at location, $\boldsymbol{x}$, on the structure and frequency, ω, to N point forces of amplitude, $\hat{F}_i(\boldsymbol{x}_i, \omega)$, $i = 1, N$ normal to the structure at locations, $\boldsymbol{x}_i$, on the surface of a structure is:

$$\hat{w}(\boldsymbol{x},\omega) = \sum_{i=1}^{N} G_s(\boldsymbol{x},\boldsymbol{x}_i,\omega)\hat{F}(\boldsymbol{x}_i,\omega) \tag{3.94}$$

In Equation (3.94), the units of w are metres, the units of F are Newtons and the units of G are metres per Newton. Thus, in this case, the Green's function is the displacement of the structure at $\boldsymbol{x}$ due to a unit point force at $\boldsymbol{x}_0$.

3.5.2 Excitation of a Structure by a Distributed Force

For a distributed force such as an incident acoustic field, the displacement amplitude response of the structure at any location, $\boldsymbol{x}$, is:

$$\hat{w}(\boldsymbol{x},\omega) = \iint\limits_{S} \hat{p}(\boldsymbol{x}_0,\omega)G_s(\boldsymbol{x},\boldsymbol{x}_0,\omega)\mathrm{d}\boldsymbol{x}_0 \tag{3.95}$$

where the integration is over the area of the source.

If the structure were a plate or shell subjected to a differential pressure, $(\hat{p}_0(\boldsymbol{x}_0,\omega)-\hat{p}_i(\boldsymbol{x}_0,\omega))$, due to different acoustic pressures on the outside and inside surfaces, then Equation (3.95) may be written as:

$$\hat{w}(\boldsymbol{x},\omega) = \iint\limits_{S} \left[\hat{p}_0(\boldsymbol{x}_0,\omega) - \hat{p}_i(\boldsymbol{x}_0,\omega)\right] G_s(\boldsymbol{x},\boldsymbol{x}_0,\omega)\mathrm{d}\boldsymbol{x}_0 \tag{3.96}$$

where the units of the distributed force, p, are N/m². Note that the direction of positive pressure is the same as that of positive structural displacement.

If both point excitation and distributed excitation exist simultaneously, then Equations (3.94) and (3.95) may be added together to give the total structural response:

$$\hat{w}(\boldsymbol{x}, \omega) = \iint_S \hat{p}(\boldsymbol{x}_0, \omega) G_s(\boldsymbol{x}, \boldsymbol{x}_0, \omega) \mathrm{d}\boldsymbol{x}_0 + \sum_{i=1}^{N} \hat{F}(\boldsymbol{x}_i, \omega) G_s(\boldsymbol{x}, \boldsymbol{x}_i, \omega) \qquad (3.97)$$

This equation is based on the assumption that the structural response is not affected by any sound field that it radiates (not valid for radiation into liquids) and that excitation forces are normal to the structure.

3.5.3 Excitation of an Acoustic Medium by a Number of Point Acoustic Sources

When the source of excitation is acoustic, the acoustic pressure amplitude at any point, $\boldsymbol{r}$, in the acoustic medium due to N point sources with a volume velocity amplitude of $\omega \hat{w}(\boldsymbol{r}_i, \omega) \, \mathrm{d}S_i$ is:

$$\hat{p}(\boldsymbol{r}, \omega) = -\omega^2 \rho \sum_{i=1}^{N} G(\boldsymbol{r}, \boldsymbol{r}_i, \omega) \hat{w}(\boldsymbol{r}_i, \omega) \mathrm{d}S_i \qquad (3.98)$$

where $\mathrm{d}S_i$ is the surface area of the ith point source boundary. It is worth noting that Equation (3.98) has a similar form to the time domain Equation (3.9).

If each of the acoustic sources were distributed sources rather than point sources then each term in the sum would become a triple integral over the volume of each source, or if the sources were distributed over a boundary surface, then the integral would be a double integral over the boundary surface.

3.5.4 Excitation of an Acoustic Medium by a Vibrating Structure

When an acoustic medium is excited by a vibrating structure with a normal displacement amplitude distribution, $\hat{w}(\boldsymbol{x}, \omega)$, the acoustic pressure at any point, $\boldsymbol{r}$, in the acoustic medium is:

$$\hat{p}(\boldsymbol{r}, \omega) = \mathrm{j}\omega^2 \rho \iint_S \hat{w}(\boldsymbol{x}, \omega) G(\boldsymbol{r}, \boldsymbol{x}, \omega) \mathrm{d}\boldsymbol{x} \qquad (3.99)$$

where the units of the acoustic Green's function, G, are L^{-1}. Note that the points, $\boldsymbol{x}$, lie on the surface, S, of the structure. For radiation from a plane surface surrounded by a large baffle, this equation reduces to the well known Rayleigh integral of Equation (3.25), where the complex normal surface velocity amplitude, $\hat{u}_n(\boldsymbol{x}, \omega) = \mathrm{j}\omega \hat{w}(\boldsymbol{x}, \omega)$.

The complex normal displacement amplitude, $\hat{w}(\boldsymbol{x}, \omega)$, of the radiating surface at frequency, ω, may be determined by summing the contributions (amplitude and phase) due to each structural mode at location, $\boldsymbol{x}$.

It can be seen from Equation (3.99) that the acoustical Green's function relates the acoustic pressure, at some point, $\boldsymbol{r}$, in the acoustic medium, to the volume velocity of the source. Physically, this means that the total acoustic pressure at a point in space is determined by summing the complex contributions (amplitude and phase) from all points on the radiating surface. Although the vibration modes describing the motion of the surface are orthogonal in terms of structural vibration (see Chapter 5), they are not orthogonal in terms of their individual contributions to the radiated sound field. This means that the radiated sound pressure squared or radiated sound power cannot be calculated simply by adding together all of the contributions from each mode. This is because when the quantity, $w(\boldsymbol{x}, \omega)$, in Equation (3.99) is squared to allow the pressure squared at $\boldsymbol{r}$ to be calculated, the result is made up of products of mode shapes squared as well as products of each mode shape with all of the others, the latter being referred to as cross-coupling terms. If the space averaged surface vibration amplitude squared were of interest, then the cross-coupling terms would integrate to zero, as the modes are orthogonal. However, the cross-coupling term contributions to the squared radiated pressure field do not, in general, go to zero (except for the case of radiation from a uniform spherical shell), when integrated over an imaginary surface in space to give the radiated sound power, because they are multiplied by the acoustical Green's function prior to integration. This is an extremely important concept from the viewpoint of active control of sound radiated by structures, as it illustrates that attempting to control one or more structural modes individually may not necessarily lead to a reduction in overall radiated sound power, even if the controlled modes would be the most efficient radiators if present in isolation.

The idea that modal sound powers cannot be added to give the total radiated sound power may be easier to understand if one imagines that the sound radiation efficiency of a surface is a function of the overall velocity distribution over it, and is independent of whether the velocity distribution is described in terms of modes or individual amplitudes and phases as a function of surface location. Generally, the more complicated the surface vibration pattern (or the greater the number of in-phase and out-of-phase areas), the less efficient will be the sound radiation. This is because adjacent areas of the surface that are out of phase effectively cancel the sound radiation from one another (provided that the separation of their mid-points is much less than a wavelength of sound in the adjacent acoustic medium). This results in a much less efficiently radiating surface at low frequencies, although at high frequencies where the separation between the mid-points of adjacent areas is much greater than a wavelength of sound in the acoustic medium, there will be no noticeable decrease in efficiency.

Another important concept involves the difference in sound fields radiated by a structure or plates excited by an incident acoustic wave and one excited by a mechanical localised force. In the former case the structure will be forced to respond in modes that are characterised by bending waves having wavelengths equal to the trace wavelengths of the incident acoustic field. Thus, at excitation frequencies below the structure critical frequency, the modes that are excited will not be resonant because the structural wavelength of the resonant modes will always be smaller than the wavelength in the acoustic medium. Thus, lower order modes will be excited at frequencies above their resonance frequencies. As these lower order modes are more efficient than the higher order modes that would have been resonant at the excitation frequencies, the radiated sound level will be higher than it would be for a resonantly excited structure having the same mean square velocity levels at the same excitation frequencies. As excitation of a structure by a mechanical force results in resonant structural response, then it can be concluded that sound radiation from an acoustically excited structure will be greater than that radiated by a structure excited mechanically to the same vibration level (McGary, 1988). A useful item of information that follows from this conclusion is that structural damping will only be effective for controlling mechanically excited structures because it is only the resonant structural response that is significantly influenced by damping.

3.6   Structural Sound Radiation and Wavenumber Transforms

While on the topic of sound radiation, it is of interest to examine another technique for characterising it. This is a method known as wavenumber transforms. This technique is useful because it provides some insight into the physical mechanisms associated with structural sound radiation.

A wavenumber transform essentially transforms a quantity expressed as a function of spatial coordinates (for example, surface vibration velocity) into a quantity expressed as a function of wavenumber variables; the inverse transform does vice versa. It is essentially the same Fourier transform operation used in transforming a signal from the time domain to the frequency domain, as will be discussed in Section 7.3. Thus, just as a discrete number of frequency bins is obtained in the frequency domain, a discrete number of wavenumber bins is obtained in the wavenumber domain. Remember that the acoustic wavenumber k_a is defined as ω/c, where c is the speed of sound in the acoustic medium of interest.

The wavenumber transform, as indicated by its name, essentially describes the response of a structure in terms of waves (rather than in terms of modes, as was done earlier in this section) and the acoustic radiation in terms of waves coupling with the acoustic medium rather than modes coupling with it. An important property of the wavenumber transform is that each of the wavenumbers so determined corresponds to a particular angle of sound radiation that happens to coincide with the matching of the trace wavelength of the acoustic field with the wavelength of the structural vibration. The trace wavelength of the acoustic field radiated from a plane surface at some angle is defined as the wavelength that it would effectively 'project' on the surface, as shown in Figure 3.1.

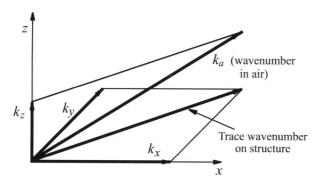

FIGURE 3.1 Relationship between wave vectors, k_a, k_x, k_y and k_z, for a radiating plane surface located in the $x - y$ plane.

Only wavenumbers corresponding to structural wavelengths greater than the wavelength in the acoustic medium surrounding the structure will radiate sound. These wavenumbers are referred to as 'supersonic', because they are characterised by a higher phase speed in the structure than the speed of sound in the acoustic medium at the same frequency. Except at structural boundaries and discontinuities, it is not possible for structural waves with wavelengths shorter than the corresponding waves in the acoustic medium to radiate sound because there is no angle of incidence at which the acoustic wavelength will be equal to the trace wavelength on the structure.

Thus, the purpose of using wavenumber transforms is to describe the structural vibration in terms of waves of different wavenumbers, which is really an alternative to describing it in terms of modes. The main advantage in using this method is that it is easy to identify those wavenumbers that are responsible for sound radiation. Another advantage of wavenumber transforms is that they enable the acoustic radiation integral expression to be transformed to a partial differential equation, which is usually more easily solved to obtain the structural radiation efficiency.

The basic description of wavenumber transforms applied to waves in a general sense, as well as to bending waves on structures radiating sound, has been discussed in detail by Fahy and Gardonio (2007). Here, a brief review of wavenumber transforms applied to bending waves in structures is given with the aim of formulating the general comments made above into quantitative equations that can be usefully applied to radiating structures.

As shown in Chapter 7, the Fourier transform, which is used to obtain the frequency spectrum of a time domain signal, is given as:

$$X(f) = \int\!\!\!\int\limits_{-\infty}^{\infty} x(t)\mathrm{e}^{-\mathrm{j}2\pi ft}\mathrm{d}t \tag{3.100}$$

where $x(t)$ is a time varying signal and f is the frequency in Hz of the Fourier component of the signal.

The two-dimensional spatial equivalent of this equation (or wavenumber transform) for a flat surface in the $x - y$ plane, vibrating in the z-direction (normal to the surface) is:

$$U_z(k_x, k_y) = \int\limits_{-\infty}^{\infty}\int\limits_{-\infty}^{\infty} \hat{u}_z(x, y)\mathrm{e}^{\mathrm{j}(k_x x + k_y y)}\mathrm{d}x\,\mathrm{d}y \tag{3.101}$$

where $\hat{u}_z(x, y)$ is the surface normal velocity amplitude distribution, and k_x, k_y are bending wavenumbers on the surface in the x- and y-directions, respectively, so that:

$$k_x = \frac{\omega}{c_{bx}} \;; \qquad k_y = \frac{\omega}{c_{by}} \tag{3.102}$$

where $\omega = 2\pi f$. Note that the sign on the exponent in Equation (3.101) is opposite to that used in Equation (3.100), which represents the transform from the time domain to the frequency domain. This is because positive time dependence is used here and negative spatial dependence. This means that the term, $\mathrm{e}^{\mathrm{j}\omega t}$, represents positive or increasing time and the term, $\mathrm{e}^{-\mathrm{j}k_a x}$, represents a wave travelling in the positive x-direction.

The inverse transform is expressed in a similar way to the inverse Fourier transform discussed in Chapter 7 and may be written as follows:

$$\hat{u}_z(x, y) = \frac{1}{(2\pi)^2} \int\limits_{-\infty}^{\infty}\int\limits_{-\infty}^{\infty} U_z(k_x, k_y)\mathrm{e}^{-\mathrm{j}(k_x x + k_y y)}\,\mathrm{d}k_x\,\mathrm{d}k_y \tag{3.103}$$

The normal surface vibration velocity amplitude field, $\hat{u}_z(x, y)$, on the plane surface can be thought of as made up of an infinite number of sinusoidal travelling waves, each of which is described by:

$$\hat{u}_{z,k}(x, y) = U_z(k_x, k_y)\mathrm{e}^{-\mathrm{j}(k_x x + k_y y)} \tag{3.104}$$

in a similar way to the transient time signal $x(t)$ of Equation (3.100) being thought of as consisting of an infinite number of pure tones.

Although the surface velocity normal amplitude distribution, $\hat{u}_z(x, y)$, has been selected as the variable of interest because it is directly related to sound power radiation, the transform equations are equally valid for surface displacement, and indeed also for a two-dimensional acoustic wave in air with acoustic pressure used instead of surface velocity as the transform variable.

For a plane surface radiating sound into the surrounding medium, the normal velocity of the fluid at the surface is equal to the normal velocity of the surface. Thus, following a similar line of reasoning to that used in Section 1.4.2, the acoustic pressure gradient normal to the surface at the surface is related to the surface normal velocity by:

$$\mathrm{j}\omega\rho\hat{u}_z(x, y) = -\frac{\partial \hat{p}(x, y, 0)}{\partial z} \tag{3.105}$$

where z represents the axis normal to the surface lying in the $x - y$ plane. Equation (3.105) applies to either instantaneous values or amplitudes of pressure and velocity, provided that the same descriptor is used for each at any one time. The boundary condition of Equation (3.104) can also be expressed in terms of the transformed variables, P and U_z, as follows:

$$j\omega\rho U_z(k_x, k_y) = -\left.\frac{\partial P(k_x, k_y, z)}{\partial z}\right|_{z=0} \tag{3.106}$$

The acoustic pressure field amplitude must also satisfy the transformed Helmholtz (wave) equation. That is:

$$\int_{-\infty}^{\infty} \int_{-\infty}^{\infty} \left(\frac{\partial^2}{\partial x^2} + \frac{\partial^2}{\partial y^2} + \frac{\partial^2}{\partial z^2} + k_a^2\right) \hat{p}(x, y, z) e^{j(k_x x + k_y y)}\, dx\, dy = 0 \tag{3.107}$$

which can be written in terms of the pressure transform as (Junger and Feit, 1986):

$$\left(k_a^2 - k_x^2 - k_y^2 + \frac{\partial^2}{\partial z^2}\right) P(k_x, k_y, z) = 0 \tag{3.108}$$

A solution to Equation (3.108) is:

$$P(k_x, k_y, z) = A e^{-jk_z z} \tag{3.109}$$

where:

$$k_z^2 = k_a^2 - k_x^2 - k_y^2 \tag{3.110}$$

Another solution would be the same as Equation (3.109) with a positive exponent, but this would imply waves converging on the radiating surface from infinity, thus not satisfying the Sommerfeld radiation condition (Junger and Feit, 1986), and so it is not allowed. Substitution of Equation (3.109) into Equation (3.106) gives:

$$A = \frac{\omega\rho U_z(k_x, k_y)}{k_z} \tag{3.111}$$

Thus, substituting Equation (3.111) into Equation (3.109), the following is obtained:

$$P(k_x, k_y, z) = \frac{\omega\rho U_z(k_x, k_y) e^{-jk_z z}}{k_z} \tag{3.112}$$

Taking the inverse transform gives an expression for the sound pressure amplitude at any location in the near or far field of the vibrating surface as follows:

$$\hat{p}(x, y, z) = \frac{\omega\rho}{(2\pi)^2} \int_{-\infty}^{\infty} \int_{-\infty}^{\infty} \frac{U_z(k_x, k_y) e^{-j(k_x x + k_y y + k_z z)}}{k_z}\, dk_x\, dk_y \tag{3.113}$$

This type of integral is almost always analytically intractable in the near field of vibrating surfaces, although it is generally possible to find an analytical solution for the far field sound pressure. However, in most cases (both near and far field), it is usually the simplest to use fast Fourier transform techniques to evaluate the integral.

It is of interest to examine the physical significance of the wavenumber, k_z, or wavevector as it is sometimes called (given its vector nature). It should be pointed out that although wavenumbers may be thought of as vector quantities, the related quantity, wavelength, is always a scalar. The subscript, z, on k_z indicates that the wavevector quantity k_z represents the component of the wavevector in the z-direction. The actual radiated wave characterised by k_a does not necessarily radiate normally to the surface, and in fact it hardly ever does. The wavevectors,

k_x and k_y, on the structure correspond to the x and y components of the acoustic wavevector, k_a, which explains the physical significance of Equation (3.109). Thus each discrete wavevector pair, (k_x, k_y), corresponds to a wave radiating at a particular angle from a vibrating surface. On any particular surface where more than one vibration mode is excited, there will always be more than one direction in which waves will radiate from the surface.

From Equation (3.110), it can be seen that if $k_a^2 < (k_x^2 + k_y^2)$, the wave vector k_z will be imaginary and will correspond to a wave that decays exponentially with distance from the surface and does not contribute to far field sound radiation. As wavenumber is inversely proportional to wave phase speed, this condition corresponds to waves in the structure with speeds less than the speed of the wave in the acoustic medium, and are thus referred to as subsonic structural waves. Such waves can only radiate from parts of the plate near the edges, where there is no out-of-phase plate motion to provide the cancelling sound pressure. Such sound radiation is thus quite inefficient.

The condition, $k_a^2 > (k_x^2 + k_y^2)$, corresponds to supersonic structural waves that radiate far field sound well, as there is always some angle of radiation for which the trace wavelength on the radiating surface of the acoustic wave matches the structural wavelength (see Figure 3.1 where the relationship between k_a, k_x, k_y and k_z is illustrated).

Wavenumber transforms are also useful for allowing the radiated acoustic power to be described in terms of quantities that can be measured on the surface of the radiating structure.

The sound power radiated to the far field by a harmonically vibrating surface with a complex normal surface velocity distribution of $u_z(x, y)$ is given by (see Section 4.2.3):

$$W = \frac{1}{2}\mathrm{Re}\left\{ \iint_S \hat{p}(x, y, 0)\hat{u}_z^*(x, y)\mathrm{d}x\mathrm{d}y \right\} \tag{3.114}$$

where $p(x, y, 0)$ is the complex acoustic pressure in the fluid adjacent to the vibrating surface.

Following an argument similar to that used by Fahy and Gardonio (2007) for the one-dimensional problem, it can be shown that the equivalent expression, in terms of transformed pressure and velocity, is:

$$W = \frac{1}{8\pi^2}\mathrm{Re}\left\{ \int_{-\infty}^{\infty} \int_{-\infty}^{\infty} P(k_x, k_y)U_z^*(k_x, k_y)\mathrm{d}k_x\mathrm{d}k_y \right\} \tag{3.115}$$

Setting Equations (3.114) and (3.115) equal is really a way of expressing Parseval's theorem in two-dimensional form (see Jenkins and Watts (1968), for the equivalent one-dimensional expression for Fourier analysis).

Substituting Equation (3.112) for P and then Equation (3.110) for k_z into Equation (3.115) gives:

$$W = \frac{\omega\rho}{8\pi^2}\mathrm{Re}\left\{ \int_{-\infty}^{\infty} \int_{-\infty}^{\infty} \frac{|U_z(k_x, k_y)|^2}{\sqrt{k_a^2 - k_x^2 - k_y^2}}\, \mathrm{d}k_x\, \mathrm{d}k_y \right\} \tag{3.116}$$

Only wavenumber components that satisfy, $k_a \geq \sqrt{k_x^2 + k_y^2}$ contribute to the real part of Equation (3.116); thus, the equation can be rewritten as:

$$W = \frac{\omega\rho}{8\pi^2}\mathrm{Re}\left\{ \iint_{k_x^2 + k_y^2 \leq k_a^2} \frac{|U_z(k_x, k_y)|^2}{\sqrt{k_a^2 - k_x^2 - k_y^2}}\, \mathrm{d}k_x\, \mathrm{d}k_y \right\} \tag{3.117}$$

The sound power can also be written in terms of the wavenumber transform of the acoustic pressure immediately adjacent to the radiating surface as:

$$W = \frac{\omega\rho}{8\pi^2}\text{Re}\left\{\iint\limits_{k_x^2+k_y^2\leq k^2} |P(k_x,k_y,0)|^2 \sqrt{k_a^2 - k_x^2 - k_y^2}\ \mathrm{d}k_x\,\mathrm{d}k_y\right\} \qquad (3.118)$$

Using the principles of acoustic holography (Veronesi and Maynard, 1987), the transform of the surface acoustic pressure (or the pressure at any other plane away from the surface) can be derived from the transform of acoustic pressure measurements taken at an array of points on a plane parallel to the radiating surface and a distance, z_m, from it as follows:

$$P(k_x,k_y,z) = P(k_x,k_y,z_m)\mathrm{e}^{-\mathrm{j}k_z(z-z_m)} \qquad (3.119)$$

where $z = 0$ if the pressure transform on the surface is desired.

4

Acoustic and Structural Impedance and Intensity

4.1 Introduction

Impedance and intensity are common terms used both in sound radiation from vibrating structures and in sound propagation. In this chapter, the various types of impedance are discussed in detail, followed by a discussion of both sound intensity for sound propagation in a gas (usually air) and structural intensity for sound waves propagating in structures. A number of similarities and differences may be observed for the two different types of intensity. Fundamentally, sound and structural intensity are both a measure of the energy that is propagating in the respective medium.

4.2 Impedance

There are four types of impedance commonly used in acoustics. Each type is directly related to the other three and can be derived from any one of the other three. Impedances are generally complex quantities (characterised by an amplitude and a phase), which are defined as a function of frequency. Thus, the quantities used in the equations to follow are complex amplitudes defined at specific frequencies, allowing the time dependent term, $e^{j\omega t}$, to be omitted.

Impedances are generally associated with acoustic sources or acoustic propagation and a number of specific examples are discussed later on in this section. However, before discussing any specific examples, it is useful to differentiate between the different types of impedance.

4.2.1 Specific Acoustic Impedance, Z_s

The specific acoustic impedance is defined as the ratio of the acoustic pressure to particle velocity, u, in the direction of wave propagation, anywhere in an acoustic medium, including the surface of a noise source or at an air/material interface. If the acoustic medium is infinite in extent such that waves are travelling in only one direction, the specific acoustic impedance is equal to the characteristic impedance, ρc, of the medium. Thus:

$$Z_s = \frac{p}{u} \tag{4.1}$$

4.2.2   Acoustic Impedance, Z_A

The acoustic impedance is particularly useful for describing sound propagation in ducts, and for a plane wave in a duct of cross-sectional area, S, it is defined as the ratio of the acoustic pressure to the volume velocity at a duct cross section. Thus:

$$Z_A = \frac{p}{uS} = \frac{Z_s}{S} \tag{4.2}$$

4.2.3   Mechanical Impedance, Z_m

The mechanical impedance is defined as the ratio of the force, F, acting on a surface or system to the velocity of the system at the point of application of the force. If the system is a vibrating surface, then the surface velocity, u, is equal to the acoustic particle velocity, u, at an adjacent point in the surrounding fluid. If the vibrating surface of area, S, is subject to a uniform acoustic pressure, p, and is vibrating with a uniform normal velocity, then the mechanical impedance is:

$$Z_m = \frac{F}{u} = \frac{pS}{u} = Z_s S = Z_A S^2 \tag{4.3}$$

For a more generally vibrating surface, the mechanical impedance is:

$$Z_m = \frac{1}{\langle u^2 \rangle} \iint_S p(\boldsymbol{x}) u^*(\boldsymbol{x}) \mathrm{d}\boldsymbol{x} \tag{4.4}$$

where p is the acoustic pressure, u^* is the complex conjugate of the acoustic particle velocity adjacent to and normal to the vibrating surface at $\boldsymbol{x}$, $\langle u^2 \rangle$ is the mean square particle velocity at the vibrating surface and averaged over the surface area, S, and $\boldsymbol{x}$ is a vector location of a point on the surface.

Note that in Equations (4.1) to (4.4), all quantities except $\langle u^2 \rangle$ may be instantaneous, peak or root mean square, provided that only one type is used at any one time.

4.2.4   Radiation Impedance and Radiation Efficiency

The radiation impedance of a surface or acoustic source is a measure of the reaction of the acoustic medium against the motion of the surface or source. It is really a special case of a mechanical impedance applied to a source of sound. Thus, Equations (4.3) and (4.4) may also be used to define the radiation impedance.

When the radiation impedance is normalised by the characteristic impedance (ρc) of the acoustic medium and the surface area, S, of the noise source, the resulting quantity is known as the radiation efficiency, σ (sometimes called radiation ratio, as it can sometimes exceed unity). Physically, the radiation efficiency is the ratio of the sound power radiated by a particular sound source of surface area, S, to the power that would be carried in one direction by a plane wave of area, S.

For a monopole source (pulsating sphere) of radius, a, and surface area, S, the radiation efficiency, σ_m, is:

$$\sigma_m = \frac{pS}{uS\rho c} = \frac{p}{u\rho c} = \frac{Z_s}{\rho c} \tag{4.5}$$

It can be shown (Bies et al., 2017) that:

$$\sigma_m = \frac{\mathrm{j}k_a a}{1 + \mathrm{j}k_a a} = \cos \beta \mathrm{e}^{\mathrm{j}\beta} \tag{4.6}$$

where:

$$\beta = \tan^{-1}(1/k_a a) \tag{4.7}$$

and where $k_a = 2\pi/\lambda$ is the wavenumber of the radiated sound.

The real and imaginary parts of σ_m for a pulsating sphere are shown in Figure 4.1 as a function of $k_a r$, where:

$$\sigma_m = \sigma_R + j\sigma_I \tag{4.8}$$

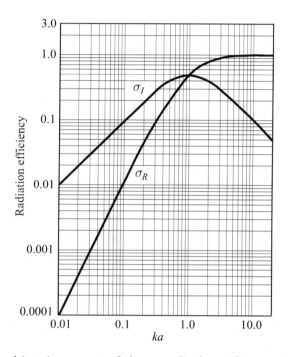

FIGURE 4.1 Real and imaginary parts of the normalised specific acoustic impedance $Z_s/(\rho c)$ of the air load on a pulsating sphere of radius, a, located in free space. Frequency is plotted on a normalised scale, where $k_a a = 2\pi f a/c = 2\pi a/\lambda$. The ordinate is equal to $Z_M/(\rho c S)$, where Z_M is the mechanical impedance; and to $Z_A S/(\rho c)$, where Z_A is the acoustic impedance. The quantity, S, is the area for which the impedance is being determined, and ρc is the characteristic impedance of the medium.

The real part of the radiation efficiency is associated with the radiation of acoustic energy to the far field, while the imaginary part is associated with energy storage in the near field of the vibrating surface (Bies et al., 2017).

The real and imaginary parts of the radiation efficiency of a vibrating structure can be determined by measuring the space averaged complex acoustical intensity in the acoustic medium at the surface of the structure. Thus:

$$\sigma = \frac{\langle I \rangle}{\langle u^2 \rangle \rho c} \tag{4.9}$$

where $\langle u^2 \rangle$ is the space and time averaged normal velocity of the radiating surface and $\langle I \rangle$ is the space averaged complex sound intensity in a direction normal to the surface.

If only the real part of the radiation efficiency (relevant to far field sound radiation) is needed, then the radiation efficiency can be determined by measuring the sound power, W, radiated by the surface or structure and using the following relation:

$$\sigma = \frac{W}{\langle u^2 \rangle S \rho c} \tag{4.10}$$

where S is the area of the radiating surface.

Another useful example is that of a plane circular piston, that is, a uniformly vibrating surface radiating into free space. Three cases will be considered: radiation from one side of the piston mounted in an infinite rigid baffle; radiation from the piston mounted in the end of a long tube; and radiation from both sides of a piston in free space.

For the piston mounted in the infinite rigid baffle, the expression for the complex radiation efficiency is (Kinsler et al., 1982):

$$\sigma_B = \sigma_R + j\sigma_I = R(2k_a a) + jX(2k_a a) \tag{4.11}$$

where:

$$R(x) = \frac{x^2}{2 \times 4} - \frac{x^4}{2 \times 4^2 \times 6} + \frac{x^6}{2 \times 4^2 \times 6^2 \times 8} - \dots \tag{4.12}$$

and:

$$X(x) = \frac{4}{\pi}\left[\frac{x}{3} - \frac{x^3}{3^2 \times 5} + \frac{x^5}{3^2 \times 5^2 \times 7} - \dots\right] \tag{4.13}$$

where a is the radius of the circular piston and $k_a = 2\pi/\lambda$ = wavenumber in the adjacent fluid. The real and imaginary parts of σ_B are plotted as a function of $k_a a$ and shown as solid lines in Figure 4.2.

The case of a plane, circular piston in the end of a pipe has been analysed in detail by Levine and Schwinger (1948) and the results are shown as dashed lines in Figure 4.2.

The case of a piston radiating from both sides into free space was analysed in detail by Wiener (1951) and the results are shown as dotted lines in Figure 4.2. Note that at high frequencies, the real part of the radiation efficiency asymptotes to 2 rather than unity, as sound is radiated from both sides.

Another example that will now be considered is the radiation efficiency of a plane circular plate vibrating in one of its resonant modes (not necessarily at the resonance frequency), mounted in the plane of an infinite rigid baffle and radiating into free space. Three specific cases will be considered: sound radiation from a simply supported circular plate vibrating in one of its first seven low order modes; sound radiation for a clamped edge circular plate; and sound radiation from a simply supported rectangular plate, vibrating in its first few low order modes. For the latter case, only the real part of the radiation efficiency will be given.

The analysis used to derive the results for the circular plates is complicated, making use of the oblate spheroidal coordinate system, and is discussed in detail elsewhere (Hansen, 1980). Results are given for the clamped edge and simply supported edge circular plates, both real and imaginary components, in Figures 4.3(a) to (h).

The concept of modes of vibration as a means of representing the motion of a vibrating surface is discussed more thoroughly in Chapter 5.

If only the real part of the radiation efficiency (associated with the radiation of energy to the far field) is of interest, then the following expression may be used to calculate just the real part:

$$\sigma = \frac{1}{\langle u^2\rangle S\rho c}\int_0^\pi\int_0^{2\pi}\frac{|\hat{p}^2(\mathbf{r})|}{2\rho c}r^2 d\theta d\phi \tag{4.14}$$

where $|\hat{p}(\mathbf{r})|^2$ is the square of the modulus of the acoustic pressure amplitude at some location, $\mathbf{r} = (r, \theta, \phi)$, in the far field of the vibrating surface and $\langle u^2\rangle$ is the space and time averaged normal acoustic particle velocity, u, averaged over the surface (equal to the normal surface velocity) and given by:

$$\langle u^2\rangle = \frac{1}{S}\iint_S\left[\frac{1}{T}\int_0^T u^2(\mathbf{x},t)dt\right]d\mathbf{x} \tag{4.15}$$

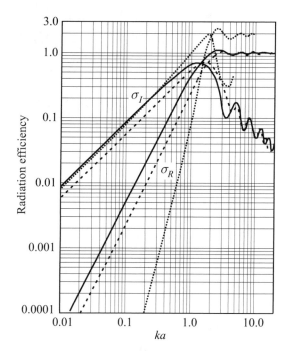

FIGURE 4.2 Real and imaginary parts of the normalised mechanical impedance, $Z_M/(\pi a^2 \rho c)$, or radiation efficiency for a plane piston of radius, a. Frequency is plotted on a normalised scale, where $k_a a = 2\pi f a/c = 2\pi a/\lambda$. Note also that the ordinate is equal to $Z_A \pi a^2/(\rho c)$, where Z_A is the acoustic impedance.

—————— Piston mounted in an infinite flat baffle and radiating from one side.
— — — — — Piston mounted in the end of a long tube and radiating from one side.
· · · · · · · · · · · · · Piston in free space and radiating from both sides.

where S is the area of plane vibrating surface, $\boldsymbol{x}$ is a vector location (x, y) on the surface and T is a suitable time period over which to estimate the mean square velocity of the surface.

The integration in Equation (4.14) is over a hemisphere at a distance, r, from the centre of the vibrating surface. Equations (4.10) and (4.14) can be combined to give an expression for the radiated sound power in terms of the far field radiated sound pressure.

The acoustic pressure, $p(\boldsymbol{r}, t)$, at location, $\boldsymbol{r}$, in space, radiated by a plane, infinitely baffled surface vibrating arbitrarily at angular frequency, ω, may be calculated using the following well-known integral formulation first introduced by Lord Rayleigh in 1896 and discussed in Chapter 3:

$$p(\boldsymbol{r},t) = \frac{\mathrm{j}\omega\rho}{2\pi} \iint\limits_{S} \frac{u(\boldsymbol{x},t)\mathrm{e}^{-\mathrm{j}k_a r}}{r}\mathrm{d}\boldsymbol{x} \qquad (4.16)$$

where $u(\boldsymbol{x}, t)$ is the normal surface velocity amplitude at location, $\boldsymbol{x}$, and r is the distance from the location, $\boldsymbol{x}$, on the surface to location, $\boldsymbol{r}$, in space and for a fixed point in space, $\boldsymbol{r}$ varies with location, $\boldsymbol{x}$, on the surface. At first, it may seem that an integral expression for calculating sound radiation from an infinitely baffled plane surface for a single frequency is a bit restrictive. In practice, the baffle need be only a few wavelengths in size, provided that radiation from the back of the surface is prevented from interfering with that radiated from the front (for example the wall of an enclosure). In fact, at low frequencies (where the radiating surface is small compared with a wavelength of sound) no baffle is needed at all, and very good predictions for the radiated field can be obtained.

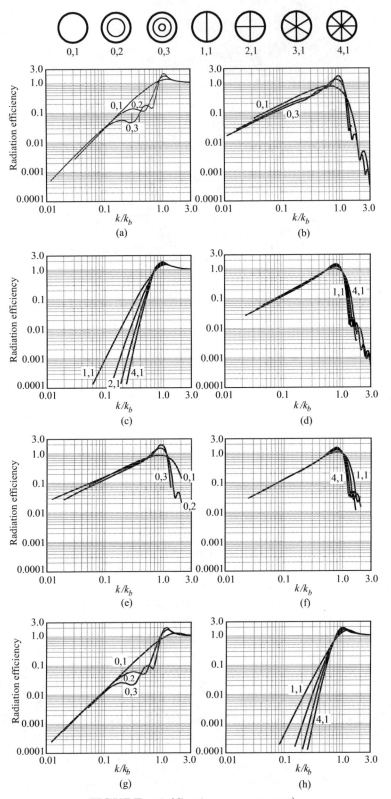

FIGURE 4.3 (Caption on next page.)

FIGURE 4.3 (a) Resistive radiation efficiency for modes with circular nodes and plates with clamped edges calculated using classical plate theory. (b) Reactive radiation efficiency for modes with circular nodes and plates with clamped edges calculated using classical plate theory. (c) Resistive radiation efficiency for modes with diametral nodes and plates with clamped edges calculated using classical plate theory. (d) Reactive radiation efficiency for modes with diametral nodes and plates with clamped edges calculated using classical plate theory. (e) Reactive radiation efficiency for modes with circular nodes and plates with simply supported edges calculated using classical plate theory. (f) Reactive radiation efficiency for modes with diametral nodes and plates with simply supported edges calculated using classical plate theory. (g) Resistive radiation efficiency for modes with circular nodes and plates with simply supported edges calculated using classical plate theory. (h) Resistive radiation efficiency for modes with diametral nodes and plates with simply supported edges calculated using classical plate theory.

However, for sound radiation from only one side of the plane surface, the 2 in the denominator of Equation (4.16) is replaced with 4 to account for radiation into a spherical rather than a hemispherical space.

Equation (4.16), in practice, can also be generalised to apply to a narrow band of noise with a centre frequency, ω. If multiple frequencies or frequency bands are considered, the sound field due to each can be combined by adding pressures squared to give the total sound pressure squared, from which the sound pressure level can be obtained using Equation (1.7).

Given the normal surface velocity amplitude distribution, $\hat{u}(\boldsymbol{x})$, over a plane vibrating surface, the previous expressions may be used to calculate the real part of the surface radiation impedance or radiation efficiency. Wallace (1972) presented the analysis for a simply supported rectangular plate where he used the following modal velocity distribution for the m, nth mode (m, n are integers and for the lowest order mode, $m = n = 1$). Note that the assumption of light structural damping is implicit in this formulation:

$$\hat{u}_{mn}(\boldsymbol{x}) = \hat{u}_{mn}(x, y) = A_{mn} \sin(m\pi x/L_x) \sin(n\pi y/L_y) \tag{4.17}$$

where A_{mn} is the modal velocity amplitude and L_x and L_y are the dimensions of the plate. Thus the following result for the far field radiated sound pressure amplitude at frequency, ω, at a point (r, θ, ϑ) in space is obtained:

$$\hat{p}(\boldsymbol{r}) = \frac{jA_{mn}k_a\rho c}{2\pi r} e^{-jk_a r} \frac{L_x L_y}{mn\pi^2} \left[\frac{(-1)^m e^{-j\alpha} - 1}{(\alpha/m\pi)^2 - 1} \right] \left[\frac{(-1)^n e^{-j\beta} - 1}{(\beta/n\pi)^2 - 1} \right] \tag{4.18}$$

where:

$$\alpha = k_a L_x \sin\theta \cos\vartheta \quad \text{and} \quad \beta = k_a L_y \sin\theta \sin\vartheta \tag{4.19}$$

and where r is the distance of the point in space from the corner of the plate where $x = y = 0$.

The following result is then obtained for the radiation efficiency of the mnth mode:

$$\sigma_{mn} = \frac{64k_a^2 L_x L_y}{\pi^6 m^2 n^2} \int_0^{\pi/2} \int_0^{\pi/2} \left[\frac{\begin{matrix}\cos\\\sin\end{matrix}\left(\dfrac{\alpha}{2}\right) \begin{matrix}\cos\\\sin\end{matrix}\left(\dfrac{\beta}{2}\right)}{[(\alpha/m\pi)^2 - 1][(\beta/n\pi)^2 - 1]} \right]^2 \sin\theta \, d\theta \, d\vartheta \tag{4.20}$$

where $\cos(\alpha/2)$ is used when m is odd and $\sin(\alpha/2)$ is used when m is even. Similarly $\cos(\beta/2)$ is used when n is odd and $\sin(\beta/2)$ is used when n is even.

The mode order, $(m - 1, n - 1)$, refers to the number of nodal lines across the plate in the vertical and horizontal directions, respectively. That is, number of vertical nodal lines $= (m-1)$.

Results for the radiation efficiency of some low order vibration modes of a simply supported rectangular plate are given in Figures 4.4(a) to (d) (Wallace, 1972).

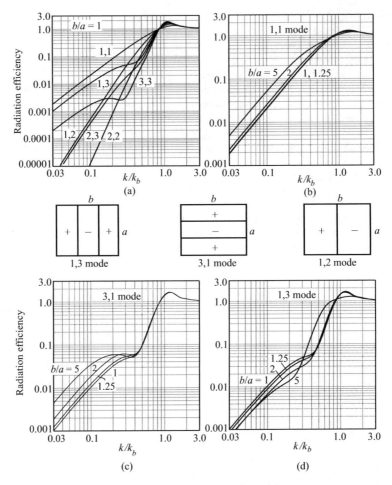

FIGURE 4.4 Radiation efficiencies for various modes and dimensional aspect ratios for a simply supported rectangular plate (Wallace, 1972).

When a plate is excited "off-resonance", several modes are likely to contribute to the overall plate response and to the overall sound radiation. In this case, the far field sound pressure may be calculated as follows:

1. The amplitude and relative phase of the plate response can be calculated or measured at a large number of points on the plate.

2. The resulting far field sound pressure at vector location, r, is then calculated by replacing the integral in Equation (4.16) by a sum over all of the calculated displacements (which are all complex, being represented by an amplitude and a phase).

It is not possible to obtain good results by determining the amplitudes and relative phases of the vibration modes contributing to the sound pressure, and then combining the sound pressures or sound pressures squared due to each vibration mode, to obtain the total sound pressure. This is because the vibration modes are not orthogonal with respect to sound radiation; they are only orthogonal with respect to vibration on the plate.

The radiation efficiency can then be calculated as before by integrating the calculated squared sound pressure over a hemispherical surface surrounding the plate to obtain the sound power radiated by the plate, and then dividing the result by the plate surface area, S, the characteristic impedance, ρc, and the mean square surface velocity of the plate. Note that the relative amplitudes and phases of the modes excited on the plate will be dependent on the location and type of exciting force.

Sometimes, the sound radiation efficiency is dominated by modes other than those that dominate the structural response, even if the response is close to resonance. This is because some very efficient modes could be excited at a much lower level than one or more modes which are not very efficient radiators. This phenomenon often occurs when the structure is excited with an acoustic wave, but not when it is excited mechanically.

4.2.4.1 Structural Input Impedance

The input impedance (sometimes called the mechanical impedance) of a structure is a quantity that allows the vibrational input power or energy transfer to the structure to be calculated for a defined force or moment acting at a defined point or points. It also allows the energy transfer from one structure to another to be expressed in fairly simple terms, and is a function of excitation frequency and location on the structure. Structural input impedance essentially relates the motion of a structure to a disturbance applied at the same location. The reciprocal of structural input impedance is called the point mobility, and is more commonly used in the discipline of structural dynamics. If the structural response is desired at a different location to the applied disturbance, then the related quantities are the transfer impedance or its reciprocal, transfer mobility.

Structural input impedance is an important concept used in the application of active vibration control to reduce vibratory power transmission in structures, as it enables the power input to the structure generated by the control sources to be calculated, and the type and necessary strength of the control source to be quantified. Although the expressions to follow refer to a point force or point moment, they can be applied in practice to forces or moments that act over a small area. The concept of structural input impedance also facilitates the analysis of vibratory power transmission through a complex structure, which is an essential part of the design of active systems to control this power transmission. It is also very useful in the analysis of the effectiveness of active vibration isolation systems (as well as passive isolation systems).

There are three types of structural input impedance that are important: force impedance, Z_F, moment impedance, Z_M, and wave impedance, Z_W. The first two are defined as follows:

$$Z_F = F/u \tag{4.21}$$

$$Z_M = M/\dot{\theta} \tag{4.22}$$

where F is the excitation force, u is the velocity of the structure in the direction of the force at the point of application of the force, M is the excitation moment and $\dot{\theta}$ is the angular velocity of the structure in the direction of the moment at the point of application of the moment. Thus, for the concept of structural input impedance to be applied, it is assumed that the force or moment excitation is localised in a region that is small compared to a wavelength of sound. Impedance is usually defined as a function of frequency, and it is a complex quantity characterised by an amplitude and phase. Thus the quantities, F, u, M and $\dot{\theta}$ in the preceding equations are usually complex quantities evaluated at the frequency of interest.

If a force and moment act together on a beam, for example, the above results for independent force and moment impedances cannot be simply added together, as there will be coupling between the force and moment response. This arises because the point force will result in a rotation as well as a lateral displacement of the beam section and the moment will result in a displacement as well as a rotation of the beam section. Similarly, if multiple forces or moments act at different locations, the impedance matrix will have coupling terms containing transfer impedances from one point to another.

The complex (real and imaginary) power transmission into a structure, generated by a harmonic (single frequency) point excitation force of complex amplitude, $\hat{F}$, is:

$$W = \frac{1}{2}\hat{F}\,\hat{u}_0^*$$ (4.23)

where the * indicates the complex conjugate and $\hat{u}_0$ is the complex velocity amplitude of the structure at the point of application of the force, $\hat{F}$. In terms of impedance the complex power is:

$$W = \frac{\hat{F}\hat{F}^*}{2Z_F} = \frac{1}{2}\hat{u}_0\hat{u}_0^* Z_F$$ (4.24)

The time averaged propagating part of the complex power is proportional to the product of the exciting force and the in-phase component of the structural velocity at the point of application of the force, and is:

$$\mathrm{Re}\{W\} = \frac{1}{2}\mathrm{Re}\{\hat{F}^*\hat{u}_0\} = \frac{|\hat{F}|^2}{2\,\mathrm{Re}\{Z_F\}} = \frac{1}{2}|\hat{u}_0|^2\mathrm{Re}\{Z_F\}$$ (4.25)

The part that represents the amplitude of the non-propagating stored energy is proportional to the product of the excitation force and the in quadrature component of the structural velocity at the point of application of the force, and is:

$$\mathrm{Im}\{W\} = \frac{1}{2}\mathrm{Im}\{\hat{F}^*\hat{u}_0\} = \frac{|\hat{F}^2|}{2\,\mathrm{Im}\{Z_F\}} = \frac{1}{2}|\hat{u}_0|^2\mathrm{Im}\{Z_F\}$$ (4.26)

Similarly, the power injected by a moment amplitude excitation, $\hat{M}$, is:

$$\mathrm{Re}\{W\} = \frac{1}{2}\mathrm{Re}\{\hat{M}^*\hat{\dot{\theta}}_0\} = \frac{|\hat{M}|^2}{2\mathrm{Re}\{Z_M\}} = \frac{1}{2}|\hat{\dot{\theta}}_0|^2\,\mathrm{Re}\{Z_M\}$$ (4.27)

where $\hat{\dot{\theta}}_0$ is the complex angular velocity amplitude of the structure in the direction of the moment at the point of application of the moment, $\hat{M}$.

The reciprocal of the force impedance, Z_F, is referred to as the mobility of a structure and this latter quantity is often referred to in the literature. Force and moment impedances of simple structures can be calculated analytically using the wave equation; however, for more complex structures it is necessary to determine these quantities by measurement, or from the results of a modal analysis (see Chapter 5). Force impedance is usually measured by exciting the structure with a shaker (electrodynamic, hydraulic, piezoelectric, etc.), then measuring the force input (with a piezoelectric crystal) and the acceleration (with an accelerometer) at the point of interest on the structure. The structural velocity signal is obtained by integrating the signal from the accelerometer. The force to structural velocity ratio at the location of the input force is known as the point impedance. If the velocity is measured at a location different to the force input location, then the ratio of the force to velocity is known as the transfer impedance.

Sometimes the force and acceleration measurements are made using an impedance head, which is a single transducer containing two piezoelectric crystals, one for measuring force and the other for measuring acceleration. When measuring the force impedance or mobility of a structure, it is important that the shaker is connected to the structure using a ball joint or a length of thin wire so that bending moments are not transmitted to the structure (see, for example, Ewins (2000)). It is also important to mount the force transducer at the structure end of the shaker attachment. When an impedance head is used, the end containing the force crystal should be attached to the structure. When using an impedance head it is also possible for the stiffness of the joint between the force crystal and the acceleration crystal to cause errors in the phase of the acceleration measurement at higher frequencies, so it is generally better to use separate force and acceleration transducers and mount both directly on to the structure (preferably with a screwed mounting stud).

The measurement of the moment impedance is more difficult and requires the use of two shakers exciting the structure at two locations as close together as possible. The phases and amplitudes of the input forces and accelerations at each location are measured and these measurements, together with the distance between the excitation points, are used to obtain the input moment and the angular acceleration of the structure midway between the two excitation points.

The derivation of theoretical force and moment impedances can be tedious, even for simple structures. However, an example derivation will be given here for the bending wave point force impedance at the centre of a thin, infinitely long beam. In practice, a beam of finite length may be considered infinitely long if either there is a large amount of damping at its ends, or if its internal loss factor, η, is sufficiently large that waves reflected from the ends have a sufficiently diminished amplitude, by the time they arrive back at their source, that they may be ignored. For a beam of length, L, excited in the centre and with negligible damping at the ends, the required value of the internal loss factor (for bending waves) for the beam to be considered infinite is such that (Fahy and Gardonio, 2007):

$$\eta >> 2/k_b L \qquad (4.28)$$

where the bending (or flexural) wavenumber, k_b is:

$$k_b = (\omega^2 \rho_m S / EJ)^{1/4} \qquad (4.29)$$

where ω is the excitation frequency (radians/s), S is the beam cross-sectional area, ρ_m is the density of the beam material, E is Young's modulus of elasticity, and J is the second moment of area of the beam cross section about an axis perpendicular to the beam axis and also perpendicular to the bending wave displacement.

4.2.5 Force Impedance of an Infinite Beam (bending Waves)

In Section 2.3.3.3, the classical wave equation for bending waves in a beam was found to be:

$$EJ \frac{\partial^4 w(x,t)}{\partial x^4} + \rho_m S \frac{\partial^2 w(x,t)}{\partial t^2} = 0 \qquad (4.30)$$

where w, the bending wave displacement at axial location, x, along the beam, is positive in the positive z-direction (see Figure 2.12). Note that this equation is only valid if the bending wavelength, $\lambda = 2\pi/k_b$, is much larger than the thickness, h, of the beam (beam dimension in the direction of the bending wave displacement). In general, Equation (4.30) is valid if $\lambda > 6h$. For beams that are wide with respect to their thickness, E must be replaced with $E(1 - \nu^2)$. Even with this adjustment, the beam equation will only hold at frequencies below which bending waves begin to propagate across the width.

Note that the bending wave equation for the angular displacement, θ, has the same form as that for the normal displacement, w.

For a beam excited by a normal point force, $F = \hat{F} e^{j\omega t}$, at a location $x = a$, Equation (4.30) becomes:

$$\frac{\partial^4 w(x,t)}{\partial x^4} + \rho_m S \frac{\partial^2 w(x,t)}{\partial t^2} = \hat{F} \delta(x - a) e^{j\omega t} \qquad (4.31)$$

where $\delta(x - a)$ is the Dirac delta function discussed in detail in Chapter 3.

Equation (4.30) is equivalent to Equation (4.31) at every point on the beam except where the force is applied. The solution to Equation (4.30) was given in Section 2.3.3 as:

$$w(x,t) = \left[A_1 e^{-jk_b x} + A_2 e^{jk_b x} + A_3 e^{-k_b x} + A_4 e^{k_b x} \right] e^{j\omega t} = \hat{w}(x) e^{j\omega t} \qquad (4.32)$$

Here, it will be assumed that the centre of the beam is at $x = 0$ and the beam extends to infinity in each direction. Thus, in the region, $x < 0$, B and D must be zero (no reflected waves from the left end) and where $x > 0$, A and C are zero. The coefficients, A, B, C and D, may be found by satisfying equilibrium conditions immediately to the left $(x = 0^-)$ and right $(x = 0^+)$ of $x = 0$.

As shown in Section 2.3.3, the elastic shear stresses produce an upward force of magnitude:

$$\frac{F}{2} = EJ\frac{\partial^3 \hat{w}(x)}{\partial x^3} \tag{4.33}$$

at the left end of an elemental beam section and a downward force of equal magnitude at the right-hand end of the section (see Figure 4.5). As the applied force is considered positive in the positive w-direction, then at $x = 0^+$:

$$\frac{F}{2} - EJ(\mathrm{j}k_b^3 A_1 - k_b^3 A_3) = 0 \tag{4.34}$$

and at $x = 0^-$:

$$\frac{F}{2} + EJ(-\mathrm{j}k_b^3 A_2 + k_b^3 A_4) = 0 \tag{4.35}$$

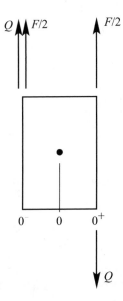

FIGURE 4.5 External forces, F, and internal shear forces, Q, acting on an infinitesimally small cross-sectional element at the centre of a beam.

Because of symmetry, the slope of the beam at $x = 0$ is zero. Hence:

$$- \mathrm{j}k_b A_1 - k_b A_3 = \mathrm{j}k_b A_2 + k_b A_4 = 0 \tag{4.36}$$

Equations (4.34) to (4.36) give:

$$A_1 = A_2 = \mathrm{j}A_3 = \mathrm{j}A_4 \tag{4.37}$$

and:

$$A_1 = -\frac{\mathrm{j}F}{4EJk_b^3} \tag{4.38}$$

Thus, at $(x = 0^+), (x = 0^-)$, the displacement, $\hat{w}$, is given by Equation (4.32) as:

$$\hat{w}(0^+) = \hat{w}(0^-) = \left(\frac{-\mathrm{j}F}{4EJk_b^3}\right)(1 - \mathrm{j}) \tag{4.39}$$

The point impedance (for a force applied at a point) is thus:

$$Z_F = \frac{F}{\dfrac{\partial w(0)}{\partial t}} = (2EJk_b^3\omega)(1+\text{j}) = 2\rho_m S c_b(1+\text{j}) \tag{4.40}$$

where k_b is given by Equation (4.29) and c_b is the wave speed of the bending waves given by $c_b = \omega/k_b$. A more complicated expression applies for thick beams and can be derived using the wave equation for thick beams discussed in Section 2.3.3.

4.2.6 Summary of Impedance Formulae for Infinite and Semi-Infinite Isotropic Beams and Plates

The expressions given in Tables 4.1 and 4.2 to follow for the point force and point moment impedances of thin beams and plates were derived using the appropriate wave equation and boundary conditions, and the method just illustrated for an infinite beam.

TABLE 4.1 Summary of point impedance formulae, Z_F and Z_T, or Z_M, for isotropic thin beams excited by a point force or point torque (or moment), respectively (Cremer et al., 1973; Lyon and DeJong, 1995), where $k_b = \omega/c_b$

Wave type	Structural element	Z_F	Z_T or Z_M
Longit-udinal	Infinite thin beam	$2S\sqrt{E\rho_m}$ $= 2\rho_m S c_L$	—
Longit-udinal	Semi-infinite thin beam	$S\sqrt{E\rho_m}$ $= \rho_m S c_L$	—
Torsional	Infinite shaft	—	$2J\sqrt{G\rho_m} = 2\rho_m J c_T$
Torsional	Semi-infinite shaft	—	$J\sqrt{G\rho_m} = \rho_m J c_T$
Bending	Infinite thin beam	$2\rho_m S c_b(1+\text{j})$	$2\rho_m S c_b(1-\text{j})/k_b^2$
Bending	Semi-infinite thin beam	$\frac{1}{2}\rho_m S c_b(1+\text{j})$	$\frac{1}{2}\rho_m S c_b(1-\text{j})/k_b^2$

In all the formulae given, it is assumed that the source impedance, Z_s, is sufficiently small not to contribute to the total impedance of the beam or plate. Where this is not true, the impedance of the source can simply be added to that of the beam or plate provided that there are no reflected waves present (that is, the ends of the beam or edges of the plate not adjacent to the source must be an infinite distance from the source or else they must be absorptive). Where waves are reflected back to the driving point and Z_s cannot be neglected, the reflection coefficients of both the propagating and non-propagating waves at the driving point must be used to determine the point impedance of the driving source and beam or plate in combination.

The point force impedance function is the ratio of input force to structural vibration velocity at the same location on the structure. On the other hand, the transfer impedance is the ratio of the input force at one location to the vibration velocity at another location. The point moment impedance is the ratio of the input moment to the angular velocity. The point moment impedance, Z_M, for a beam is related to the point force impedance, Z_F, by:

$$Z_M = \frac{-jZ_F}{k_b^2} \tag{4.41}$$

It is of interest to note that the point force impedances of infinite systems such as those listed in Tables 4.1 and 4.2 correspond to frequency averaged impedances of equivalent finite systems, as illustrated in Figure 4.6. The only difference is that the finiteness of the finite systems add the peaks and troughs that span the line for the infinite system (Skudrzyk, 1980). Thus the infinite system impedances provide a good estimate of the average impedance of finite systems with the modal peaks being smaller as structural damping is increased. An excellent treatment of this topic, which includes tables of impedance formulae for many infinite and semi-infinite systems, is given by Pinnington (1988).

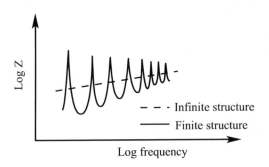

FIGURE 4.6 Infinite system impedance approximation to the impedance of a similar finite system.

4.2.7   Point Force Impedance of Finite Systems

As discussed in Chapter 5, the motion of a finite vibrating structure can be characterised in terms of its modes of vibration, and the normal surface velocity amplitude, $\hat{u}(\boldsymbol{x})$, at location, $\boldsymbol{x}$, due to an applied force amplitude, $\hat{F}(\boldsymbol{x}_F)$, at location, $\boldsymbol{x}_F$, and frequency, ω, is given by Fahy and Gardonio (2007):

$$\hat{u}(\boldsymbol{x}) = \hat{F}(\boldsymbol{x}_F) \sum_{n=1}^{\infty} \frac{j\omega \psi_n(\boldsymbol{x}_F)\psi_n(\boldsymbol{x})}{m_n(\omega_n^2 - \omega^2 + j\eta\omega_n^2)} \tag{4.43}$$

from which the force impedance at location, $\boldsymbol{x}_F$, is:

$$Z_F = \frac{\hat{F}(\boldsymbol{x}_F)}{\hat{u}(\boldsymbol{x}_F)} = \left[\sum_{n=1}^{\infty} \frac{j\omega \psi_n^2(\boldsymbol{x}_F)}{m_n(\omega_n^2 - \omega^2 + j\eta_n\omega_n^2)} \right]^{-1} \tag{4.44}$$

TABLE 4.2 Summary of point impedance formulae, Z_F and Z_M, for isotropic thin plates excited by a point force or a point moment (Cremer et al., 1973; Lyon and DeJong, 1995), where $k_b = \omega/c_b$

Wave type	Structural element	Z_F	Z_M
In-plane	Infinite, thin isotropic plate	$8\pi m_s f r^2 \left(1 - \dfrac{\mathrm{j}c_L}{2\pi f r}\right)$	—
Bending	Infinite, thin isotropic plate	$\begin{aligned} &8\omega m_s/k_b^2 \\ &= 2.3 m_s c_L h \\ &= 8 m_s \kappa_b c_L \end{aligned}$	$\dfrac{16\omega m_s}{k_b^4}\left(1 - \dfrac{4\mathrm{j}}{\pi}\log_e(0.9 k_b a)\right)^{-1}$
Bending	Semi-infinite, thin isotropic plate	$3.5\omega m_s/k_b^2$	$\dfrac{5.3\omega m_s}{k_b^4}\left(1 - 1.46\mathrm{j}\log_e(0.9 k_b a)\right)^{-1}$

The quantities in Tables 4.1 and 4.2 are defined as follows:

S = cross-sectional area of the beam;

J = second moment of area of the beam cross section (sometimes referred to as the area moment of inertia) $= \pi d^4/64$ for a circular-section solid shaft, where d is the shaft diameter;

E = Young's modulus of elasticity beam or plate material;

G = shear modulus of the beam or plate material $= E/[2(1+\nu)]$;

h = plate thickness;

ν = Poisson ratio for the plate material;

ρ_m = density of the beam or plate material;

m_s = mass per unit area of the plate ($= \rho_m h$);

$\kappa_b = \sqrt{J/S} = h/\sqrt{12}$ = radius of gyration of plate section for bending, where J and S are the second moment of area and area, respectively, of the plate cross section;

k_b = bending wavenumber (given by Equation (4.29) for a beam and Equation (4.42) for a plate);

$$k_b = \left[\frac{12\omega^2 m_s(1-\nu^2)}{Eh^3}\right]^{1/4} \tag{4.42}$$

c_b = bending wave speed ($= \omega/k_b = c_g/2$);

c_L = longitudinal wave speed ($=\sqrt{E/\rho} = c_g$ for a beam and $= \sqrt{E/[\rho_m(1-\nu^2)]} = c_g$ for a plate);

f ($= \omega/(2\pi)$) = excitation frequency in Hz;

$2r$ = diameter of the excitation footprint on the plate;

a = moment arm, or distance between the two forces used to generate the moment on the plate.

The modal mass, m_n, is defined as:

$$m_n = \iint\limits_{S} m_s(\boldsymbol{x})\psi_n^2(\boldsymbol{x})d\boldsymbol{x} \qquad (4.45)$$

where $m_s(\boldsymbol{x})$ is the surface density (mass per unit area) at vector location, $\boldsymbol{x}$, on the surface, $\psi(\boldsymbol{x}_F)$ is the value of the mode shape function at location, $\boldsymbol{x}_F$, η_n is the structural loss factor (hysteretic, not viscous) for mode, n, and $\boldsymbol{x}_F$ is the vector location on the surface at which the point force, F, is applied.

To apply the preceding equations to a practical structure, the resonance frequencies, mode shapes and modal damping must be determined. Modal damping cannot be calculated, but it can be estimated from experience with similar structures, or it can be measured (if the structure exists) using the methods outlined in Chapter 5. Structural resonance frequencies and mode shapes can be calculated from first principles for simple structures and by using finite element analysis (FEA) or the boundary element method (BEM) for more complex structures.

As an example for a simple structure, it will be shown briefly how the mode shapes and resonance frequencies may be derived for a beam of finite length and free ends. The solution to the wave equation for a thin beam is given by Equation (4.32). One end of the beam is at $x = 0$ and the other is at $x = L$. Considering the boundary conditions at $x = 0$, as the beam is free at this end, the bending moment and shear force vanish. Thus:

$$\frac{\partial^2 \hat{w}(x)}{\partial x^2} = 0; \quad \text{and} \quad \frac{\partial^3 \hat{w}(x)}{\partial x^3} = 0 \qquad (4.46)$$

The solution to the beam wave equation may also be written in terms of hyperbolic and transcendental functions as follows:

$$w(x,t) = (A\cos k_b x + B\sin k_b x + C\cosh k_b x + D\sinh k_b x)\mathrm{e}^{j\omega t} \qquad (4.47)$$

and in fact this solution form is more commonly used to determine the beam resonance frequencies and mode shapes as it simplifies the analysis.

The boundary conditions of Equation (4.46) applied to the end, $x = 0$, lead to the following relations:

$$A = C \quad \text{and} \quad B = D \qquad (4.48)$$

Substituting Equation (4.48) into Equation (4.47) and ignoring the time dependence gives:

$$\hat{w}(x) = A(\cos k_b x + \cosh k_b x) + B(\sin k_b x + \sinh k_b x) \qquad (4.49)$$

Applying the first of the boundary conditions of Equations (4.46) to (4.49) evaluated at $x = L$ gives:

$$B = -A\frac{\cosh k_b L - \cos k_b L}{\sinh k_b L - \sin k_b L} \qquad (4.50)$$

Substituting Equation (4.50) into Equation (4.49) and eliminating the constant, A, gives the following result for the beam mode shape function:

$$\psi(x) = \frac{\hat{w}(x)}{A} = \left[\cosh k_b x + \cos k_b x - \frac{\cosh k_b L - \cos k_b L}{\sinh k_b L - \sin k_b L}(\sinh k_b x + \sin k_b x)\right] \qquad (4.51)$$

Applying the second of the boundary conditions of Equations (4.46) to (4.51) at $x = L$ allows an expression for the eigen solutions, k_b, to be obtained as follows:

$$\cos(k_b L)\cosh(k_b L) = 1 \qquad (4.52)$$

TABLE 4.3 Mode shape and resonance frequency equations for beams and plates

End conditions	Frequency equation	Mode shape, $\psi(x)$
$\longleftarrow L \longrightarrow$	$\cos k_b L \cosh k_b L = 1$	$\psi(x) = (\sin k_b x - \sinh k_b x) + A(\cos k_b x - \cosh k_b x)$ $A = -\dfrac{\sin k_b L - \sinh k_b L}{\cos k_b L - \cosh k_b L}$
	$\tan k_b L = \tanh k_b L$	$\psi(x) = (\sin k_b x - \sinh k_b x) + A(\cos k_b x - \cosh k_b x)$ $A = -\dfrac{\sin k_b L + \sinh k_b L}{\cos k_b L + \cosh k_b L}$
	$\sin k_b L = 0$	$\psi(x) = \sin k_b x$
	$\cos k_b L \cosh k_b L = -1$	$\psi(x) = (\sin k_b x - \sinh k_b x) + A(\cos k_b x - \cosh k_b x)$ $A = -\dfrac{\sin k_b L + \sinh k_b L}{\sin k_b L - \sinh k_b L}$
	$\omega_{n,m} = \left[\dfrac{Eh^3}{12 m_s (1 - \nu^2)} \left\{ \left(\dfrac{n\pi}{L_1}\right)^2 + \left(\dfrac{m\pi}{L_2}\right)^2 \right\} \right]$	$\psi(x,y) = \sin\left(\dfrac{n\pi x}{L_1}\right) \sin\left(\dfrac{m\pi y}{L_2}\right)$

The solutions of Equation (4.52) correspond to the resonance frequencies of the modes of vibration of the beam and when substituted into Equation (4.51), they allow the corresponding mode shape functions to be calculated. Frequency equations and mode shapes corresponding to various beam end conditions and for a simply supported plate are given in Table 4.3. Note that for the beam, the resonance frequency is related to the solution, k_b, of the frequency equation, by Equation (4.29). For the simply supported plate in Table 4.3, n and m are integer numbers which correspond to mode order (n, m).

4.2.8   Point Force Impedance of Cylinders

4.2.8.1   Infinite cylinder

In this case, as there are no waves reflected from the ends of the cylinder, the modal displacements are independent of axial location and the mode shapes for radial and circumferential vibration are:

$$\psi_{wn} = \cos n\theta \qquad (4.53)$$

$$\psi_{\xi_\theta n} = \sin n\theta \qquad (4.54)$$

The resonance frequencies (corresponding to values of n from $n = 0$ to $n = \infty$) are:

$$\Omega_n^2 = \begin{cases} \frac{1}{2}\left[(1+n^2)(1+k_t n^2)\right] \mp \frac{1}{2}\left[(1+n^2)^2 - 2k_t n^2(1 - 6n^2 + n^4)\right]^{1/2} \\ \text{(Goldenveizer theory, radial and circumferential modes)} \\ \\ \frac{1}{2}\left[1+n^2+k_t n^4\right] \mp \frac{1}{2}\left[(1+n^2)^2 - 2k_t n^6\right]^{1/2} \\ \text{(Flügge theory, radial and circumferential modes)} \end{cases} \qquad (4.55)$$

where:

$$k_t = h^2/12a^2 \qquad (4.56)$$

and where Ω_n is the non-dimensional frequency, defined as:

$$\Omega_n = \omega_n a \sqrt{\frac{\rho_m(1 - \nu^2)}{E}} \qquad (4.57)$$

Equations (4.53) to (4.57) may be substituted into Equations (4.43) to (4.45) to give the point force impedance of an infinite cylinder. In cases involving sound radiation, only the radial vibration component, w, of mode shape Equation (4.53) is of interest.

4.2.8.2   Finite Cylinder — Shear Diaphragm Ends

For this case, it is assumed that the ends of the cylinder are closed with a thin plate, which is supported so that it cannot move normal to the cylinder surface. That is:

$$w = M_x = N_x = \xi_\theta = 0 \qquad (4.58)$$

The mode shape functions are:

$$\psi_{w,m,n} = \sin\frac{m\pi x}{L}\cos n\theta \qquad (4.59)$$

$$\psi_{\xi_\theta,m,n} = \sin\frac{m\pi x}{L}\sin n\theta \qquad (4.60)$$

$$\psi_{\xi_x,m,n} = \cos\frac{m\pi x}{L}\cos n\theta \qquad (4.61)$$

The non-dimensional resonance frequencies are solutions of:

$$\Omega^6 - (K_2 + k_t \Delta K_2)\Omega^4 + (K_1 + k_t \Delta K_1)\Omega^2 + (K_0 + \Delta K_0) = 0 \qquad (4.62)$$

where:

$$K_2 = 1 + \frac{1}{2}(3 - \nu)(n^2 + \lambda_x^2) + k_t(n^2 + \lambda_x^2)^2 \qquad (4.63)$$

$$K_1 = \frac{1}{2}(1 - \nu)\left[(3 + 2\nu)\lambda_x^2 + n^2 + (n^2 + \lambda_x^2)^2 + \frac{(3 - \nu)}{(1 - \nu)}k_t(n^2 + \lambda_x^2)\right] \qquad (4.64)$$

$$K_0 = \frac{1}{2}(1 - \nu)\left[(1 - \nu^2)\lambda_x^4 + k_t(n^2 + \lambda^2)^4\right] \qquad (4.65)$$

$$\text{where } \lambda_x = m\pi x/L \qquad (4.66)$$

For the Goldenveiser–Novozhilov theory:

$$\Delta K_2 = 2(1 - \nu)\lambda_x^2 + n^2 \qquad (4.67)$$

$$\Delta K_1 = 2(1 - \nu)\lambda_x^2 + n^2 + 2(1 - \nu)\lambda_x^4 - (2 - \nu)\lambda_x^2 n^2 - \frac{1}{2}(3 + \nu)n^4 \qquad (4.68)$$

$$\Delta K_0 = \frac{1}{2}(1 - \nu)[4(1 - \nu^2)\lambda_x^4 + 4\lambda_x^2 n^2 + n^4 - 2(2 - \nu)(2 + \nu)\lambda_x^4 n^2 - 8\lambda_x^2 n^4 - 2n^6] \qquad (4.69)$$

For the Flügge theory:

$$\Delta K_2 = \Delta K_1 = 0 \qquad (4.70)$$

$$\Delta K_0 = \frac{1}{2}(1 - \nu)[2(2 - \nu)\lambda_x^2 n^2 + n^4 - 2\nu\lambda_x^6 - 6\lambda_x^4 n^2 - 2(4 - \nu)\lambda_x^2 n^4 - 2n^6] \qquad (4.71)$$

Mode shape functions and modal resonance frequency equations for various other cylinder end conditions are given by Leissa (1973). The mode shape functions and solutions to the resonance frequency equations may be used together with Equations (4.43) to (4.45) to obtain the point force impedance. Note that for $n \geq 2$, the mode shape functions and resonance frequencies are not very dependent upon the boundary conditions at the cylinder ends. Also, for all but the lowest order modes, the modal masses are equal to one-quarter of the total cylinder mass.

4.2.9 Wave Impedance of Finite Structures

When a structure is excited by an incident acoustic field, its response is governed by its wave impedance, just as its response to a point force is governed by its point force input impedance. The wave impedance is defined as the ratio of the complex force per unit area, p (or pressure), to the complex velocity, u, at a point on the structure (Fahy and Gardonio, 2007). Thus:

$$Z_{ws} = \frac{p_i}{u} \qquad (4.72)$$

The wave impedance of a structure can be associated with a specific wavenumber in the acoustic medium adjacent to the structure, or frequency and phase speed combination, $k_a = \omega/c$. It is evaluated mathematically by applying a force to a structure in the form of a sinusoidal travelling wave and using the structure equation of motion to derive the structural response. It can be applied in practice to a random or multiple frequency sound field by using Fourier analysis to separate the signal into its frequency components and superposition to obtain the total structural response.

As an example, the wave impedance for an infinite undamped isotropic plate of thickness, h, subject to a transverse force in the form of a plane travelling wave characterised by a wavenumber, k_a, will be derived. Note that only bending waves will be considered, as in practice when structures are excited by acoustic waves in a fluid medium, other waves cannot be generated because the fluid is not capable of supporting shear forces.

The equation of motion for bending waves in a plate subjected to an applied normal force amplitude of q/unit area is given by Equation (2.186). This equation can be written for bending waves propagating in the x-direction only as:

$$D\frac{\partial^4 w(x,t)}{\partial x^4} + \rho_m h \frac{\partial^2 w(x,t)}{\partial t^2} = q e^{j(\omega t - kx)} \qquad (4.73)$$

with a solution of the form:

$$w(x,t) = A e^{j(\omega t - kx)} = \hat{w}(x) e^{j\omega t} \qquad (4.74)$$

where A is the complex displacement amplitude, q is the complex amplitude of the applied force per unit area and the stiffness, D, is:

$$D = E h^3 / 12(1 - \nu^2) \qquad (4.75)$$

Substituting Equation (4.74) into Equation (4.73) and making use of the complex modulus, $E(1 + j\eta)$, as a replacement for E in Equation (4.75) (to allow the inclusion of damping), the following is obtained:

$$\left[\frac{E(1 + j\eta)h^3}{12(1 - \nu^2)} k_a^4 - \rho_m h \omega^2 \right] A = q \qquad (4.76)$$

The wave impedance is defined by Equation (4.72), which is equivalent to:

$$Z_{ws} = \frac{q}{j\omega A} \qquad (4.77)$$

Substituting for q/A from Equation (4.76) into Equation (4.77) gives:

$$Z_{ws} = \frac{E h^3 k_a^4 \eta}{12(1 - \nu^2)\omega} - j\left[\frac{E h^3 k_a^4}{12(1 - \nu^2)\omega} - \rho_m h \omega \right] \qquad (4.78)$$

For structures radiating into dense fluids such as water, the effect of fluid loading must also be taken into account using a fluid wave impedance as discussed by Fahy and Gardonio (2007). However, as the loading is usually negligible for radiation into air, it is not considered further here.

4.3   Sound Intensity

Sound intensity is defined as a measure of the rate of local acoustic energy flow in an acoustic medium. It is a vector quantity characterised by a magnitude, a direction and a specific point location in the acoustic medium. The intensity vector can be expressed as scalar components acting along each axis in a cartesian coordinate system. The sound power being transmitted through an imaginary surface can be obtained by integrating the real component of sound intensity normal to the surface over the area of the surface. More specifically, sound intensity is commonly defined as the long time average rate of transmission of sound energy through unit area of acoustic fluid. However, sound intensity is a complex vector, with each element having both real and imaginary components. The real (or active) component is what is commonly used and has just been defined. The imaginary (or reactive) component is a measure of the energy stored in the sound field.

The active component of sound intensity has a time averaged non-zero value, corresponding to a net transport of energy, while the reactive component has a zero time averaged value corresponding to local oscillatory transport of energy. As the reactive intensity is zero when averaged over time, it is generally expressed as an amplitude. It is associated with potential energy storage and does not propagate anywhere but oscillates between the sound source and adjacent fluid. At any single frequency, these active and reactive intensity components are associated with components of the acoustic particle velocity which are, respectively, in phase and in quadrature (90° out of phase) with the local acoustic pressure. In quadrature components occur in the near field (within half a wavelength) of sound sources or sound reflecting objects, or in any part of the field where sound waves are travelling in more than one direction simultaneously. Reactive intensity amplitude only has meaning for a single frequency or a narrow band sound field. When averaged over a wide frequency band the result is not the sum of the reactive intensity for the individual frequencies as it is for the active component. In fact, the reactive intensity amplitude is zero when averaged over a wide frequency band.

As sound intensity is an energy based quantity, its measurement or calculation requires the determination of two independent quantities; namely, the acoustic pressure and the acoustic particle velocity, together with the relative phase between them.

The instantaneous sound intensity, $I_i(r, t)$, in an acoustic field at a location given by the field vector, r, is a vector quantity describing the instantaneous acoustic power transmission per unit area in the direction of the particle velocity, $u(r, t)$. The general expression for the instantaneous sound intensity is:

$$I_i(r, t) = p(r, t)u(r, t) \qquad (4.79)$$

A general expression for the active sound intensity in a three-dimensional field, $I(r)$, is the time average of the instantaneous intensity given by (4.79), which may be written as follows:

$$I(r) = \langle p(r, t)u(r, t) \rangle = \lim_{T_A \to \infty} \frac{1}{T_A} \int_0^{T_A} p(r, t)u(r, t)dt \qquad (4.80)$$

For the special case of single frequency sound, the sound pressure at a location in three-dimensional space may be represented as:

$$p(r, t) = \hat{A}_p(r)e^{j(\omega t + \theta_p(r))} = \hat{p}(r)e^{j\omega t} \qquad (4.81)$$

where both the amplitude, $\hat{A}_p(r)$, and the phase, $\theta_p(r)$, are real, space dependent quantities and $\hat{p}(r)$ is a complex quantity with real and imaginary parts. The phase term, $\theta_p(r)$, includes the term, $-k_a r$. A similar expression may also be written for the acoustic particle velocity, $u(r, t)$:

$$u(r, t) = \hat{A}_u(r)e^{j(\omega t + \theta_u(r))} = \hat{u}(r)e^{j\omega t} \qquad (4.82)$$

Integration with respect to time of Equation (1.48) and introducing the unit vector $n = r/r$, taking the gradient in the direction, n, and use of Equations (4.81) and (4.82) gives the following result for the acoustic particle velocity in direction, n:

$$u(r, t) = \frac{n j}{\omega \rho} \nabla p(r, t) = \frac{n}{\omega \rho} \left[-\hat{A}_p(r)\frac{\partial \theta_p}{\partial r} + j\frac{\partial \hat{A}_p(r)}{\partial r} \right] e^{j(\omega t + \theta_p(r))} \qquad (4.83)$$

The instantaneous intensity cannot be determined simply by multiplying Equations (4.81) and (4.83) together. This is because the complex notation formulation, $e^{j(\omega t + \theta_p)}$, can only be used for linear quantities. Thus the product of two quantities represented in complex notation is given by the product of their real components only (Skudrzyk, 1971). Thus the instantaneous sound intensity in direction, n, is:

$$I_i(r, t) = \text{Re}\{p(r, t)\}\text{Re}\{u(r, t)\}$$

$$= -\frac{n}{\omega \rho} \left[\hat{A}_p^2(r)\frac{\partial \theta_p}{\partial r} \cos^2(\omega t + \theta_p) + \hat{A}_p(r)\frac{\partial \hat{A}_p(r)}{\partial r} \sin(\omega t + \theta_p)\cos(\omega t + \theta_p) \right] \quad (\text{W/m}^2)$$

$$(4.84)$$

The first term in brackets in Equation (4.84) is the product of the real part of the pressure with the real part of the acoustic particle velocity that is in phase with the pressure (active intensity), while the second term is the product of the real part of the pressure with the real part of the velocity that is in quadrature with the pressure (reactive intensity).

Using well-known trigonometric identities (Abramowitz and Stegun, 1965), Equation (4.84) may be rewritten as:

$$I_i(\boldsymbol{r},t) = -\frac{\boldsymbol{n}}{2\omega\rho}\left\{ \hat{A}_p^2(\boldsymbol{r})\frac{\partial\theta_p}{\partial r}\left[1+\cos2(\omega t+\theta_p)\right] + \hat{A}_p(\boldsymbol{r})\frac{\partial\hat{A}_p(\boldsymbol{r})}{\partial r}\sin2(\omega t+\theta_p)\right\} \quad (\text{W/m}^2)$$

(4.85)

The time average of Equation (4.85) is the active intensity, which is thus:

$$I(\boldsymbol{r}) = -\frac{\boldsymbol{n}}{2\rho\omega}\hat{A}_p^2(\boldsymbol{r})\frac{\partial\theta_p}{\partial r} \tag{4.86}$$

Equation (4.86) is a measure of the acoustic power transmission in the direction of the acoustic intensity vector. Alternatively substitution of the real parts of Equations (4.81) and (4.82) into Equation (4.79) gives the instantaneous intensity in direction, $\boldsymbol{n}$as:

$$I_i(\boldsymbol{r},t) = \boldsymbol{n}\hat{A}_p(\boldsymbol{r})\hat{A}_u(\boldsymbol{r})\cos(\omega t+\theta_p)\cos(\omega t+\theta_u) \quad (\text{W/m}^2) \tag{4.87}$$

Using well-known trigonometric identities (Abramowitz and Stegun, 1965), Equation (4.87) may be rewritten as:

$$I_i(\boldsymbol{r},t) = \frac{\boldsymbol{n}\hat{A}_p(\boldsymbol{r})\hat{A}_u(\boldsymbol{r})}{2}\left\{[1+\cos2(\omega t+\theta_p)]\cos(\theta_p-\theta_u)+\sin2(\omega t+\theta_p)\sin(\theta_p-\theta_u)\right\} \quad (\text{W/m}^2)$$

(4.88)

Equation (4.88) is an alternative form of Equation (4.79). The first term on the right-hand side of the equation is the active intensity, which has a mean value given by:

$$I(\boldsymbol{r}) = \frac{\boldsymbol{n}\hat{A}_p(\boldsymbol{r})\hat{A}_u(\boldsymbol{r})}{2}\cos(\theta_p-\theta_u) = \frac{\boldsymbol{n}}{2}\text{Re}\{\hat{p}(\boldsymbol{r})\hat{u}^*(\boldsymbol{r})\} \quad (\text{W/m}^2) \tag{4.89}$$

where the * indicates the complex conjugate.

Although the time averaged reactive intensity is zero (as it is an oscillating quantity), its amplitude, $Q(\boldsymbol{r})$, is given by the amplitude of the second term in curly brackets in Equation (4.85) as:

$$Q(\boldsymbol{r}) = -\frac{1}{2\rho\omega}\hat{A}_p(\boldsymbol{r})\frac{\partial\hat{A}_p(\boldsymbol{r})}{\partial r} = -\frac{1}{4\rho\omega}\frac{\partial\hat{A}_p^2(\boldsymbol{r})}{\partial r} \tag{4.90}$$

Also in Equation (4.88), the second term in brackets is the reactive intensity, which has an amplitude:

$$Q(\boldsymbol{r}) = \frac{\boldsymbol{n}\hat{A}_p(\boldsymbol{r})\hat{A}_u(\boldsymbol{r})}{2}\sin(\theta_p-\theta_u) = \frac{\boldsymbol{n}}{2}\text{Im}\{\hat{p}(\boldsymbol{r})\hat{u}^*(\boldsymbol{r})\} \quad (\text{W/m}^2) \tag{4.91}$$

4.3.1 Plane Wave and Far Field Intensity

Waves radiating outward, away from any source, tend to become planar. Consequently, the equations derived in this section also apply in the far field of any source. For this purpose, the radius of curvature of an acoustic wave should be greater than about ten times the radiated wavelength. For a point source, this would imply a distance from the sound source of 10 times the acoustic wavelength. For a propagating plane wave, the characteristic impedance, ρc, is a real quantity; thus, according to Equation (1.94), the acoustic pressure and particle velocity are in phase and consequently acoustic power is transmitted. The intensity is a vector quantity but where direction is understood the magnitude is of greater interest. Consequently, the intensity will be written in scalar form as a magnitude. If Equation (1.94) is used to replace $\boldsymbol{u}$ in Equation (4.80) the expression for the scalar plane wave sound intensity at location $\boldsymbol{r}$ becomes:

$$I = \langle p^2(\boldsymbol{r},t)\rangle / \rho c \quad (\text{W/m}^2) \tag{4.92}$$

In Equation (4.92), the intensity has been written in terms of the mean square pressure. If Equation (1.94) is used to replace p in the expression for intensity, the following alternative form of the expression for the scalar plane wave sound intensity is obtained:

$$I = \rho c \langle u^2(r,t)\rangle \quad (\text{W/m}^2) \tag{4.93}$$

where again the intensity has been written in scalar form as a magnitude in terms of the scalar particle velocity, u. The mean square particle velocity is defined in a similar way as the mean square sound pressure.

For single frequency sound, of amplitude $\hat{p}(\boldsymbol{r})$, Equation (4.92) may be written as:

$$I = \frac{1}{2}\text{Re}\{\hat{p}(\boldsymbol{r})\hat{u}^*(\boldsymbol{r})\} = \langle \hat{p}^2(\boldsymbol{r},t)\rangle / 2\rho c \quad (\text{W/m}^2) \tag{4.94}$$

where $\langle\ \rangle$ denotes the time average.

4.3.2 Spherical Wave Intensity

If Equations (1.117) and (1.118) are substituted into Equation (4.80) and use is made of the following equation:

$$\lim_{T_A \to \infty} \frac{1}{T_A} \int_0^{T_A} \text{f f}' \text{d}t = 0 \tag{4.95}$$

then Equation (4.92) is obtained, showing that the latter equation also holds for a spherical wave at any distance, r, from the source. Alternatively, similar reasoning shows that Equation (4.95) is only true of a spherical wave at distances r from the source, which are large (see Section 1.4.11). To simplify the notation to follow, the $\boldsymbol{r}$ dependence (dependence on location) and time dependence, t, of the quantities p and u will be assumed, and specific reference to these will be omitted. It is convenient to rewrite Equation (1.124) in terms of its magnitude and phase. Carrying out the indicated algebra gives:

$$\frac{p}{u} = \rho c\, \text{e}^{\text{j}\beta}\cos\beta \quad (\text{W/m}^2) \tag{4.96}$$

where $\beta = (\theta_p - \theta_u)$ is the phase angle by which the acoustic pressure leads the particle velocity and is defined as:

$$\beta = \tan^{-1}[1/(k_a r)] \tag{4.97}$$

Equation (4.88) gives the instantaneous intensity for the case considered here in terms of the real pressure amplitude, $\hat{A}_p$, and real particle velocity amplitude, $\hat{A}_u$. Solving Equation (4.96) for the particle velocity in terms of the pressure shows that $\hat{A}_u = \hat{A}_p/(\rho c \cos\beta)$. Substitution of this expression and Equation (4.97) into Equation (4.88) gives the following expression for the scalar instantaneous intensity of a spherical wave:

$$I_{si}(\boldsymbol{r},t) = \frac{\hat{A}_p^2}{2\rho c}\left\{[1 + \cos 2(\omega t + \theta_p)] + \frac{1}{k_a r}\sin 2(\omega t + \theta_p)\right\} \quad (\text{W/m}^2) \tag{4.98}$$

Consideration of Equation (4.98) shows that the time average of the first term on the right-hand side is non-zero and is the same as that of a plane wave given by Equation (4.92), while the time average of the second term is zero and thus the second term is associated with the non-propagating reactive intensity. The second term tends to zero as the distance, r, from the source to observation point becomes large; that is, the second term is negligible in the far field of the source. On the other hand, the reactive intensity becomes quite large close to the source; this is a near field effect. Integration over time of Equation (4.98), taking note that the integral of the second term is zero, gives the same expression for the intensity of a spherical wave as was obtained previously for a plane wave (see Equation (4.92)).

4.3.3   Sound Power

When sound propagates, transmission of acoustic power is implied. The intensity, as a measure of the energy passing through a unit area of the acoustic medium per unit time, was defined for plane and spherical waves and found to be the same. It will be assumed that the expression given by Equation (4.92) holds in general for sources that radiate more complicated acoustic waves, at least at sufficient distance from the source so that, in general, the power, W, measured in units of watts (W) radiated by any acoustic source is:

$$W = \iint\limits_{S} \boldsymbol{I} \cdot \boldsymbol{n} \, \mathrm{d}S \quad \text{(W)} \tag{4.99}$$

where $\boldsymbol{n}$ is the unit vector normal to the surface of area S. For the cases of the plane wave and spherical wave, the mean square pressure, $\langle p^2 \rangle$, is a function of a single spatial variable in the direction of propagation. The meaning is now extended to include, for example, variations with angular direction, as is the case for sources that radiate more power in some directions than in others. A loudspeaker that radiates most power on-axis to the front would be such a source. According to Equation (4.99), the sound power, W, radiated by a source is defined as the integral of the sound intensity over a surface surrounding the source. Most often, a convenient surface is an encompassing sphere or spherical section, but sometimes other surfaces are chosen, as dictated by the circumstances of the particular case considered. For a sound source producing uniformly spherical waves (or radiating equally in all directions), a spherical surface is most convenient, and in this case Equation (4.99) leads to the following expression:

$$W = 4\pi r^2 I \quad \text{(W)} \tag{4.100}$$

where the magnitude of the active sound intensity, I, normal to the imaginary surface surrounding the source, is measured at a distance, r, from the source. In this case, the source has been treated as though it radiates uniformly in all directions. Consideration is given to sources that do not radiate uniformly in all directions in Bies et al. (2017).

In practice, the accuracy of sound intensity measurement is affected by the magnitude of the reactive intensity (see Equation (4.91)), which is why it is best not to take sound intensity measurements too close to the sound source or in highly reverberant environments.

4.3.4   Measurement of Sound Intensity

The measurement of sound intensity in a complex sound field consisting of many waves travelling in different directions provides a means for directly determining the magnitude and direction of the net acoustic power flow at any location in space. Measuring and averaging the sound intensity over an imaginary surface surrounding a machine allows determination of the total acoustic power radiated by the machine.

Theoretically, measurements can be conducted in the near field of a machine, in the presence of reflecting surfaces and near other noisy machinery. However, if the reactive field associated with reflecting surfaces, or the near field of the sound source, is greater than the active field by 10 dB or more, or if the contributions of other nearby sound sources is 10 dB or more greater than the contributions due to the source under investigation, then in practice, reliable sound intensity measurements cannot be made due to limitations in the accuracy achievable in measuring the phase relationship between the two transducers used to determine sound intensity.

The measurement of sound intensity requires the simultaneous determination of sound pressure and particle velocity. The determination of sound pressure is straightforward, but the determination of particle velocity presents some difficulties; thus, there are two principal techniques for the determination of particle velocity and consequently the measurement of sound intensity. Either the acoustic pressure and particle velocity are measured directly ($p - u$ method) or the acoustic pressure is measured simultaneously at two closely spaced points and the mean pressure and particle velocity are calculated ($p - p$ method). In either case, the pressure is multiplied by the particle velocity to produce the instantaneous intensity and the time average intensity. In the following analysis, the vector location, r, will be omitted from the notation, as it will be assumed that the measurement applies to a particular location that is defined by the location of the transducers. Errors inherent in intensity measurements and limitations of instrumentation are discussed in Sections 4.3.4.4 and 4.3.4.4, and by Fahy (1995).

4.3.4.1 Sound Intensity Measurement by the $p - u$ Method

For the $p - u$ method, the acoustic particle velocity is measured directly. One method involves using two parallel ultrasonic beams travelling from source to receiver in opposite directions. Any particle movement in the direction of the beams will cause a phase difference in the received signals at the two receivers. The phase difference is related to the acoustic particle velocity in the space between the two receivers and may be used to calculate an estimate of the particle velocity up to a frequency of 6 kHz. The ultrasound technique is not used very much any more as it has been overtaken by a much more effective technology known as the "Microflown" particle velocity sensor (de Bree et al., 1996; Druyvesteyn and de Bree, 2000), which uses a measure of the temperature difference between two resistive sensors spaced 40 µm apart to estimate the acoustic particle velocity. The temperature difference between the two sensors is caused by the transfer of heat from one sensor to the other by convection as a result of the acoustic particle motion. This, in turn, leads to a variation in resistance of the sensor, which can be detected electronically. To get a temperature difference that is sufficiently high to be detected, the sensors are heated with a d.c. current to about 500 Kelvin. The sensor consists of two cantilevers of silicon nitride (dimensions $800 \times 40 \times 1$ µm) with an electrically conducting platinum pattern, used as both the sensor and heater, placed on them. The base of the sensor is silicon, which allows it to be manufactured using the same wafer technology as used to make integrated circuit chips. Up to 1000 sensors can be manufactured in a single wafer. The sensitivity of the sensors (and signal to noise ratio) can be increased by packaging of the sensor in such a way that the packaging increases the particle velocity near the sensor.

The spacing of 40 µm between the two resistive sensors making up the particle velocity sensor is an optimal compromise. Smaller spacing reduces the heat loss to the surroundings so that more of the heat from one sensor is convected to the other, making the device more sensitive. On the other hand, as the sensors come closer together, conductive heat flow between the two of them becomes an important source of heat loss and thus measurement error.

It has been shown (Jacobsen and Liu, 2005) that measurement of sound intensity using the "Microflown" sensor to determine the particle velocity directly is much more accurate than the indirect method involving the measurement of the pressure difference between two closely spaced microphones, which is described in Section 4.3.4.3. Jacobsen and Liu (2005) also showed that the "Microflown" sensor was more accurate than microphones for acoustic holography which involves using the measurement of acoustic pressure OR particle velocity in a plane front of a noise source to predict the acoustic pressure AND particle velocity in another plane.

Particle velocity sensors such as the "Microflown" are more useful than a pair of microphones for measuring particle velocity, as they are directional, which makes them less susceptible to background noise, and they are more sensitive. When used close to a reasonably stiff noise emitting structure their superiority over microphones is even more apparent, as the pressure associated with any background noise is approximately doubled at the surface, whereas the particle velocity associated with the background noise will be close to zero where it is reflected. It is possible to purchase a small sound intensity probe (12 mm diameter), which includes a microphone and a "microflown", and this can be used to directly measure sound intensity.

4.3.4.2   Accuracy of the $p - u$ Method

With sound intensity measurements, there are systematic errors, random errors and calibration uncertainty. The calibration uncertainty for a typical $p - u$ probe is approximately $\pm$ 0.5 dB. However, the systematic error arises from the inaccuracy in the model that compensates for the phase mismatch between the velocity sensor and the pressure sensor in the probe. This model is derived by exposing the $p - u$ probe to a sound field where the relative amplitudes and phases between the acoustic particle velocity and pressure are well known (Jacobsen and de Bree, 2005).

Although the phase error is not as important for a $p - u$ probe as it is for a $p - p$ probe (as the $p - p$ probe also relies on the phase between the two microphones to estimate the particle velocity), it is still the major cause of uncertainty in the intensity measurement, especially for reactive fields where the reactive intensity amplitude is larger or comparable with the active intensity. Reactive intensity fields occur close to noise radiating structures. For such fields, the errors in sound intensity measurement using a $p - u$ probe increase as the phase between the acoustic pressure and particle velocity increases. The Microflown handbook gives the systematic error for intensity measurements using a $p - u$ probe at a particular frequency as:

$$\text{Error(dB)} = 10 \log_{10} \left[1 + \left| \frac{\hat{I} - I}{I} \right| \right] = 10 \log_{10}(1 + \beta_e \tan \beta_f) \qquad (4.101)$$

where I is the actual active intensity at the $p - u$ probe location, $\hat{I}$ is the intensity measured by the probe, β_e (in radians) is the phase calibration error, which for a typical $p - u$ probe is of the order of $2.5°$ or 0.044 radians, and β_f is the phase between the acoustic pressure and particle velocity in the sound field. It can be seen that this systematic error will exceed 1 dB if the phase between the acoustic pressure and particle velocity exceeds $80°$. In practice, if the sound field reactivity is too high, the probe can be moved further from the noise source and the reactivity will decrease. Also as reflected sound and sound from sources other than the one being measured do not usually increase the reactivity of the sound field, they will not influence the accuracy of the intensity measurement made with the $p - u$ probe, in contrast to their significant influence on intensity measurements made with the $p - p$ probe (see Section 4.3.4.4). Note that very reactive fields are unlikely to occur except at low frequencies, whereas the phase error in the $p - p$ probe affects the accuracy of the intensity measurement over the entire audio frequency range. Thus the $p - u$ probe usually gives a more accurate value of sound intensity at high frequencies and in fact, is useful up to 20 kHz, whereas the limit of commercially available $p - p$ probes is 10 kHz.

Estimations of the magnitude of random errors for the $p - u$ method are very difficult to make and no estimates have been reported in the literature.

4.3.4.3   Sound Intensity Measurement by the $p - p$ Method

For the $p - p$ method, the determinations of acoustic pressure and acoustic particle velocity are both made using a pair of high quality condenser microphones. The microphones are generally mounted side by side or facing one another and separated by a fixed distance (6 mm to 50 mm) depending upon the frequency range to be investigated. A signal proportional to the particle velocity at a point midway between the two microphones and along the line joining their acoustic centres is obtained, using the finite difference in measured pressures to approximate the pressure gradient, while the mean is taken as the pressure at the midpoint.

The useful frequency range of $p - p$ intensity meters is largely determined by the selected spacing between the microphones used for the measurement of pressure gradient, from which the particle velocity may be determined by integration. The spacing must be sufficiently small to be much less than a wavelength at the upper frequency bound so that the measured pressure difference approximates the pressure gradient. On the other hand, the spacing must be sufficiently large for the phase difference in the measured pressures to be determined at the lower frequency bound with sufficient precision to determine the pressure gradient with sufficient accuracy. Clearly, the microphone spacing must be a compromise and a range of spacings is usually offered to the user. The assumed positive sense of the determined intensity is in the direction of the centre-line from microphone 1 to microphone 2. For convenience, where appropriate in the following discussion, the positive direction of intensity will be indicated by unit vector, n, and this is in the direction from microphone 1 to microphone 2.

Taking the gradient in the direction of unit vector, n, and using Equation (1.47) gives the equation of motion relating the pressure gradient to the particle acceleration. That is:

$$n \frac{\partial p}{\partial n} = -\rho \frac{\partial u_n}{\partial t} \tag{4.102}$$

where u_n is the component in direction, n, of particle velocity, u; p and u are both functions of vector location r and time t; and ρ is the density of the acoustic medium. The normal component of particle velocity, u_n, is obtained by integration of Equation (4.102) where the assumption is implicit that the particle velocity is zero at time $t = -\infty$:

$$u_n(t) = -\frac{n}{\rho} \int_{-\infty}^{t} \frac{\partial p(\tau)}{\partial n} \, \mathrm{d}\tau \tag{4.103}$$

The integrand of Equation (4.103) is approximated using the finite difference between the pressure signals, p_1 and p_2, from microphones 1 and 2, respectively, and Δ is the separation distance between them:

$$u_n(t) = -\frac{n}{\rho \Delta} \int_{-\infty}^{t} [p_2(\tau) - p_1(\tau)] \, \mathrm{d}\tau \tag{4.104}$$

The pressure midway between the two microphones is approximated as the mean:

$$p(t) = \frac{1}{2} [p_1(t) + p_2(t)] \tag{4.105}$$

Thus, the instantaneous intensity in direction, n, at time, t, is approximated as:

$$I_{n,i}(t) = \frac{n}{2\rho\Delta} [p_1(t) + p_2(t)] \int_{-\infty}^{t} [p_1(\tau) - p_2(\tau)] \, \mathrm{d}\tau \tag{4.106}$$

and the active intensity in direction, n, is then:

$$I = \frac{n}{2\rho\Delta} \left\langle [p_1(t) + p_2(t)] \int_{-\infty}^{t} [p_1(\tau) - p_2(\tau)] \, \mathrm{d}\tau \right\rangle \tag{4.107}$$

where $\langle \ \rangle$ represents a time average.

For stationary sound fields the instantaneous intensity can be obtained from the product of the signal from one microphone and the integrated signal from a second microphone in close proximity to the first (Fahy, 1995):

$$I_{n,i}(t) = \frac{n}{\rho\Delta} p_2(t) \int\limits_{-\infty}^{t} p_1(\tau)\mathrm{d}\tau \tag{4.108}$$

The time average of Equation (4.108) gives the following expression for the time average intensity in direction n (where n is the unit vector):

$$I_n = \frac{n}{\rho\Delta} \lim_{T_A\to\infty} \frac{1}{T_A} \int\limits_{0}^{T_A} \left[p_2(t) \int\limits_{-\infty}^{t} p_1(\tau)\mathrm{d}\tau \right] \mathrm{d}t = \frac{n}{\rho\Delta} \left\langle p_2(t) \int p_1(\tau)\mathrm{d}\tau \right\rangle \tag{4.109}$$

Commercial instruments with digital filtering (one third octave or octave) are available to implement Equation (4.109).

As an example, consider two harmonic pressure signals from two closely spaced microphones:

$$p_i(t) = \hat{A}_{p,i}\mathrm{e}^{\mathrm{j}(\omega t+\theta_i)} = \hat{A}_{p,i}\cos(\omega t + \theta_i) + \mathrm{j}\hat{A}_{p,i}\sin(\omega t + \theta_i); \qquad i = 1, 2 \tag{4.110}$$

Substitution of the real components of these quantities in Equation (4.106) gives, for the instantaneous intensity in direction, n, the following result:

$$\begin{aligned}
I_i(t) &= \frac{n}{2\rho\omega\Delta} \left[\hat{A}_{p,1}\cos(\omega t + \theta_1) + \hat{A}_{p,2}\cos(\omega t + \theta_2) \right] \times \left[\frac{\hat{A}_{p,1}}{\omega}\sin(\omega t + \theta_1) - \frac{\hat{A}_{p,2}}{\omega}\sin(\omega t + \theta_2) \right] \\
&= \frac{n}{4\rho\omega\Delta} \left[\hat{A}_{p,1}^2\sin(2\omega t + 2\theta_1) - \hat{A}_{p,2}^2\sin(2\omega t + 2\theta_2) + 2\hat{A}_{p,1}\hat{A}_{p,2}\sin(\theta_1 - \theta_2) \right]
\end{aligned} \tag{4.111}$$

Taking the time average of Equation (4.111) gives the following expression for the active intensity:

$$I_n = \frac{n\hat{A}_{p,1}\hat{A}_{p,2}}{2\rho\omega\Delta}\sin(\theta_1 - \theta_2) \tag{4.112}$$

If the argument of the sine is a small quantity then Equation (4.112) becomes approximately:

$$I_n = \frac{n\hat{A}_{p,1}\hat{A}_{p,2}}{2\rho\omega\Delta}(\theta_1 - \theta_2); \qquad \theta_1 - \theta_2 \ll 1 \tag{4.113}$$

This equation also follows directly from Equation (4.86), where the finite difference approximation is used to replace $\partial\theta_p/\partial r$ with $-(\theta_1 - \theta_2)/\Delta$ and $\hat{A}_p^2$ is approximated by $\hat{A}_{p,1} \times \hat{A}_{p,2}$.

The first two terms of the right-hand side of Equation (4.111) describe the reactive part of the intensity. If the phase angles θ_1 and θ_2 are not greatly different; for example, the real sound pressure amplitudes, $\hat{A}_{p,1}$ and $\hat{A}_{p,2}$, are measured at points that are closely spaced compared to a wavelength, the magnitude of the reactive component of the intensity is approximately:

$$Q = \frac{1}{4\omega\rho\Delta} \left[\hat{A}_{p,1}^2 - \hat{A}_{p,2}^2 \right] \tag{4.114}$$

Equation (4.114) also follows directly from Equation (4.90) where $\hat{A}_{p,1}$ is replaced by $(\hat{A}_{p,1} + \hat{A}_{p,2})/2$ and $\partial\hat{A}_p(r)/\partial r$ is replaced with the finite difference approximation, $-(\hat{A}_{p,1} - \hat{A}_{p,2})/\Delta$.

Measurement of the intensity in a harmonic stationary sound field can be made with only one microphone, a phase meter and a stable reference signal if the microphone can be located sequentially at two suitably spaced points. The single reference signal can be the driving signal for the acoustic or vibration source (such as the input to the amplifier driving a loudspeaker) that is generating the sound field for which the intensity is to be determined. Indeed, the 3-D sound intensity field can be measured by automatically traversing, stepwise, a single microphone over an area of interest, measuring the acoustic pressure at each step and also the phase between the microphone signal and the reference signal.

Use of a single microphone for intensity measurements eliminates problems associated with microphone, amplifier and integrator phase mismatch as well as enormously reducing diffraction problems encountered during the measurements. Although useful in the laboratory, this technique is difficult to implement in the field as it is often difficult to obtain the stable reference signal of exactly the same frequency as the sound or vibration field.

In general, the determination of the total instantaneous intensity vector, $\boldsymbol{I}_i(t)$, requires the simultaneous determination of three orthogonal components of particle velocity. Current instrumentation is available to do this using either a single $p - u$ probe or single $p - p$ probe, with three separate measurements in three orthogonal directions or with a 3-D probe.

Note that with a $p - p$ probe, the finite difference approximation used to obtain the acoustic particle velocity has problems at both low and high frequencies, which means that two or three different microphone spacings are needed to cover the audio frequency range. At low frequencies, the instantaneous pressure signals at the two microphones are very close in amplitude and a point is reached where the precision in the microphone phase matching is insufficient to accurately resolve the difference. At high frequencies the assumption that the pressure varies linearly between the two microphones is no longer valid. The $p - u$ probe suffers from neither of these problems, as it does not use an approximation to the sound pressure gradient to determine the acoustic particle velocity.

4.3.4.4 Accuracy of the $p - p$ Method

The accuracy of the $p - p$ method is affected by both systematic and random errors. The systematic error stems from the amplitude sensitivity difference and phase mismatch between the microphones and is a result of the approximations inherent in the finite difference estimation of particle velocity from pressure measurements at two closely spaced microphones.

The error due to microphone phase mismatch and the associated finite difference approximation can be expressed in terms of the difference, δ_{pI}, between sound pressure and intensity levels measured in the sound field being evaluated and the difference, δ_{pIO}, between sound pressure and intensity levels measured by the instrumentation in a specially controlled uniform pressure field in which the phase at each of the microphone locations is the same and for which the intensity is zero.

The quantity, δ_{pIO}, known as the "Residual Pressure–Intensity Index", is a measure of the accuracy of the phase matching between the two microphones making up the sound intensity probe and the higher its value, the higher is the quality of the instrumentation (that is, the better is the microphone phase matching).

The quantity, δ_{pI}, known as the "Pressure–Intensity Index" for a particular sound field, is defined as the difference in dB between the measured intensity and the intensity that would characterise a plane wave having the measured sound pressure level, corrected by the term, $10 \log_{10}(\rho c/400)$. Thus:

$$\delta_{pI} = L_p - L_I = 10 \log_{10} \left(\frac{\rho c}{400} \right) - 10 \log_{10} \left(\frac{\beta_f \lambda}{2\pi \Delta} \right) \tag{4.115}$$

where β_f is the actual phase difference between the sound field at the two microphone locations, Δ is the microphone spacing and λ is the wavelength of the sound. The first term on the right of Equation (4.115) accounts for the difference in reference levels for sound intensity and sound pressure. If noise is coming from sources other than the one being measured or there are reflecting surfaces in the vicinity of the noise source, the Pressure–Intensity Index will increase and so will the error in the intensity measurement. Other sources or reflected sound do not affect the accuracy of intensity measurements taken with a $p - u$ probe, but the $p - u$ probe is more sensitive than the $p - p$ probe in reactive sound fields (such as the near field of a source, see Jacobsen and de Bree (2005)).

The normalised systematic error in intensity due to microphone phase mismatch is a function of the actual phase difference, β_f, between the two microphone locations in the sound field and the phase mismatch error, β_s and is given by Fahy (1995) as:

$$e_\beta(I) = \beta_s/\beta_f \tag{4.116}$$

The difference between the Residual Pressure–Intensity Index and the Pressure–Intensity Index may be written in terms of this error as:

$$\delta_{pIO} - \delta_{pI} = 10\log_{10}|1 + (1/e_\beta(I))| = 10\log_{10}|1 + (\beta_f/\beta_s)| \tag{4.117}$$

A normalised error of $\beta_s/\beta_f = 0.25$ corresponds to a sound intensity error of approximately 1 dB ($= 10\log_{10}(1 + 0.25)$) and this corresponds to a difference, $\delta_{pIO} - \delta_{pI} = 10\log_{10}(1 + 1/0.25) = 7$ dB. A normalised error of 0.12 corresponds to an sound intensity error of approximately 0.5 dB and a difference, $\delta_{pIO} - \delta_{pI} = 10$ dB.

The Pressure–Intensity Index will be large (leading to relatively large errors in intensity estimates) in near fields and reverberant fields and this can extend over the entire audio frequency range.

The phase mismatch between the microphones in the $p - p$ probe is related to the Residual Pressure–Intensity Index (which is often supplied by the $p - p$ probe suppliers) by:

$$\beta_s = k_a\Delta 10^{-\delta_{pIO}/10} \tag{4.118}$$

Phase mismatch also distorts the directional sensitivity of the $p - p$ probe so that the null in response of the probe (often used to locate noise sources) is changed from the 90° direction to β_m, as given by (Fahy, 1995):

$$\beta_m = \cos^{-1}(\beta_s/k_a\Delta) \tag{4.119}$$

where k_a is the wavenumber at the frequency of interest and Δ is the spacing between the two microphones in the $p - p$ probe.

In state of the art instrumentation, microphones are available that have a phase mismatch of less than 0.05°. In cases where the phase mismatch is larger than this, the instrumentation sometimes employs phase mismatch compensation in the signal processing path.

The error due to amplitude mismatch is zero for perfectly phase matched microphones. However, for imperfectly phase matched microphones, the error is quite complicated to quantify and depends on the characteristics of the sound field being measured.

Fahy (1995) shows that random errors in intensity measurements using the $p - p$ method add to the uncertainty due to systematic errors and in most sound fields where the signals received by the two microphones are random and have a coherence close to unity, the normalised random error is given by $(BT_A)^{-1/2}$, corresponding to an intensity error of:

$$e_r(I) = 10\log_{10}\left(1 + (BT_A)^{-1/2}\right) \tag{4.120}$$

where B is the bandwidth of the measurement in Hz and T_A is the effective averaging time, which may be less than the measurement time unless real-time processing is performed. The coherence of the two microphone signals will be less than unity in high frequency diffuse fields, or where the microphone signals are contaminated by electrical noise, unsteady flow or cable vibration. In this case the random error will be greater than that indicated by Equation (4.120).

Finally, there is an error due to instrument calibration that adds to the random errors. This is approximately ±0.2 dB for one particular manufacturer but the reader is advised to consult calibration charts that are supplied with the instrumentation.

4.3.5 Frequency Decomposition of the Intensity

It is often necessary to decompose the intensity signal into its frequency components. This may be done either directly or indirectly.

4.3.5.1 Direct Frequency Decomposition

$p - u$ Probe

For this probe type, the frequency distribution of the mean intensity may be obtained by passing the two output signals (p and u) through identical bandpass filters prior to performing the time averaging.

$p - p$ Probe

For this probe type, the frequency distribution may be determined by passing the two signals through appropriate identical bandpass filters, either before or after performing the sum, difference and integration operations of Equation (4.109) and then time averaging the resulting outputs.

4.3.5.2 Indirect Frequency Decomposition

Determination of the intensity using the indirect frequency decomposition method is based on Fourier analysis of the two probe signals (either the $p - u$ signals or the $p - p$ signals).

$p - u$ Probe

Fahy (1995) (pp. 95-97) shows that for a $p - u$ probe, the intensity as a function of frequency, in the same direction as the measured particle velocity, is given by the single-sided cross-spectral density, G_{Dpu}, between the two signals as:

$$I_D(f_n) = \mathrm{Re}\{G_{Dpu}(f_n)\} \tag{4.121}$$

$$Q_D(f_n) = -\mathrm{Im}\{G_{Dpu}(f_n)\} \tag{4.122}$$

where $I_D(f_n)$ represents the real (or active) time averaged intensity spectral density at frequency f_n and $Q_D(f_n)$ represents the amplitude of the spectral density of the reactive component.

The cross-spectral density of two signals is defined as the product of the complex instantaneous spectral density of one signal with the complex conjugate of the complex instantaneous spectral density of the other signal (see Section 7.3.11). Thus, if $G_{Dp}(f_n)$ and $G_{Du}(f_n)$ represent the complex single-sided spectral densities of the pressure and velocity signals, respectively, then the associated cross-spectral density is:

$$G_{Dpu}(f_n) = G_{Dp}^*(f_n)G_{Du}(f_n) \tag{4.123}$$

where the * represents the complex conjugate.

For single frequency signals and harmonics, the cross-spectral density function, $G_{Dpu}(f_n)$, may be replaced with the cross spectrum, $G_{pu}(f_n)$, obtained by multiplying the cross-spectral density by the bandwidth of each FFT filter (or the frequency resolution) used to obtain the cross spectrum.

$p - p$ Probe

For the $p - p$ probe, Fahy (1995) shows that the mean active intensity spectral density, $I_D(f_n)$, and spectral density of the amplitude, $Q_D(f_n)$, of the reactive intensity spectral density in the direction from p_1 to p_2, at frequency, f_n, are:

$$I_D(f_n) = -\frac{1}{\rho f_n \Delta}\mathrm{Im}\{G_{Dp1p2}(f_n)\} \tag{4.124}$$

and:

$$Q_D(f_n) = -\text{Im}\{G_{Dpu}(f_n)\} = \frac{1}{2\rho f_n \Delta}\left[G_{Dp1p1}(f_n) - G_{Dp2p2}(f_n)\right] \qquad (4.125)$$

where G_{Dp1p2} is the cross-spectral density of the two pressure signals, and G_{Dp1p1} and G_{Dp2p2} represent the auto-spectral densities (see Section 7.3.11). The intensity spectrum is obtained using the same equations with the subscript, D, removed, so that cross spectra and auto spectra are used in place of cross- and auto-spectral densities, respectively.

For the case of a stationary harmonic sound field, it is possible to determine the sound intensity by using a single microphone and the indirect frequency decomposition method just described by taking the cross spectrum between the microphone signal and a stable reference signal (referred to as A) of the same frequency, for two locations p_1 and p_2 of the microphone. Thus, the effective cross spectrum, G_{p1p2}, for use in the preceding equations can be calculated as:

$$G_{p1p2}(f_n) = \frac{G_{p1}^*(f_n)G_A(f_n)G_{p2}^*(f_n)G_{p2}(f_n)}{G_{p2}(f_n)^*G_A(f_n)} = \frac{G_{p1A}(f_n)G_{p2p2}(f_n)}{G_{p2A}(f_n)} \qquad (4.126)$$

Equations (4.124) to (4.126) are easily implemented using a standard commercially available FFT spectrum analyser. Both the two-channel real-time and parallel filter analysers are commercially available and offer the possibility of making a direct measurement of the sound intensity of propagating sound waves in a sound field using two carefully matched microphones and band limited filters. The filter bandwidth is generally not in excess of one octave. If the microphones are not well matched, measurements are still possible (Fahy, 1995).

Most spectrum analysers measure a cross spectrum as well as a cross-spectral density, with the cross-spectral density being useful for calculating the intensity for broadband noise sources and the cross spectrum being useful for calculating the intensity of tonal or harmonic noise sources.

4.4   Structural Intensity and Structural Power Transmission

Structural intensity allows quantification of the rate of local vibratory power transmission in a structure in a similar way that sound intensity allows the quantification of the acoustic power transmission in a fluid medium. The structural power transmission through an imaginary structural cross section is obtained by integrating the component of structural intensity normal to the cross section over the area of the cross section. Structural intensity is commonly defined as the long time rate of vibratory energy flow through unit area of a solid structure. An alternative definition of structural intensity, which is often used for thin plates, is that it is the vibrational power transmission per unit width of structure in a given direction. Like sound intensity, structural intensity is a complex vector, with each scalar element having both real and imaginary components, with similar definitions as given for sound intensity in Section 4.3. However, unlike sound intensity, which only applies to longitudinal wave propagation in fluids, structural intensity can apply to longitudinal, shear (or torsional) and bending waves in solid structures. Note that the symbol used for power transmission in structures is Π, whereas the symbol used for power transmission in fluids is W, whereas in both cases the symbol representing intensity is I. This is consistent with existing conventions.

Measuring the structural intensity allows the determination of the residual vibratory power transmission through a particular structural section, and can be used as a cost function for an active control system. The measurement of structural intensity requires the determination of the local vibratory stress and particle velocity, and the relative phase between them. Real structures are characterised by both active and reactive structural intensity fields, and the instantaneous structural intensity is defined as:

$$\boldsymbol{I} = -\boldsymbol{u} \cdot \mathfrak{S} \qquad (4.127)$$

where $\boldsymbol{u}$ is the structural velocity vector and $\mathfrak{S}$ is the stress tensor. For a three-dimensional coordinate system, Equation (4.127) may be written as:

$$\boldsymbol{I} = \begin{bmatrix} I_1 \\ I_2 \\ I_3 \end{bmatrix} = - \begin{bmatrix} \sigma_{11} & \sigma_{12} & \sigma_{13} \\ \sigma_{21} & \sigma_{22} & \sigma_{23} \\ \sigma_{31} & \sigma_{32} & \sigma_{33} \end{bmatrix} \begin{bmatrix} u_1 \\ u_2 \\ u_3 \end{bmatrix} \tag{4.128}$$

where the σ_{ij}, $i, j = 1, 3$ $(i \neq j)$ represent shear stresses, the σ_{ii} represent tensile or compressive stresses and $\boldsymbol{I}$ is the instantaneous intensity vector, which is a time dependent quantity.

For longitudinal waves, the particle displacements, velocities and accelerations are in the same and opposite directions as the wave propagation; that is, they move back and forth about a mean position, cycling in the direction of wave propagation and then in the opposite direction. For bending waves and shear waves the displacements, velocities and accelerations are in a direction normal to the direction of wave propagation. For bending waves the direction is normal to the structure surface, and for shear waves the direction is in the plane of the surface.

For a Cartesian coordinate system, the subscripts in Equation (4.128) have the equivalence: $x = 1$, $y = 2$, $z = 3$. For a cylindrical system, $r = 1$, $\theta = 2$, $x = 3$ where r is the radial coordinate, θ is the angular coordinate and x is the axial coordinate. The negative sign in Equations (4.127) and (4.128) appears because the stress is directly related to the derivative of the displacement or the displacement gradient. If this gradient is negative, then the intensity must be oriented in the positive direction.

The intensity in direction 1 can be derived simply from Equation (4.128) and is:

$$I_1 = -(\sigma_{11}u_1 + \sigma_{12}u_2 + \sigma_{13}u_3) = -\sum_{k=1}^{3} \sigma_{1k}u_k \tag{4.129}$$

The instantaneous power transmission per unit width through any beam, plate or shell cross section can be found by averaging the intensity over the cross-sectional area as follows. For Cartesian coordinates, the power transmission per unit width in the axial x-direction for a shell of thickness, h, is:

$$\Pi_x = \int_{-h/2}^{h/2} I_x \, dz \tag{4.130}$$

where z is the thickness coordinate, having the value of zero at the centre of the cross section. Power transmission in the y-direction is described by a similar relationship. However, for a curved surface described by a cylindrical coordinate system, the axial power transmission (in the non-curved direction) is described by a slightly more complex expression, which takes into account the curvature of the cross section through which the power is being transmitted (see Figure 4.7), as follows (Romano et al., 1990):

$$\Pi_x = \int_{-h/2}^{h/2} I_x \left(1 + \frac{z}{a}\right) dz \tag{4.131}$$

where a is the radius of curvature of the shell. The term $(1 + z/a)$ in the integral arises because the arc length of lines parallel to the $z = 0$ line shown in Figure 4.7 differs from the arc length at $z = 0$ by this factor. Thus, this must be included in the integral to provide the correct area weighting. The nomenclature and sign convention used here differ from that used by Pavic (1976); Cremer et al. (1973). However, they are consistent with those used by Romano et al. (1990); Leissa (1969, 1973).

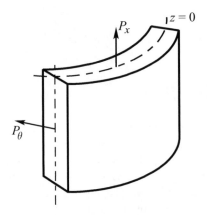

FIGURE 4.7 Shell power flows in axial and tangential directions.

For a flat plate, $a = \infty$, and Equation (4.131) becomes the same as Equation (4.130). As the section surface in the θ-direction in Figure 4.7 is the same shape as for a flat plate, the expression for power transmission in this direction is obtained from Equation (4.130) by substituting x for θ.

In the equations to follow, the x and t dependencies, which would normally appear in brackets following the symbols for displacements, rotations, moments and forces, have been omitted to reduce the complexity of the notation; for example, $w(x, t)$ has been replaced simply by w and $w(x)$ by $\hat{w}$. Also a dot over a variable indicates differentiation with respect to time and a double dot indicates double differentiation with respect to time.

Substituting Equation (4.129) into Equation (4.131), and using the cylindrical coordinate system, the following is obtained for the power transmission in the axial x-direction per unit length of cylinder circumference:

$$\Pi_x = -\int_{-h/2}^{h/2} \left(\sigma_{xz}\dot{w} + \sigma_{x\theta}\dot{\xi}_\theta + \sigma_{xx}\dot{\xi}_x \right) (1 + z/a)\, \mathrm{d}z \tag{4.132}$$

and for the power transmission in the circumferential θ-direction per unit length of cylinder:

$$\Pi_\theta = -\int_{-h/2}^{h/2} \left(\sigma_{\theta\theta}\dot{\xi}_\theta + \sigma_{\theta x}\dot{\xi}_x + \sigma_{\theta z}\dot{w} \right) \mathrm{d}z \tag{4.133}$$

Note that the displacements ξ_θ, ξ_x and w refer to the displacement of the elemental thickness $\mathrm{d}z$ (at location $a + z$) in the θ-, x- and z-directions, respectively, and are not the same as the displacements of the section centre-line. To express Equations (4.132) and (4.133) in terms of centre-line displacements, a series expansion is used as follows (Romano et al., 1990):

$$\begin{pmatrix} w(a+z) \\ \xi_x(a+z) \\ \xi_\theta(a+z) \end{pmatrix} = \begin{pmatrix} w(a) \\ \xi_x(a) \\ \xi_\theta(a) \end{pmatrix} + \begin{pmatrix} w^{(1)}(a) \\ \xi_x^{(1)}(a) \\ \xi_\theta^{(1)}(a) \end{pmatrix} \frac{z}{1!} + \begin{pmatrix} w^{(2)}(a) \\ \xi_x^{(2)}(a) \\ \xi_\theta^{(2)}(a) \end{pmatrix} \frac{z^2}{2!} + \dots \tag{4.134}$$

where the superscript (1) denotes differentiation with respect to x, the superscript (2) denotes double differentiation with respect to x and "!" represents "factorial".

The Kirchhoff assumptions discussed in Chapter 2 for a cylinder allow Equation (4.134) to be written as (Romano et al., 1990):

$$w(a + z) = w(a) \tag{4.135}$$

$$\xi_x(a+z) = \xi_x(a) - z\frac{\partial w(a)}{\partial x} \tag{4.136}$$

$$\xi_\theta(a+z) = \xi_\theta(a) + \left(\frac{\xi_\theta(a)}{a} - \frac{1}{a}\frac{\partial w(a)}{\partial \theta}\right)z \tag{4.137}$$

If Equations (4.135) to (4.137) are substituted into Equations (4.132) and (4.133), and Equations (2.318) to (2.329) from Chapter 2 are used, the following equations are obtained:

$$\Pi_x = -\left[\dot{w}Q_x + \dot{\xi}_\theta N_{x\theta} + \left(\frac{\dot{\xi}_\theta}{a} - \frac{1}{a}\frac{\partial \dot{w}}{\partial \theta}\right)M_{x\theta} + \dot{\xi}_x N_x - \frac{\partial \dot{w}}{\partial x}M_x\right] \tag{4.138}$$

$$\Pi_\theta = -\left[\dot{w}Q_\theta + \dot{\xi}_\theta N_\theta + \left(\frac{\dot{\xi}_\theta}{a} - \frac{1}{a}\frac{\partial \dot{w}}{\partial \theta}\right)M_\theta + \dot{\xi}_x N_{\theta x} - \frac{\partial \dot{w}}{\partial x}M_{\theta x}\right] \tag{4.139}$$

Note that the same coordinate system, sign conventions and definitions of moments and forces as used in Chapter 2 are used here.

In Equation (4.138), shear waves are associated with the second term, longitudinal waves with the fourth term and bending waves with the remaining terms.

The equation for power transmission per unit width (structural intensity) in a plate in the x-direction can be obtained directly from Equations (4.138) and (4.139) by replacing $\frac{1}{a}\frac{\partial}{\partial \theta}$ with $\frac{\partial}{\partial y}$, θ with y and ξ_θ/a with 0 to obtain:

$$\Pi_x = -\left[\dot{w}Q_x - \frac{\partial \dot{w}}{\partial y}M_{xy} - \frac{\partial \dot{w}}{\partial x}M_x + \dot{\xi}_y N_{xy} + \dot{\xi}_x N_x\right] \tag{4.140}$$

The expression for the y component of power transmission is obtained by interchanging the x and y subscripts in Equation (4.140).

The first three terms in Equation (4.140) correspond to power transmission associated with bending wave propagation. The first of these is the shear force contribution, the second is the twisting contribution and the third term is the bending contribution. The fourth term corresponds to shear wave propagation while the fifth corresponds to longitudinal wave propagation.

For wave propagation in a beam, Equation (4.140) becomes:

$$\Pi_x = -\left[-\dot{w}Q_x - \dot{\theta}_x M_{xy} - \frac{\partial \dot{w}}{\partial x}M_x + \dot{\xi}_x N_x\right] \tag{4.141}$$

where the first and third terms represent the bending wave power transmission, the second term represents the torsional wave power transmission and the last term represents the longitudinal wave power transmission.

Note the difference in sign between the beam and plate for the $\dot{w}Q_x$ term. This is because the sign convention for positive Q_x on the plate is different to that for a beam (see Figures 2.13 and 2.17(b)).

The quantity θ_x is the angle of rotation about the x-axis caused by a torsional wave. It is defined as:

$$\theta_x = \frac{\partial w}{\partial y} = -\frac{\partial w}{\partial z} \tag{4.142}$$

The force and moment variables are defined and derived in Chapter 2.

For a rectangular-section beam, the bending wave power transmission can be separated into two components corresponding to transverse displacement along two orthogonal axes. For the rectangular-section beam shown in Figure 2.8, these would be the y- and z-axes, and in Equation (4.141) the displacement, w, would be replaced with w_y or w_z, depending upon the wave of interest. If both wave components were present simultaneously, then the total power transmission would simply be the arithmetic sum of the two components.

The quantities of Equations (4.138) to (4.141) are functions of time as well as location on the cylinder, plate or beam. To enable us to express quantities in terms of net energy flow or time averaged power transmission, it is useful to find expressions for time averaged intensity. To do this, the same approach as that adopted for sound intensity may be used. Thus, for a harmonic vibration on a beam at frequency ω, the following is obtained:

$$\langle \Pi_x \rangle_t = -\frac{1}{2} \mathrm{Re} \left\{ -\dot{w}^* Q_x - \dot{\theta}_x^* M_{xy} - \frac{\partial \dot{w}^*}{\partial x} M_x + \dot{\xi}_x^* N_x \right\} \tag{4.143}$$

where $\langle \Pi_x \rangle_t$ is the time averaged power transmission and the $*$ denotes the complex conjugate.

Similar expressions can be obtained for plates and cylinders. In these latter two cases the left side of the equation will be time averaged power transmission per unit width. Also, as for sound intensity, the amplitude of the imaginary component of structural power transmission can be obtained by replacing "Re" in Equation (4.143) with "Im". The measurement of structural intensity or structural power transmission is really only practical on simple structures such as beams, plates and shells. Even on these simple structures, it is extremely difficult to measure reactive intensity, due to the need for a minimum of four measurement locations and stringent requirements on the relative phase calibration between transducers.

In beams, plates and shells, the determination of structural intensity is possible from measurements on the surface of these elements, because relatively simple relationships have been derived between the variables that govern the energy flow through the structure and the vibration on the surface, thus enabling the determination of the transmission of energy in a beam plate or shell by measurement of surface vibrations only. These simple relationships, however, only hold at lower frequencies where the motion of the interior of the structure is uniquely related to the motion of the surface.

Here, expressions will be presented for the structural intensity in beams, plates and cylindrical shells in terms of the normal and in-plane surface displacements and their spatial derivatives. Experimental measurement of these quantities thus allows the determination of both active and reactive intensity components throughout a structural cross section without the need to measure the stress directly. It will be shown how, in special cases, it is possible to use just two accelerometers to measure single wave intensities in beams and plates. The measurement of structural intensity on more general structures is extremely difficult, although it is possible to measure surface intensity by using strain gauges together with vibration transducers (Pavic, 1987).

Before proceeding with the analysis for specific examples, some general observations about determining the product of two transducer signals will be summarised. Consider two time varying signals, $x(t)$ and $y(t)$. The following notation will be used to denote the time averaged product in the time domain:

$$\langle xy \rangle_t = \langle x(t)\, y(t) \rangle_t \tag{4.144}$$

$$\left\langle x \int y \right\rangle_t = \left\langle x(t) \int_0^t y(\tau) \mathrm{d}\tau \right\rangle_t \tag{4.145}$$

For sinusoidal signals, the notation:

$$x(t) = \hat{x} \mathrm{e}^{\mathrm{j}\omega t} \text{ and } y(t) = \hat{y} \mathrm{e}^{\mathrm{j}\omega t} \tag{4.146}$$

where $\hat{x}$ and $\hat{y}$ may be complex, will be used.

Table 4.4 contains some time and frequency domain relationships that will be used in the intensity measurement procedures to be described later. The expressions in the right-hand column

TABLE 4.4 Signal processing relationships in time and frequency domains

Time domain	Frequency domain	
	Sinusoidal	Broadband
$\langle xy \rangle_t = \langle yx \rangle_t$ Amplitude of active intensity	$\dfrac{1}{2}\mathrm{Re}\{\hat{x}^*\hat{y}\}$	$\int\limits_0^\infty \mathrm{Re}\{G_{xy}(\omega)\}\mathrm{d}\omega$
Amplitude of reactive intensity	$\dfrac{1}{2}\mathrm{Im}\{\hat{x}^*\hat{y}\}$	$\int\limits_0^\infty \mathrm{Im}\{G_{xy}(\omega)\}\mathrm{d}\omega$
$\langle x \int y \rangle_t = -\langle y \int x \rangle_t$	$\dfrac{1}{2\omega}\mathrm{Im}\{\hat{x}^*\hat{y}\}$	$\int\limits_0^\infty \dfrac{\mathrm{Im}\{G_{xy}(\omega)\}\mathrm{d}\omega}{\omega}$
$\langle x \iint y \rangle_t = \langle y \iint x \rangle_t$	$-\dfrac{1}{2\omega^2}\mathrm{Re}\{\hat{x}^*\hat{y}\}$	$-\int\limits_0^\infty \dfrac{\mathrm{Re}\{G_{xy}(\omega)\}\mathrm{d}\omega}{\omega^2}$
$\langle \int x \iint y \rangle_t = -\langle \int y \iint x \rangle_t$	$\dfrac{1}{2\omega^3}\mathrm{Im}\{\hat{x}^*\hat{y}\}$	$\int\limits_0^\infty \dfrac{\mathrm{Im}\{G_{xy}(\omega)\}\mathrm{d}\omega}{\omega^3}$

are written for harmonic signals in terms of the cross spectrum, $G_{xy}(\omega)$ (see Section 7.3.11). Here the frequency variable used is ω radians/sec in place of f_n in Hz used in Chapter 7. Due to the complexity of the notation used in this section, the cross spectrum is sometimes written as $G_{x,y}(\omega)$ or $G(x,y,\omega)$. When broadband noise is considered the cross spectrum, $G_{xy}(\omega)$, is replaced with the cross-spectral density, $G_{Dxy}(\omega)$, $G_{D,x,y}(\omega)$ or $G(D,x,y,\omega)$.

4.4.1 Intensity and Power Transmission Measurement in Beams

The vibratory power propagating in a beam is given by the integral of the structural intensity due to all wave types over the beam cross section at the point of interest. The following analysis for beams will express the results in terms of total power transmission. Some authors refer to this power transmission quantity as intensity, but as the units are actually power units, and to avoid confusion with the actual intensity defined in Equation (4.127), the quantity will be referred to here as power. As accelerometers measure linear rather than angular accelerations, the Cartesian coordinate system will be used, even for circular-section beams. The beam displacements will be denoted w_y, w_z and ξ_x, corresponding to lateral displacement in the y-direction, lateral displacement in the z-direction and axial displacement along the length of the beam (see Figure 2.7). In the following analysis, classical (or Bernoulli–Euler) beam theory will be used.

   To avoid confusion between beam power transmission and beam bending displacement in the following analysis, the symbol used for structural power transmission will be Π. (Note that the symbol, W, is used for power transmission in an acoustic medium.)

   The analysis will begin with Equation (4.140) and proceed to derive power transmission expressions for each wave type (longitudinal, torsional and bending) in terms of the in-plane and normal beam displacements.

4.4.1.1 Longitudinal Waves

From Equation (4.140) the following may be written for the power transmission:

$$\Pi_L(t) = -\dot{\xi}_x N_x = -\dot{\xi}_x \sigma_{xx} S = -\dot{\xi}_x E \epsilon_{xx} S = -SE \frac{\partial \xi_x}{\partial t} \frac{\partial \xi_x}{\partial x} \qquad (4.147)$$

where S is the beam cross-sectional area and E is Young's modulus of elasticity. If other waves are present simultaneously, then ξ_x is the longitudinal displacement of the centre of the beam and will be written as ξ_{xo} to indicate this.

In the remainder of this section, amplitudes of harmonically varying quantities as well as instantaneous values of these quantities will be discussed. The instantaneous values are related to the amplitude as follows:

$$\xi_x = \hat{\xi}_x e^{j\omega t} \qquad (4.148)$$

Equation (4.147) can be rewritten in terms of velocity and acceleration as follows:

$$\Pi_L(t) = -SE \frac{\partial \xi_x}{\partial t} \frac{\partial \xi_x}{\partial x} = -SE u_x \int \frac{\partial u_x}{\partial x} = -SE \int a_x \iint \frac{\partial a_x}{\partial x} \qquad (4.149)$$

where:

$$\xi_x = \iint a_x(t)\mathrm{d}t\,\mathrm{d}t = \iint a_x = \int u_x \qquad (4.150)$$

$$\frac{\partial \xi_x}{\partial t} = \int a_x(t)\mathrm{d}t = \int a_x = u_x \qquad (4.151)$$

To determine the gradient, $\dfrac{\partial a_x}{\partial x}$, it is necessary to take simultaneous measurements at two points closely spaced a distance, Δ, apart and then use a finite difference approximation. Thus, the acceleration and acceleration gradient at a point midway between them are given by:

$$a_x = (a_{x1} + a_{x2})/2 \qquad (4.152)$$

$$\frac{\partial a_x}{\partial x} = \frac{a_{x2} - a_{x1}}{\Delta} \qquad (4.153)$$

where Δ is the spacing between the two accelerometers. A recommended value for Δ is between one-fifteenth and one-twentieth of a wavelength (Hayek et al., 1990), although considerations outlined in Section 4.4.4 suggest that $\lambda/10$ may be more appropriate, where λ is the structural longitudinal wavelength. The best value is dependent upon the structure characteristics, such as thickness and lateral dimensions, but is probably in the range stated above for most structures encountered in practice.

The accelerometers are numbered such that number 2 corresponds to a larger x-coordinate than number 1. Thus, positive intensity (in the positive x-direction) is energy transmission from position 1 to position 2.

Substituting Equations (4.152) and (4.153) into Equation (4.149), the following is obtained:

$$\Pi_L(t) = -\frac{SE}{2\Delta} \left[\int (a_{x1} + a_{x2}) \iint (a_{x2} - a_{x1}) \right] \qquad (4.154)$$

The time average vibratory power transmission can thus be written as:

$$\Pi_{La} = \langle \Pi_L(t) \rangle_t = -\frac{SE}{2\Delta} \left\langle \int (a_{x1} + a_{x2}) \iint (a_{x2} - a_{x1}) \right\rangle_t \qquad (4.155)$$

For harmonic excitation, $\iint (a_{x2} - a_{x1}) = -(a_{x2} - a_{x1})/\omega^2$ and Equation (4.154) can be rewritten as:

$$\Pi_L(t) = \frac{SE}{2\omega^2 \Delta} \left[\left(\int (a_{x1} + a_{x2}) \right) (a_{x2} - a_{x1}) \right] \qquad (4.156)$$

Thus, the corresponding time averaged power transmission can be written as:

$$\Pi_{La}(\omega) = \frac{SE}{2\omega^2\Delta} \left\langle (a_{x1} + a_{x2}) \int (a_{x1} - a_{x2}) d\tau \right\rangle_t \qquad (4.157)$$

or:

$$\Pi_{La}(\omega) = \frac{SE}{2\omega^2\Delta} \left\langle a_{x2} \int a_{x1} \right\rangle_t \qquad (4.158)$$

where the following properties of the two harmonic signals have been used:

$$\left\langle a_{x1} \int a_{x1} \right\rangle_t = \left\langle a_{x2} \int a_{x2} \right\rangle_t = 0 \qquad (4.159)$$

and:

$$\left\langle a_{x1} \int a_{x2} \right\rangle_t = - \left\langle a_{x2} \int a_{x1} \right\rangle_t \qquad (4.160)$$

From the similarity between Equation (4.157) and the corresponding Equation (4.107) for acoustic intensity it may be deduced that a sound intensity analyser may be used to determine the structural intensity of a harmonic wave field on a beam in the far field of any sources or reflections, by replacing the pressure signals with accelerometer signals in Equations (4.108) and (4.109) and using a different pre-multiplier.

For harmonic excitation, it is possible to derive a more convenient form of Equation (4.158) by beginning with Equation (4.147) and immediately assuming harmonic excitation as follows:

$$\hat{\xi}_x = A_1 e^{-jk_L x} + A_2 e^{jk_L x} = \hat{A}_1 e^{-j(k_L x - \theta_1)} + \hat{A}_2 e^{j(k_L x + \theta_2)} \qquad (4.161)$$

where $\hat{A}_1$ and $\hat{A}_2$ are real amplitudes with the units of displacement and θ_1 and θ_2 represent the phases of the waves at $x = 0$. Using Equation (4.161), Equation (4.147) can be rewritten as:

$$\Pi_{La}(\omega) = -\mathrm{Re}\left\{ \frac{SE}{2} \left(j\omega\hat{\xi}_x \right)^* \frac{\partial \hat{\xi}_x}{\partial x} \right\} = -\mathrm{Re}\left\{ \frac{j\omega SE}{2} \hat{\xi}_x^* \frac{\partial \hat{\xi}_x}{\partial x} \right\} = \frac{\omega^2 SE}{2c_L} (\hat{A}_1^2 - \hat{A}_2^2) \qquad (4.162)$$

where $k_L = \omega/c_L$ is the wavenumber for the longitudinal wave and c_L is the wave speed.

The amplitude of the fluctuating reactive power is:

$$\Pi_{Lr}(\omega) = \mathrm{Im}\left\{ \frac{j\omega SE}{2} \hat{\xi}_x^* \frac{\partial \hat{\xi}_x}{\partial x} \right\} = -\frac{\hat{A}_1 \hat{A}_2 \omega^2 SE}{c_L} \sin(2k_L x + \theta_2 - \theta_1) \qquad (4.163)$$

In terms of velocity (if a laser Doppler velocimeter is used to determine the beam response), Equation (4.162) may be written as:

$$\Pi_{La}(\omega) = \mathrm{Re}\left\{ \frac{jSE}{2\omega} \hat{u}_x^* \frac{\partial \hat{u}_x}{\partial x} \right\} \qquad (4.164)$$

where $\hat{u}_x$ is the longitudinal velocity amplitude of the centre of the beam section (the time derivative of the displacement) at location, x, and the hat above a variable denotes the complex amplitude of the time varying signal.

If accelerometers were used and mounted to measure axial (or longitudinal) acceleration, then Equation (4.162) could be written as:

$$\Pi_{La}(\omega) = \mathrm{Re}\left\{ \frac{jSE}{2\omega^3} \hat{a}_x^* \frac{\partial \hat{a}_x}{\partial x} \right\} \qquad (4.165)$$

and Equation (4.163) could be written as:

$$\Pi_{Lr}(\omega) = \text{Im} \left\{ \frac{\text{j}SE}{2\omega^3} \hat{a}_x^* \frac{\partial \hat{a}_x}{\partial x} \right\} \tag{4.166}$$

where a_x is the longitudinal acceleration of the centre of the beam section.

Substituting Equations (4.152) and (4.153) into Equation (4.165) gives for the active power:

$$\Pi_{La}(\omega) = -\text{Im} \left\{ \frac{SE}{4\omega^3 \Delta} (\hat{a}_{x1} + \hat{a}_{x2})^* (\hat{a}_{x2} - \hat{a}_{x1}) \right\} = \frac{SE}{2\omega^3 \Delta} |\hat{a}_{x1}||\hat{a}_{x2}| \sin(\theta_1 - \theta_2) \tag{4.167}$$

and substituting Equations (4.152) and (4.153) into Equation (4.166) gives for the reactive power:

$$\Pi_{Lr}(\omega) = \text{Re} \left\{ \frac{SE}{4\omega^3 \Delta} (\hat{a}_{x1} + \hat{a}_{x2})^* (\hat{a}_{x2} - \hat{a}_{x1}) \right\} = \frac{SE}{4\omega^3 \Delta} \left(|\hat{a}_{x2}|^2 - |\hat{a}_{x1}|^2 \right) \tag{4.168}$$

Thus the time averaged longitudinal wave active power transmission for harmonic waves can be measured in practice by multiplying the amplitudes of two closely spaced accelerometers by the sine of the phase difference between the two. The direction of positive power transmission is from accelerometer 1 to 2. It can be shown easily that Equation (4.167) is equivalent to Equation (4.158); however, in many experimental situations it is probably easier to measure the amplitudes of and the phase difference between two sinusoidal signals than it is to accurately perform analogue integrations and multiplications. Especially for broadband signals (but also for harmonic wave fields), often the most convenient way to determine the vibratory power is to use a measurement of the cross spectrum between the two accelerometer signals, as will now be explained.

Taking the time average of Equation (4.147) gives for the active power transmission:

$$\langle \Pi_L(t) \rangle_t = \Pi_{La} = -SE \left\langle \frac{\partial \xi_x}{\partial t} \frac{\partial \xi_x}{\partial x} \right\rangle_t = -SE \left\langle \int a_x \iint \frac{\partial a_x}{\partial x} \right\rangle_t \tag{4.169}$$

Using the relationships in Table 4.4, the frequency dependent active power for a harmonic signal can be written as:

$$\Pi_{La}(\omega) = \frac{SE}{\omega^3} \text{Im} \left\{ G \left(\frac{\partial a_x}{\partial x}, a_x, \omega \right) \right\} \tag{4.170}$$

Note that $\text{Im}[G(a_1, a_2, \omega) = -\text{Im}\{G(a_2, a_1, \omega)\}$, and the cross spectrum is denoted as $G_{xy}(\omega)$ or $G(x, y, \omega)$. The cross spectrum, $G_{xy}(\omega)$, can be determined by inputting the two acceleration signals into a spectrum analyser. The auto spectrum is discussed in Section 7.3.1 and the cross spectrum is discussed in Section 7.3.11. Here, for convenience with the rest of the analysis in this chapter, frequency is expressed as ω (radians/s) rather than f_n (Hz) used in Chapter 7.

Substituting Equations (4.152) and (4.153) into Equation (4.170) gives:

$$\Pi_{La}(\omega) = -\frac{SE}{\omega^3 \Delta} \text{Im} \left\{ G_{a_{x_1}, a_{x_2}}(\omega) \right\} \tag{4.171}$$

In short hand notation, Equation (4.171) may be written as:

$$\Pi_{La}(\omega) = \frac{SE}{\omega^3 \Delta} \text{Im}\{ G_{21}(\omega) \} \tag{4.172}$$

If the accelerometer numbering convention is reversed, then G_{21} is replaced by G_{12}.

In a similar way as demonstrated by Fahy (1995) for sound intensity, the amplitude of the fluctuating reactive power can be shown to be:

$$\Pi_{Lr}(\omega) = -\frac{SE}{\omega^3 \Delta} [G_{a_{x2}, a_{x2}}(\omega) - G_{a_{x1}, a_{x1}}(\omega)] = -\frac{SE}{\omega^3 \Delta} [G_{22}(\omega) - G_{11}(\omega)] \tag{4.173}$$

Note that for random noise, the functions, $G_{xy}(\omega)$, may be interpreted as cross-spectral density functions, $G_{Dxy}(\omega)$, so that the powers on the left of Equations (4.171) to (4.173) become power per Hertz.

For non-harmonic excitation, the concept of reactive power is meaningless because, unlike active structural intensity, the quantity determined over a wide frequency band will not simply be the sum of the values determined for any set of narrower frequency bands, which together make up the wide frequency band.

When measuring structural intensity using the two accelerometer method, the error associated with the finite spacing, Δ, of the accelerometers can be minimised by adjusting the measured structural intensity by a factor to obtain the actual structural intensity as follows (Kim and Tichy, 2000):

$$\Pi_{\text{actual}} = \Pi_{\text{measured}} \left[\frac{k_L \Delta}{\sin k_l \Delta} \right] \tag{4.174}$$

where k_L is the wavenumber for longitudinal waves.

4.4.1.2 Torsional Waves

From Equation (4.140):

$$\Pi_T(t) = \dot{\theta}_x M_{xy} = C \frac{\partial \theta_x}{\partial t} \frac{\partial \theta_x}{\partial x} \tag{4.175}$$

where C is the torsional stiffness of the beam defined in Table 2.1 and below Equation (2.72). Equation (4.175) is in a similar form to Equation (4.147). Thus, expressions for the active and reactive powers can be obtained from Section 4.4.1.1 simply by replacing ξ_x with θ_x, SE with $-C$ and a_{x1}, a_{x2} with $\ddot{\theta}_{x1}$, $\ddot{\theta}_{x2}$, as appropriate, in all of the equations. The determination of $\ddot{\theta}_x$ from accelerometer measurements will be discussed later in Section 4.4.1.5.

4.4.1.3 Bending Waves

From Equation (4.140), the power transmission expression for bending waves characterised by deflections, w_z, in the z-direction is:

$$\Pi_{Bz}(t) = - \left[-\dot{w}_z Q_x - \frac{\partial \dot{w}_z}{\partial x} M_x \right] \tag{4.176}$$

Using the definitions of Q_x and M_x from Section 2.3.3.3 for classical beam theory, the following is obtained:

$$\Pi_{Bz}(t) = E J_{yy} \left[\frac{\partial w_z}{\partial t} \frac{\partial^3 w_z}{\partial x^3} - \frac{\partial^2 w_z}{\partial t \partial x} \frac{\partial^2 w_z}{\partial x^2} \right] \tag{4.177}$$

For bending waves travelling in the x-direction and characterised by a normal displacement in the y-direction, the power transmission equation is found by substituting w_y for w_z and J_{zz} for J_{yy} in Equation (4.177). Note that J_{yy} is the second moment of area of the beam cross section about the y-axis. If more than one wave type is present at one time, w_z would be replaced by w_{z0}, where the subscript, 0, denotes displacement of the centre of the beam section.

Determination of the intensity or power transmission associated with bending waves is more complicated than for torsional and longitudinal waves due to the higher order derivatives involved (see Equation (4.177)). In fact it is necessary to use a minimum of four accelerometers to enable the third order derivative to be evaluated.

Two types of accelerometer configuration, illustrated in Figure 4.8, have been shown to give good results (Pavic, 1976; Hayek et al., 1990) for measuring structural intensity due to bending waves.

For the configuration shown in Figure 4.8(a), the derivatives are calculated using the following relations:

$$w = w_0 = \frac{w_3 + w_2}{2} \tag{4.178}$$

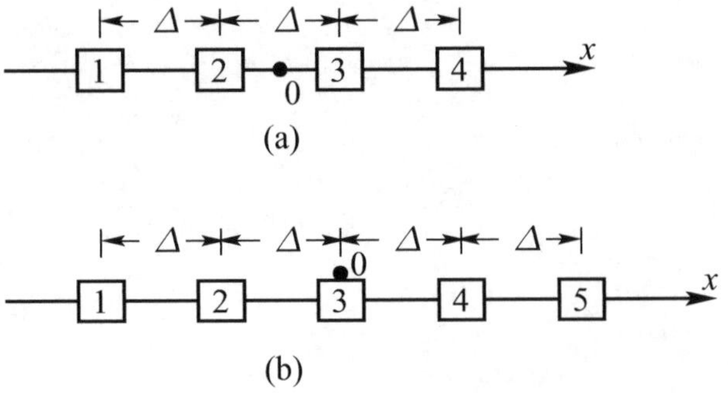

FIGURE 4.8 Accelerometer configurations for determining higher order derivatives of the beam displacement at 0: (a) Pavic (1976); (b) Hayek et al. (1990).

$$\frac{\partial w}{\partial x} = \frac{w_3 - w_2}{\Delta} \qquad (4.179)$$

$$\frac{\partial^2 w}{\partial x^2} = \frac{w_1 - w_2 - w_3 + w_4}{2\Delta^2} \qquad (4.180)$$

$$\frac{\partial^3 w}{\partial x^3} = \frac{-w_1 + 3w_3 - 3w_2 + w_4}{\Delta^3} \qquad (4.181)$$

Note that the accelerometer numbering convention adopted has been increasing the number in the direction of increasing x. Although this convention is opposite to that adopted by Pavic (1976), it is the convention almost universally used by others. Also, as mentioned previously, the moment and force convention that has been used for decades by those concerned with the dynamic analysis of plates is maintained here. Unfortunately this convention has not been followed universally by those involved in the measurement of structural intensity. Thus, some of the results presented here may look a little different to what appears elsewhere.

For the configuration in Figure 4.8(b), the following expressions apply:

$$w = w_o; \quad \text{and} \quad \frac{\partial w}{\partial x} = \frac{w_4 - w_2}{2\Delta} \qquad (4.182)$$

$$\frac{\partial^3 w}{\partial x^3} = \frac{-w_1 + 2w_2 - 2w_4 + w_5}{2\Delta^3} \qquad (4.183)$$

The latter equations (involving five accelerometers) give more accurate results than the equations for the four accelerometer case. However, Hayek et al. (1990) show that no further benefit is gained by using more measurement points, and that the optimum value for Δ is one-twentieth of a structural wavelength.

Taking the time average of Equation (4.177) gives the active component of the power transmission as:

$$\Pi_{Bza} = \langle \Pi_{Bz}(t) \rangle_t = EI_{yy} \left\langle \left(\frac{\partial w_z}{\partial t} \frac{\partial^3 w_z}{\partial x^3} - \frac{\partial}{\partial t} \left(\frac{\partial w_z}{\partial x} \right) \frac{\partial^2 w_z}{\partial x^2} \right) \right\rangle_t \qquad (4.184)$$

Using the notation introduced earlier, and assuming that the measurements are made with accelerometers:

$$w_z = \iint a_z \qquad (4.185)$$

$$\frac{\partial w_z}{\partial t} = \int a_z \qquad (4.186)$$

Thus:

$$\Pi_{Bza} = EI_{yy} \left\langle \int a_z \iint \frac{\partial^3 a_z}{\partial x^3} - \int \frac{\partial a_z}{\partial x} \iint \frac{\partial^2 a_z}{\partial x^2} \right\rangle_t \qquad (4.187)$$

Using Equations (4.178) to (4.181) (assuming four accelerometers) the following is obtained (Pavic, 1976):

$$\Pi_{Bza} = \frac{EI_{yy}}{\Delta^3}\left\langle 4\int a_{z2}\iint a_{z3} - \int a_{z2}\iint a_{z4} - \int a_{z1}\iint a_{z3}\right\rangle_t \tag{4.188}$$

Thus, using the relations in Table 4.4, the power is given in terms of the cross spectrum by:

$$\Pi_{Bza}(\omega) = \frac{EJ_{yy}}{\Delta^3\omega^3}\left[4G_{a_2,a_3}(\omega) - G_{a_2,a_4}(\omega) - G_{a_1,a_3}(\omega)\right] \tag{4.189}$$

If the measurements are performed using a laser Doppler velocimeter rather than accelerometers, corresponding cross-spectral expressions for structural power transmission can be derived using the relationships of Table 4.4. If u_z represents the measured velocity, then the relationship corresponding to Equation (4.189) for the active power transmission is:

$$\Pi_{Bza}(\omega) = \frac{EJ_{yy}}{\Delta^3\omega}\left[4G_{u_2,u_3}(\omega) - G_{u_2,u_4}(\omega) - G_{u_1,u_3}(\omega)\right] \tag{4.190}$$

A simpler expression can be obtained if the accelerometers are located in the far field of all vibration sources and reflections. In this case, it may be shown that the shear force and bending moment contributions to the power transmission are the same (Noiseux, 1970) and the instantaneous total power transmission due to bending waves may be written as:

$$\Pi_{Bz}(t) = -2EJ_{yy}\left[\frac{\partial^2 w_z}{\partial t \partial x}\frac{\partial^2 w_z}{\partial x^2}\right] \tag{4.191}$$

The time averaged (or active) power may be written as:

$$\langle\Pi_{Bz}(t)\rangle = \Pi_{Bza} = -2EJ_{yy}\left\langle\int\frac{\partial a}{\partial x}\iint\frac{\partial^2 a}{\partial x^2}\right\rangle \tag{4.192}$$

which is a similar form to Equation (4.169) for longitudinal waves with J_{yy} replaced by S.

Thus, from Table 4.4, the following is obtained for the frequency domain:

$$\Pi_{Bza}(\omega) = 2EJ_{yy}\frac{\mathrm{Im}\left\{G\left(\frac{\partial^2 a}{\partial x^2}, \frac{\partial a}{\partial x}, \omega\right)\right\}}{\omega^3} \tag{4.193}$$

In the absence of near fields, the relation between the Fourier components of $\frac{\partial^2 a}{\partial x^2}$ and a is:

$$\frac{\partial^2 a}{\partial x^2} = -k_b^2 a \tag{4.194}$$

From Section 2.3.3.3:

$$\omega(\rho_m S/B)^{1/2} = k_b^2 \tag{4.195}$$

where ρ_m is the density of the beam material, S is the beam cross-sectional area and $B = EJ_{yy}$ is the the bending stiffness. Remembering the finite difference approximations for a two-accelerometer arrangement:

$$a = (a_{z1} + a_{z2})/2 \tag{4.196}$$

$$\frac{\partial a}{\partial x} = (a_{z2} - a_{z1})/\Delta \tag{4.197}$$

the following is obtained for the active power transmission:

$$\Pi_{Bza}(\omega) = \frac{2(B\rho_m S)^{1/2}}{\omega^2\Delta}\mathrm{Im}\left\{G_{a_{z2},a_{z1}}(\omega)\right\} \tag{4.198}$$

The corresponding expression for the reactive power is:

$$\Pi_{Bzr}(\omega) = \frac{(B\rho_m S)^{1/2}}{\omega^2 \Delta} \left[G_{a_{z1},a_{z1}}(\omega) - G_{a_{z2},a_{z2}}(\omega) \right] \tag{4.199}$$

where $G_{a_{z1},a_{z1}}(\omega)$ is the auto (power) spectrum (units of acceleration squared) of the accelerometer signal, z_1.

Assuming a harmonic wave field and substituting Equations (4.194) and (4.195) into Equation (4.192) gives for the time average active power:

$$\Pi_{Bza} = -\frac{2k_b^2 E J_{yy}}{\omega^2} \left\langle \left(\int \frac{\partial a}{\partial x} \right) a \right\rangle_t = -\frac{2(B\rho_m S)^{1/2}}{\omega} \left\langle \left(\int \frac{\partial a}{\partial x} \right) a \right\rangle_t \tag{4.200}$$

Using Equations (4.196) and (4.197), Equation (4.200) may be rewritten as:

$$\Pi_{Bza} = \frac{(B\rho_m S)^{1/2}}{\omega \Delta} \left\langle (a_{z1} + a_{z2}) \int (a_{z1} - a_{z2}) d\tau \right\rangle_t \tag{4.201}$$

Using Equations (4.159) and (4.160), Equation (4.201) can be rewritten as:

$$\Pi_{Bza} = \frac{2(B\rho_m S)^{1/2}}{\omega \Delta} \left\langle a_{z2} \int a_{z1} \right\rangle_t \tag{4.202}$$

From the similarity between Equation (4.202) and the corresponding Equation (4.109) for sound intensity, it may be deduced that a sound intensity analyser may be used to determine the structural intensity for a harmonic wave field on a beam in the far field of any sources or reflections, by replacing the pressure signals with accelerometer signals in Equation (4.106). Alternatively, Equation (4.202) and an analogue multiplying circuit may be used to multiply the signal from one accelerometer with the integrated signal from the other.

As discussed by Craik et al. (1995), it is possible to measure structural power transmission in the far field of the source by using a single bi-axial accelerometer, as shown in Figure 4.9.

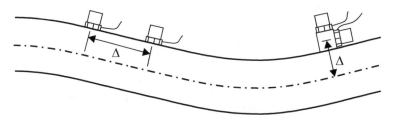

FIGURE 4.9 Definition of the spacing, Δ, for a dual accelerometer measurement of bending wave intensity, showing an array and a bi-axial accelerometer.

For the bi-axial accelerometer, the spacing, Δ, is the distance from the centre of the horizontal accelerometer to the neutral axis of the beam. This horizontal accelerometer actually measures the beam rotation about an axis normal to the page. The same equations as derived previously for the dual accelerometer array apply to the bi-axial configuration (Craik et al., 1995).

A comparison of the accuracy associated with various methods of field measurement of structural intensity on a beam was reported by Bauman (1994).

As was done for the calculation of longitudinal wave power transmission, an alternative relationship can be derived for bending wave power by beginning with Equation (4.177) and assuming harmonic excitation immediately. Thus, from Equation (4.177), the time averaged active power transmission is:

$$\Pi_{Bza} = \mathrm{Re} \left\{ \frac{E J_{yy}}{2} \left((\mathrm{j}\omega \hat{w}_z(x))^* \frac{\partial^3 \hat{w}_z(x)}{\partial x^3} - \frac{\partial \left(\mathrm{j}\omega \hat{w}_z(x) \right)^*}{\partial x} \frac{\partial^2 \hat{w}_z(x)}{\partial x^2} \right) \right\} \tag{4.203}$$

or:

$$\Pi_{Bza} = \mathrm{Re}\left\{ \frac{\mathrm{j}\omega E J_{yy}}{2} \left(\hat{w}_z^*(x)\frac{\partial^3 \hat{w}_z(x)}{\partial x^3} - \frac{\partial \hat{w}_z^*(x)}{\partial x}\frac{\partial^2 \hat{w}_z(x)}{\partial x^2} \right) \right\} \tag{4.204}$$

and the amplitude of the fluctuating reactive component is:

$$\Pi_{Bzr} = \mathrm{Im}\left\{ \frac{\mathrm{j}\omega E J_{yy}}{2} \left(\hat{w}_z^*(x)\frac{\partial^3 \hat{w}_z(x)}{\partial x^3} - \frac{\partial \hat{w}_z^*(x)}{\partial x}\frac{\partial^2 \hat{w}_z(x)}{\partial x^2} \right) \right\} \tag{4.205}$$

where the * denotes the complex conjugate.

As for the general case, simpler expressions can be obtained for Π_{Bza} if the measurements are conducted at least one half of a wavelength away from any power source or sources of reflection; that is, in a region where near field effects may be neglected. Recalling the solution for a sinusoidal bending wave travelling in a beam derived in Section 2.3.3.3:

$$w_z(x,t) = \left[A_1 \mathrm{e}^{-\mathrm{j}k_b x} + A_2 \mathrm{e}^{\mathrm{j}k_b x} + A_3 \mathrm{e}^{-k_b x} + A_4 \mathrm{e}^{k_b x} \right] \mathrm{e}^{\mathrm{j}\omega t} = \hat{w}_z(x)\mathrm{e}^{\mathrm{j}\omega t} \tag{4.206}$$

The first and second terms represent the propagating vibration field in the positive and negative x-directions, respectively, while the second two terms represent the decaying near field. Thus if the near field is ignored, $A_3 = A_4 = 0$, and if the time dependent term, $\mathrm{e}^{\mathrm{j}\omega t}$, is also omitted for convenience, Equation (4.206) becomes:

$$\hat{w}_z(x) = A_1 \mathrm{e}^{-\mathrm{j}k_b x} + A_2 \mathrm{e}^{\mathrm{j}k_b x} = \hat{A}_1 \mathrm{e}^{-\mathrm{j}(k_b x - \theta_1)} + \hat{A}_2 \mathrm{e}^{\mathrm{j}(k_b x + \theta_2)} \tag{4.207}$$

where A_1 and A_2 are complex numbers, $\hat{A}_1$ and $\hat{A}_2$ are real, and θ_1 and θ_2 are the signal phases at $x = 0$. Substituting Equation (4.207) into Equation (4.204) gives for the active power:

$$\Pi_{Bza} = E J_{yy}\omega k_b^3 (\hat{A}_1^2 - \hat{A}_2^2) \tag{4.208}$$

However:

$$\mathrm{Re}\left\{ \mathrm{j}\left(\hat{w}_z^*\frac{\partial \hat{w}_z}{\partial x} - \hat{w}_z\frac{\partial \hat{w}_z^*}{\partial x} \right) \right\} = 2k_b(\hat{A}_1^2 - \hat{A}_2^2) \tag{4.209}$$

Thus:

$$\Pi_{Bza} = \mathrm{Re}\left\{ \left(\frac{\mathrm{j}\omega}{2}\right) E J_{yy}k_b^2 \left(\hat{w}_z^*\frac{\partial \hat{w}_z}{\partial x} - \hat{w}_z\frac{\partial \hat{w}_z^*}{\partial x} \right) \right\} = \omega E J_{yy}k_b^2 \mathrm{Im}\left\{ \hat{w}_z\frac{\partial \hat{w}_z^*}{\partial x} \right\} \tag{4.210}$$

That is, the contribution due to the shear force is equal to the contribution due to the bending moment in the far field. This is in contrast to the situation in the near field of sources where it has been found (Pavic, 1990) that the shear force component dominates the bending wave component.

Substituting Equation (4.207) into Equation (4.205) gives for the reactive power:

$$\Pi_{Bzr} = 2E J_{yy}\omega k_b^3 \hat{A}_1 \hat{A}_2 \sin(2k_b x + \theta_2 - \theta_1) = \omega E J_{yy}k_b^2 \mathrm{Re}\left\{ \hat{w}_z\frac{\partial \hat{w}_z^*}{\partial x} \right\} \tag{4.211}$$

If accelerometers were used and mounted to measure lateral (or bending) acceleration, then Equation (4.210) could be written as:

$$\Pi_{Bza} = -\mathrm{Im}\left\{ \frac{E J_{yy}k_b^2}{\omega^3}\hat{a}_z^*\frac{\partial \hat{a}_z}{\partial x} \right\} \tag{4.212}$$

and Equation (4.211) could be written as:

$$\Pi_{Bzr} = \mathrm{Re}\left\{ \frac{E J_{yy}k_b^2}{\omega^3}\hat{a}_z^*\frac{\partial \hat{a}_z}{\partial x} \right\} \tag{4.213}$$

where a_z is the acceleration of the centre of the beam section in the z-direction.

Substituting Equations (4.196) and (4.197) into Equation (4.212) gives for the active power:

$$\Pi_{Bza} = -\text{Im}\left\{ \frac{EJ_{yy}k_b^2}{2\omega^3\Delta}(\hat{a}_{z1} + \hat{a}_{z2})^*(\hat{a}_{z2} - \hat{a}_{z1}) \right\} = \frac{EJ_{yy}k_b^2}{\omega^3\Delta}|\hat{a}_{z1}||\hat{a}_{z2}|\sin(\theta_1 - \theta_2) \qquad (4.214)$$

and substituting Equations (4.196) and (4.197) into Equation (4.213) gives for the reactive power:

$$\Pi_{Bzr} = \text{Re}\left\{ \frac{EJ_{yy}k_b^2}{2\omega^3\Delta}(\hat{a}_{z1} + \hat{a}_{z2})^*(\hat{a}_{z2} - \hat{a}_{z1}) \right\} = \frac{EJ_{yy}k_b^2}{2\omega^3\Delta}\left(|\hat{a}_{z2}|^2 - |\hat{a}_{z1}|^2\right) \qquad (4.215)$$

Thus, the time averaged bending wave active power transmission for harmonic waves can be measured in practice by multiplying the amplitudes of two closely spaced accelerometers with the sine of the phase difference between the two. The direction of positive power transmission is from accelerometer 1 to 2. It can be shown easily that Equation (4.214) is equivalent to Equation (4.201); however, in many experimental situations it is probably easier to measure the amplitudes of and the phase difference between two sinusoidal signals than it is to accurately perform analog integrations and multiplications.

Two-element probes with the accelerometers mounted on a common base are commercially available and have been designed specifically for measuring bending wave structural intensity in the far field away from structural discontinuities or vibration sources. However, at high frequencies large errors can occur using these probes, due to the phase error introduced because of the flexibility of the base on which the accelerometers are mounted.

When measuring structural intensity using the two accelerometer method, the error associated with the finite spacing, Δ, of the accelerometers can be minimised by adjusting the measured structural intensity by a factor to obtain the actual structural intensity as follows (Kim and Tichy, 2000):

$$\Pi_{actual} = \Pi_{measured}\left[\frac{k_b\Delta}{\sin k_b\Delta} \right] \qquad (4.216)$$

4.4.1.4   Total Power Transmission

Using Equations (4.147), (4.175) and (4.177), the total instantaneous power transmission in the beam as a result of the propagation of all wave types (two orthogonal bending waves, one longitudinal wave and one torsional wave) can be written in matrix form as:

$$\Pi(t) = \frac{\partial}{\partial t}\boldsymbol{W}_0^{\text{T}}(t)\boldsymbol{\Lambda}\boldsymbol{W}_0(t) \qquad (4.217)$$

where:

$$\boldsymbol{W}_0(t) = [\xi_{x0}(t), w_{y0}(t), w_{z0}(t), \theta_x(t), \theta_y(t), \theta_z(t)]^{\text{T}} \qquad (4.218)$$

and where T denotes the transpose of a matrix or vector, ξ_{x0}, w_{y0} and w_{z0} represent the displacements of the centre of the beam section in the x-, y- and z-directions, respectively, and θ_x, θ_y and θ_z represent the rotations of the whole section about the x-, y- and z-axes, respectively. The diagonal matrix, Λ is:

$$\boldsymbol{\Lambda} = \begin{bmatrix} -ES\dfrac{\partial}{\partial x} & & & & & \\ & EJ_{zz}\dfrac{\partial^3}{\partial x^3} & & & & \\ & & EJ_{yy}\dfrac{\partial^3}{\partial x^3} & & & \\ & & & -C\dfrac{\partial}{\partial x} & & \\ & & & & -EJ_{yy}\dfrac{\partial}{\partial x} & \\ & & & & & -EJ_{zz}\dfrac{\partial}{\partial x} \end{bmatrix} \qquad (4.219)$$

where J_{yy} and J_{zz} are the second moments of area of the cross section about the y- and z-axes, respectively, E is Young's modulus of elasticity, C is the torsional stiffness and S is the area of beam cross section. The angular rotations are defined as:

$$\theta_x = \frac{\partial w_{z0}}{\partial y} = -\frac{\partial w_{y0}}{\partial z} \qquad (4.220)$$

$$\theta_y = -\frac{\partial w_{z0}}{\partial x} \qquad (4.221)$$

$$\theta_z = \frac{\partial w_{y0}}{\partial x} \qquad (4.222)$$

For single frequency beam excitation, the time averaged (or active) power transmission is:

$$\Pi_a = \frac{1}{2}\mathrm{Re}\left\{ j\omega \hat{\boldsymbol{W}}_0^{\mathrm{H}} \boldsymbol{\Lambda} \hat{\boldsymbol{W}}_0 \right\} \qquad (4.223)$$

and the amplitude of the fluctuating imaginary component is:

$$\Pi_r = \frac{1}{2}\mathrm{Im}\left\{ j\omega \hat{\boldsymbol{W}}_0^{\mathrm{H}} \boldsymbol{\Lambda} \hat{\boldsymbol{W}}_0 \right\} \qquad (4.224)$$

where the hat over a variable indicates the complex amplitude of a time varying quantity, and H is the transpose of the complex conjugate and positive power flow is in the positive directions of the particular axis under consideration or in the anti-clockwise direction for rotational power flow.

If the beam vibration is broadband as opposed to sinusoidal, the overall active intensity can be obtained by averaging the quantities in Equation (4.223) in the time domain or by using the cross-spectral density representation in the frequency domain.

4.4.1.5 Measurement of Beam Accelerations

In the previous analysis it has been assumed that the overall displacements of the beam cross section $(w_{y0},\ w_{z0},\ w_{x0})$ and θ_{x0} or corresponding accelerations, a_{y0}, a_{z0}, a_{x0} and $\ddot{\theta}_{x0}$, can be determined by simple measurements on the beam surface. In practice, the measurements are not as straightforward as they seem at first, due to the dependence on more than one wave type of the surface displacement in any one direction. However, this problem can be mostly overcome for symmetrical beams using an arrangement suggested by Verheij (1990) for measurements on circular pipes, and illustrated in Figure 4.10. Means of extracting the amplitudes corresponding to each wave type are given in the figure caption. Note that the accelerometers are oriented and their outputs combined in such a way as to isolate each of the wave types from any influence from the others. For example, the Poisson contraction and expansion associated with longitudinal wave propagation acts in opposite directions on two opposite faces of the beam; thus subtracting one from the other of the outputs of the two accelerometers shown for measuring bending waves (w_y or w_z) will null this effect.

Similarly, the longitudinal displacement of the beam surface as a result of section rotation due to bending wave propagation is nulled by adding the outputs of the two accelerometers shown for measuring longitudinal wave displacements (or ξ_x). Thus, use of the measurement scheme shown in Figure 4.10 theoretically allows measurement of the amplitude of each wave type regardless of the simultaneous presence or otherwise of other wave types. However, in practice problems can arise as a result of the non-zero cross-axis (or transverse) sensitivity of the accelerometers if all three wave types are present. Cross-axis sensitivity means that accelerometers are sensitive to motion in directions other than along their main axis as well as along their main axis. The cross-axis sensitivity is direction dependent with a maximum value usually of about 5% of the main axis sensitivity. Some manufacturers indicate the direction of least sensitivity with a mark

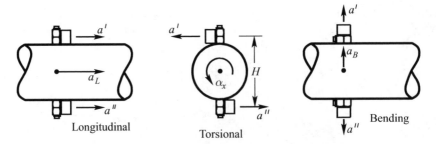

FIGURE 4.10 Simplified scheme for determining accelerations at a point on the beam, assuming that each wave exists in isolation from the others. A similar scheme would also apply to rectangular cross section beams: $a_L = (a' + a'')/2$, $\alpha_x = (a' + a'')/H$ and $a_B = (a' - a'')/2$.

on the accelerometer, and this can be two orders of magnitude less than the maximum value. Ignoring the effects of cross-axis sensitivity can lead to serious measurement errors, especially if one is attempting to measure the amplitude of a wave that is much smaller than the amplitude of the other types. For any measurement using the configuration shown in Figure 4.10, it is possible to minimise the error resulting from cross-axis sensitivity by appropriate orientation of the accelerometers used for the measurements, such that the least sensitive axis is in the direction of maximum response for wave types that are not being measured. Another problem associated with the scheme shown in Figure 4.10 is that it only applies to beams that are rectangular, circular or ellipsoidal in cross section.

An alternative scheme, which allows the amplitude of each wave type to be determined accurately, and which also allows the cross-axis sensitivity of the accelerometers to be taken into account, involves the measurement of the x-, y- and z-components of the displacement on the beam surface at three or more locations. In addition, the beam does not have to be rectangular, circular or ellipsoidal in cross section for this method to work, in contrast to the scheme shown in Figure 4.6. However, the y- and z-axes must coincide with the two principal orthogonal axes of the beam cross section, as shown in Figure 4.11. The displacement components, w_y, w_z and ξ_x, at any location on a beam cross section are described by the following matrix equation, which relates them to the displacements and rotations of the centre of the section (or section as a whole):

$$
\begin{bmatrix} \xi_x \\ w_y \\ w_z \end{bmatrix} = \begin{bmatrix} 1 & 0 & 0 & 0 & z & -y \\ 0 & 1 & 0 & -z & 0 & 0 \\ 0 & 0 & 1 & y & 0 & 0 \end{bmatrix} \begin{bmatrix} \xi_{x0} \\ w_{y0} \\ w_{z0} \\ \theta_x \\ \theta_y \\ \theta_z \end{bmatrix} \tag{4.225}
$$

where y and z are coordinate locations indicating the distance of the measurement point from the x-axis of the beam. The subscript, 0, indicates a measurement at the centre of the beam.

In theory at least, Equation (4.225) can be used to determine the displacements of the centre of the beam section from only six measurements on the surface of the beam. Thus:

$$
\begin{bmatrix} \xi_{x1} \\ w_{y1} \\ w_{z1} \\ \xi_{x2} \\ w_{y2} \\ w_{z2} \end{bmatrix} = \begin{bmatrix} 1 & 0 & 0 & 0 & z_1 & -y_1 \\ 0 & 1 & 0 & -z_1 & 0 & 0 \\ 0 & 0 & 1 & y_1 & 0 & 0 \\ 1 & 0 & 0 & 0 & z_2 & -y_2 \\ 0 & 1 & 0 & -z_2 & 0 & 0 \\ 0 & 0 & 1 & y_2 & 0 & 0 \end{bmatrix} \begin{bmatrix} \xi_{x0} \\ w_{y0} \\ w_{z0} \\ \theta_x \\ \theta_y \\ \theta_z \end{bmatrix} \tag{4.226}
$$

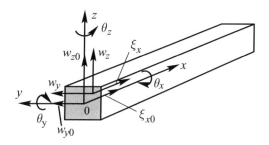

FIGURE 4.11 Coordinate system for a rectangular beam.

Unfortunately, the determinant of the coefficient matrix on the right-hand side of Equation (4.225) is zero, which indicates that more measurements are needed. If a third measurement location is used, then:

$$
\begin{bmatrix}
\xi_{x1} \\
w_{y1} \\
w_{z1} \\
\xi_{x2} \\
w_{y2} \\
w_{z2} \\
\xi_{x3} \\
w_{y3} \\
w_{z3}
\end{bmatrix}
=
\begin{bmatrix}
1 & 0 & 0 & 0 & z_1 & -y_1 \\
0 & 1 & 0 & -z_1 & 0 & 0 \\
0 & 0 & 1 & y_1 & 0 & 0 \\
1 & 0 & 0 & 0 & z_2 & -y_2 \\
0 & 1 & 0 & -z_2 & 0 & 0 \\
0 & 0 & 1 & y_2 & 0 & 0 \\
1 & 0 & 0 & 0 & z_3 & -y_3 \\
0 & 1 & 0 & -z_3 & 0 & 0 \\
0 & 0 & 1 & y_3 & 0 & 0
\end{bmatrix}
\begin{bmatrix}
\xi_{x0} \\
w_{y0} \\
w_{z0} \\
\theta_x \\
\theta_y \\
\theta_z
\end{bmatrix}
\tag{4.227}
$$

For future reference, the 9 x 6 location matrix in Equation (4.227) will be denoted $\boldsymbol{A}$. A unique solution will exist for Equation (4.227) only if the matrix, $\boldsymbol{A}^{\mathrm{T}}\boldsymbol{A}$, is non-singular. If the accelerometer locations are so selected, $\boldsymbol{W}_0$ can be calculated using:

$$
\boldsymbol{W}_0 = \left(\boldsymbol{A}^{\mathrm{T}}\boldsymbol{A}\right)^{-1}\left[\xi_{x1}, w_{y1}, w_{z1}, \xi_{x2}, w_{y2}, w_{z2}, \xi_{x3}, w_{y3}, w_{z3},\right]^{\mathrm{T}}
\tag{4.228}
$$

For a rectangular-section beam, the matrix will be non-singular if the measurement locations are in the centre of any three of the four sides of a given section. For a circular beam, the measurement locations should be on the orthogonal y- and z-axes. These axes are defined for all but perfectly circular beams and in this latter case, the measurements should be on any two cross-sectional axes, y and z, separated by 90°.

Note that Equations (4.225) to (4.227) give complex amplitudes of the displacements of the centre of the beam cross section, which allows determination of the relative phases of the various propagating wave types.

One limitation of the method just described is that the Poisson contraction effect, resulting in strain in the transverse direction as a result of longitudinal waves travelling in the x-direction, has been ignored. In most cases, this will not be a problem, as it will be too small to be of importance. The lateral displacement at location, z, on a particular cross section as a result of longitudinal wave propagation is:

$$
w_z = z\epsilon_{zz} = -\nu z\epsilon_{xx} = -\nu z\frac{\partial \xi_x}{\partial x}
\tag{4.229}
$$

Substituting Equations (4.161) into (4.229), assuming a wave travelling in only one direction and carrying out the differentiation it can be shown that for a material with a Poisson ratio of approximately 0.3, the lateral displacement due to longitudinal wave propagation is approximately:

$$
w_z = \frac{2z}{\lambda_L}\xi_x
\tag{4.230}
$$

For a rectangular beam, the maximum value of z is half the beam thickness, and for many practical beams the wavelength of longitudinal waves is much greater than the beam thickness, so the quantity in Equation (4.230) is usually very small compared to the lateral displacement due to bending waves. Nevertheless, there will be cases where longitudinal waves dominate the response, and the Poisson contraction effect must be taken into account. For beams of regular section, this is best done using the arrangements shown in Figure 4.10, where the Poisson effect is automatically cancelled by using two bending wave accelerometers mounted on opposite sides of the beam.

It is interesting to note that in a thin, wide beam, it is difficult to excite the torsional wave and the bending wave that is characterised by displacement along the width of the beam. For this case, the problem is reduced to one of identifying only one bending wave and the longitudinal wave; then the arrangement shown in Figure 4.10 can lead to quite accurate results, if the accelerometers for measuring the longitudinal waves are orientated so that their direction of minimum cross-axis sensitivity is in the direction of the bending wave displacement and if the accelerometers for measuring the bending wave are orientated so that their direction of minimum cross-axis sensitivity is in the direction of the longitudinal wave displacement. Only one accelerometer need be used for measuring the longitudinal wave if it is mounted on the thin edge of the beam.

To determine the amplitudes of waves travelling simultaneously in two different directions along the beam, at least four accelerometers are needed to resolve each pair of bending waves (provided that they are in the far field of any source or beam discontinuity), at least four accelerometers are needed to resolve the pair of longitudinal waves and at least four accelerometers are needed to resolve the pair of torsional waves. They must be mounted on at least two different beam cross sections separated by no less than one-fifteenth of a wavelength and no more than one-third of a wavelength. Use of more accelerometers and more measurement cross sections increases the accuracy of the results substantially. If more than two beam cross sections are used, then the separation between the sections furthest apart should be no more than one-third of a wavelength. Note that the same cross-sectional locations may be different for each wave type (mainly because the wavelengths corresponding to each wave type are generally very different). For lightweight structures, the number of accelerometers needed may be sufficient to significantly affect the beam dynamics, even if very small accelerometers are used.

Perhaps a more efficient way of measuring structural intensity is to use a scanning laser doppler vibrometer which can provide relative amplitude and phase information at each location on the beam. If only bending wave transmission is of interest then a 1-D vibrometer will suffice. However, for simultaneous measurement of longitudinal, torsional and bending intensity, a 3-D scanning vibrometer is necessary (Hayek et al., 1990; Pascal et al., 1993; Blotter et al., 2002; Wang et al., 2006).

4.4.1.6   Effect of Transverse Sensitivity of Accelerometers

If accelerometers are used for the measurements, they will invariably exhibit some degree of sensitivity (transverse sensitivity) along axes at right angles to their measurement axis, thus resulting in inaccuracies in the determination of the measured w_y, w_z and ξ_x. However, if the cross-axis sensitivity as a fraction of the main axis sensitivity is known or measured beforehand, then the following relation may be used to determine the actual displacements (or accelerations) at any location from the measured ones. Assuming the actual displacements are w_y, w_z and ξ_x, at a given measurement point, and the measured displacements are w_{my}, w_{mz} and ξ_{mx}, the two are related by the accelerometer cross-axis sensitivities as follows:

$$\begin{bmatrix} \xi_{mx} \\ w_{my} \\ w_{mz} \end{bmatrix} = \begin{bmatrix} 1 & \alpha_x & \alpha_x \\ \alpha_y & 1 & \alpha_y \\ \alpha_z & \alpha_z & 1 \end{bmatrix} \begin{bmatrix} \xi_x \\ w_y \\ w_z \end{bmatrix} \tag{4.231}$$

or:

$$
\left[\begin{array}{c} \xi_x \\ w_y \\ w_z \end{array}\right] = \left[\begin{array}{ccc} 1 & \alpha_x & \alpha_x \\ \alpha_y & 1 & \alpha_y \\ \alpha_z & \alpha_z & 1 \end{array}\right]^{-1} \left[\begin{array}{c} \xi_{mx} \\ w_{my} \\ w_{mz} \end{array}\right] \tag{4.232}
$$

where α_y, α_z and α_x are the cross-axis sensitivities (expressed as a fraction of the main-axis sensitivity) of the accelerometers measuring w_{my}, w_{mz} and ξ_{mx}, respectively. Note that the cross-axis sensitivity of an accelerometer is usually strongly dependent upon the direction of the axis of interest (or angular orientation of the accelerometer) and this must also be taken into account during calibration and mounting of the accelerometers. Some accelerometers are available for which the direction of minimum cross-axis sensitivity is clearly marked, and in some cases this sensitivity is negligible, thus making the corrections outlined in this section unnecessary for situations involving the propagation of only two wave types.

4.4.2 Structural Power Transmission Measurement in Plates

The instantaneous power transmission per unit width in a plate (often referred to as intensity) given by Equation (4.140) may be expressed in terms of plate displacements using Equations (2.176) to (2.178), (2.183) and (2.184) for classical plate theory and Equations (2.219) to (2.221), (2.224) and (2.225) for Mindlin–Timoshenko plate theory which should be used for thick plates and/or high frequencies.

It can be seen from Equation (4.140) that first, second and third order derivatives must be approximated to completely determine the structural power transmission in a thin plate. However, it will be shown here that under certain conditions, the two-accelerometer method (discussed for beams) may be used to determine the bending wave component of the total power transmission. It can also be seen from Equation (4.140) that in both the near and far field of sources it is necessary to measure the in-plane plate displacements and the second and third derivatives of the normal plate displacements to determine the structural power transmission in longitudinal and bending waves.

As discussed by Pavic (1976), it is necessary to use eight accelerometers to evaluate the required derivatives in Equation (4.140). The required finite difference equations may be formulated as for a beam. Here the simpler case of a harmonic sound field will be examined, for which the following expressions for the active and reactive intensity can be derived from Equation (4.140):

$$
\Pi_{xa} = \frac{1}{2}\mathrm{Re}\left\{j\omega\left[\hat{w}^*\hat{Q}_x - \frac{\partial\hat{w}^*}{\partial x}\hat{M}_x - \frac{\partial\hat{w}}{\partial y}\hat{M}_{xy} + \hat{\xi}_x^*\hat{N}_x + \xi_y^*\hat{N}_{xy}\right]\right\} \tag{4.233}
$$

$$
\Pi_{xr} = \frac{1}{2}\mathrm{Im}\left\{j\omega\left[\hat{w}^*\hat{Q}_x - \frac{\partial\hat{w}^*}{\partial x}\hat{M}_x - \frac{\partial\hat{w}^*}{\partial y}\hat{M}_{xy} + \hat{\xi}_x^*\hat{N}_x + \hat{\xi}_y^*\hat{N}_{xy}\right]\right\} \tag{4.234}
$$

where the asterisk denotes the complex conjugate and all quantities in the equations are complex (expressed as an amplitude and a relative phase). The first three terms in Equations (4.233) and (4.234) are associated with bending waves, the fourth term is associated with longitudinal waves and the last term is associated with in-plane shear waves. Simpler expressions, which can be implemented using two accelerometers, will now be derived for each of these wave types.

4.4.2.1 Longitudinal Waves

From Equation (4.140), the power transmission per unit plate width associated with longitudinal wave propagation in the x-direction is:

$$
\Pi_L(t) = -\dot{\xi}_x N_x \tag{4.235}
$$

From Equations (2.162), (2.168) to (2.171) and (2.173), the force component, N_x, may be written as:

$$N_x = \int_{-h/2}^{h/2} \sigma_{xx} \mathrm{d}z = \frac{Eh}{1-\nu^2} \left[\frac{\partial \xi_x}{\partial x} + \nu \frac{\partial \xi_y}{\partial y} \right] \tag{4.236}$$

For longitudinal wave propagation in the x-direction only, ξ_y is negligible and Equation (4.235) may be written as:

$$\Pi_L(t) = -\frac{Eh}{1-\nu^2} \frac{\partial \xi_x}{\partial t} \frac{\partial \xi_x}{\partial x} \tag{4.237}$$

which is equivalent to Equation (4.149) for a beam, where the cross-sectional area, S, for a beam has been replaced with $h/(1-\nu^2)$, where h is the plate thickness and ν is Poisson's ratio.

Thus, all of the equations and measurement techniques derived previously for longitudinal wave power transmission in a beam are valid for longitudinal wave power transmission in a plate provided that S is replaced with $h/(1-\nu^2)$. Note that the power expressions for plates are power per unit plate width whereas for beams the expressions represent the total power transmitting along the beam.

Similar arguments hold for longitudinal wave propagation in the y-direction, where the same equations may be used with the x and y subscripts interchanged.

4.4.2.2 Transverse Shear Waves

From Equation (4.140), the power transmission per unit width associated with shear waves is:

$$\Pi_s(t) = -\dot{\xi}_y N_{xy} \tag{4.238}$$

From Equations (2.154) to (2.157), the force component, N_{xy}, can be written as:

$$N_{xy} = Gh \frac{\partial \xi_y}{\partial x} \tag{4.239}$$

For shear wave only propagation, Equation (4.238) may be written as:

$$\Pi_s(t) = -Gh \frac{\partial \xi_y}{\partial t} \frac{\partial \xi_y}{\partial x} \tag{4.240}$$

which has the same form as Equation (4.149) for a beam, except that the longitudinal displacement is measured in the y-direction rather than the x-direction. Thus, all of the previously derived expressions for a beam can be used if the quantity SE is replaced with Gh and the measurement transducers are configured to measure longitudinal displacement in the y-direction (remember that only power transmission in the x-direction is being considered for now). Note that although the displacement in the y-direction is to be measured, it is the gradient of this in the x-direction which is required; thus, the two measurement transducers must be aligned with the x-axis. Similar arguments hold for shear wave propagation in the y-direction, where the same equations may be used with the x and y subscripts interchanged.

4.4.2.3 Bending Waves

From Equation (4.140), the power transmission per unit plate width associated with bending wave propagation in the x-direction is:

$$\Pi_{Bx}(t) = -\dot{w}Q_x + \frac{\partial \dot{w}}{\partial y}M_{xy} + \frac{\partial \dot{w}}{\partial x}M_x \tag{4.241}$$

Substituting Equations (2.176), (2.178) and (2.183) into Equation (2.342) gives:

$$\Pi_{Bx}(t) = D \left[\frac{\partial w}{\partial t} \left(\frac{\partial^3 w}{\partial x^3} + \frac{\partial^3 w}{\partial x\, \partial y^2} \right) - (1-\nu) \frac{\partial^2 w}{\partial t\, \partial y} \frac{\partial^2 w}{\partial x\, \partial y} - \frac{\partial^2 w}{\partial t\, \partial x} \left(\frac{\partial^2 w}{\partial x^2} + \nu \frac{\partial^2 w}{\partial y^2} \right) \right] \tag{4.242}$$

where $D = Eh^3/12(1-\nu^2)$, is the plate bending stiffness per unit width. The expression for the y-component of power transmission is obtained simply by interchanging the x and y subscripts in Equation (4.242).

Using eight accelerometers to obtain the gradients in Equation (4.242), as demonstrated by Pavic (1976), and taking the time averaged result allows an expression similar in form to Equation (4.188), but much more complex, to be obtained. Because of the many accelerometers and gradient estimates required, it is very difficult to obtain accurate results. The interested reader is referred to Pavic (1976) for more details.

For the special case of harmonic wave propagation, it is possible to simplify the expression for power transmission in the x-direction. For this case, the time average of Equation (4.242) may be written as:

$$\Pi_{Bxa} = \frac{1}{2}\mathrm{Re}\left\{ \mathrm{j}\omega D \left[\hat{w} \left(\frac{\partial^3 \hat{w}^*}{\partial x^3} + \frac{\partial^3 \hat{w}^*}{\partial x\,\partial y^2} \right) \right.\right.$$
$$\left.\left. -(1-\nu)\frac{\partial \hat{w}}{\partial y}\frac{\partial^2 \hat{w}^*}{\partial x\,\partial y} - \frac{\partial \hat{w}}{\partial x}\left(\frac{\partial^2 \hat{w}^*}{\partial x^2} + \nu\frac{\partial^2 \hat{w}^*}{\partial y^2} \right) \right] \right\} \tag{4.243}$$

If a general solution is assumed for the plate equation of motion for any arbitrary plate edge boundary conditions, and if it is further assumed that the measurements will be made in the far field of any sources, the following may be written:

$$\hat{w} = X(x)Y(y) = \left(A_1 \mathrm{e}^{-\mathrm{j}k_x x} + A_2 \mathrm{e}^{\mathrm{j}k_x x} \right)\left(B_1 \mathrm{e}^{-\mathrm{j}k_y y} + B_2 \mathrm{e}^{\mathrm{j}k_y y} \right) \tag{4.244}$$

Then:

$$\frac{\partial^2 \hat{w}}{\partial x^2} = -k_x^2 \hat{w} \tag{4.245}$$

and:

$$\frac{\partial^2 \hat{w}}{\partial y^2} = -k_y^2 \hat{w} \tag{4.246}$$

and Equation (4.243) becomes:

$$\Pi_{Bxa} = -\frac{1}{2}\mathrm{Im}\left\{ \omega D \left[\hat{w} \left(-k_x^2 \frac{\partial \hat{w}^*}{\partial x} - k_y^2 \frac{\partial \hat{w}^*}{\partial x} \right) \right.\right.$$
$$\left.\left. -(1-\nu)\frac{\partial \hat{w}}{\partial y}\frac{\partial^2 \hat{w}^*}{\partial x\,\partial y} - \frac{\partial \hat{w}}{\partial x}\left(-k_x^2 \hat{w}^* - \nu k_y^2 \hat{w}^* \right) \right] \right\} \tag{4.247}$$

Using the relationship, $\mathrm{Im}\{ab^*\} = -\mathrm{Im}\{a^*b\}$, and assuming A_1, A_2, B_1, B_2, k_x and k_y are less than unity, then it can be shown that Equation (4.247) can be written approximately as:

$$\Pi_{Bxa} = \omega D k_x^2 \mathrm{Im}\left\{ \hat{w}\frac{\partial \hat{w}^*}{\partial x} \right\} \tag{4.248}$$

which is similar to Equation (4.210) for bending wave propagation in a beam.

A similar expression can be obtained for wave propagation in the y-direction. Thus, the intensity vector for bending waves in a plate can be measured by measuring the x- and y-components, then calculating the overall magnitude and direction in the usual way.

As Equation (4.248) is similar to Equation (4.210) for beams, except for the constant multiplier ($D k_x^2$ instead of $E J_{yy} k_b^2$), all of the techniques for power transmission measurement embodied in Equations (4.198), (4.201) and (4.212) to (4.215) are also valid for any particular direction on a plate. The quantities, k_x and k_y, are dependent upon the plate boundary conditions, but in many cases may be approximated as:

$$k_x^2 = k_y^2 = \omega\sqrt{\frac{\rho_m h}{D}} \tag{4.249}$$

Indeed, this is the approximation which is implicitly assumed when the two accelerometer technique is used to determine the intensity in a plate. Although the approximation embodied in Equations (4.248) and (4.249) gives good results in many cases, in general it is necessary to use eight accelerometers and evaluate all of the gradients in Equation (4.243) directly (Pavic, 1976). This needs to be done for each of the x- and y-components of intensity to obtain the overall intensity vector. Of course it is much easier and more accurate to do these measurements with a scanning laser vibrometer or a point measuring laser vibrometer to determine the amplitude and relative phase of the normal surface velocity at the number of points required to obtain accurate gradient estimates. In all of the above equations, the normal surface acceleration, a, may be replaced with $u\omega$, where u is the normal surface velocity.

The equivalence of Equations (4.243) and (4.248) can be demonstrated numerically for particular cases (Pan and Hansen, 1994).

4.4.3   Intensity Measurement in Circular Cylinders

The power transmission per unit width in a cylinder is given by Equations (4.138) and (4.139). Again, the two accelerometer method will provide good results away from vibration sources and structural discontinuities, only if bending waves are all that are present. Unfortunately, all wave types are coupled and exist on a cylinder surface, and thus the validity of the two accelerometer method in this case is open to question.

As for the plate, the derivatives to evaluate Equations (4.138) and (4.139) can be determined using the finite difference technique and eight accelerometers (or a laser vibrometer). Note that additional accelerometers would be needed to measure the in-plane displacements, ξ_x and ξ_θ.

4.4.4   Sources of Error in the Measurement of Structural Intensity

Because of the increased complexity of structural wave fields compared to acoustic fields, it is much more difficult to obtain accurate structural intensity measurements than it is to obtain accurate sound intensity measurements. Sources of error in structural intensity measurements are associated with mass loading effects of accelerometers (which may be avoided by using very small accelerometers or by using laser doppler velocimetry), the presence of wave types other than the one that is being measured, phase matching inaccuracies between the measurement channels, inaccuracies associated with the finite difference approximation, and the presence of highly reactive fields associated with sources, sinks, discontinuities and boundaries, or with the simultaneous presence of reflected and incident wave fields.

When the reactive field is small compared to the active field and when only one wave type is present, errors associated with phase mismatch between instrumentation channels are only significant at low frequencies ($\Delta/\lambda < 0.1$, where Δ is the transducer separation and λ is the structural wavelength). For the same field situation, errors associated with the use of the finite difference approximation for the derivative of the displacement result in significant errors (approximately 20%) at frequencies above $\Delta/\lambda = 0.14$. Taylor (1990) showed that this error could be made insignificant for the four accelerometer method measurement of bending wave power if the measured power were multiplied by K_{fd}, where K_{fd} is defined as:

$$K_{fd} = \frac{1}{\sin k_b \delta} \frac{k^3 \Delta^3}{2(\cos k_b \delta - \cos 3k\delta)} \qquad (4.250)$$

and where $\delta = \Delta/2$ and k_b is the bending wavenumber. For the two accelerometer measurement of bending wave power transmission, Equation (4.216) can be used.

In view of the preceding discussion, it is probably advisable to adjust Δ to be in the region of $\lambda/10$ when using either two or four accelerometers to measure the structural intensity of only one wave type in the absence of any reactive field.

TABLE 4.5 Expressions for the wavenumber for various wave types in beams and thin plates (k_L for longitudinal waves, k_s for torsional and shear waves and k_b for bending or bending waves)

Wave type	Beam expression	Plate expression
Longitudinal, $k_L =$	$\omega\left[\rho_m/E\right]^{1/2}$	$\omega\left[\rho_m(1-\nu^2)/E\right]$
Torsional and shear, $k_s =$	$\omega\left[\rho_m/G\right]^{1/2}$ (Torsional)	$\omega\left[\rho_m/G\right]^{1/2}$ (Shear)
Flexural (or bending), $k_b =$	$\left[\omega^2\rho_m S/EJ\right]^{1/4}$	$\left[\dfrac{12\omega^2 m_s(1-\nu^2)}{Eh^3}\right]^{1/4}$

Notes: k = wavenumber, ρ_m = density of beam or plate material, E, G = Young's modulus and shear modulus for the plate material, respectively, J = second moment of area of the beam cross section of area, S, ν = Poisson's ratio for the plate material, ω = frequency of excitation (rad/s), ρ_m = density of the beam material, $m_s = \rho_m h$ = mass per unit area of the plate and h is the plate thickness.

In cases where the reactive field is significant, the effects of phase errors in the instrumentation are magnified. Such a situation can exist in the near field close to a vibration source, sink or structural discontinuity, or alternatively if incident and reflected waves exist simultaneously, causing the structure to have a partial or fully modal response. In the former cases, to minimise the effects of the near field, accelerometers or laser vibrometer measurement locations must be at least half a structural wavelength away from structural discontinuities, boundaries or vibration sources. A structural wavelength for the various wave types that can propagate in beams and plates may be calculated using the relations in Table 4.5, where:

$$\lambda = \frac{2\pi}{k_{st}} \quad \text{for all wave types} \tag{4.251}$$

and where k_{st} is the structural wavenumber for the wave of interest. Unfortunately, because of the finite size of structures, it is often not possible to make measurements in the absence of near field effects and many measurement situations are characterised by highly reactive fields, making phase matching between the two measurement channels of crucial importance. Of course if a laser vibrometer is used, the phases of all measurements are referenced to a particular location chosen by the user.

In the presence of reflected waves, the measurement accuracy is also influenced by the mode shapes of the structure at the measurement location, as measurements made near a vibration antinode will be more sensitive to instrumentation phase errors than those made near a node. This can be explained by considering the spatial variation of displacement between two measurement points separated by Δ. At an antinode this variation is very small and the phase difference between the two measurements will also be very small. Thus any relative phase error due to the instrumentation in this case would have a large effect on the measured results. This effect has been quantified by Taylor (1990) for measurements on beams in the presence of reflected waves. He showed that very large power transmission errors (over 200%) can result from instrumentation phase errors of less than 0.1° for $\Delta/\lambda \leq 0.05$. As the quantity Δ/λ (λ is the structural wavelength) increases, the error reduces but it is still 20% at $\Delta/\lambda = 0.12$. It could be concluded that in the presence of reflected waves of half the amplitude or more of the incident waves that structural power transmission measurements are so inaccurate as to be useless. In other words, structural intensity or power transmission measurements will be questionable when made at frequencies corresponding to structural resonances or on structures that are lightly damped. Note also, that if the vibration of a structure can be expressed in terms of a sum of normal modes (see Chapter 5), then there will be no active power transmission unless these modes are complex; that is, unless there is some structural damping within the structure or at the boundaries. Otherwise the power per unit width will be entirely reactive.

One way of determining whether or not the presence of a reactive field will cause significant errors in the measurement of structural intensity is to measure the residual intensity in much the same way as is done for sound intensity measurements. The residual intensity must be as small as possible but its unsigned value must be well below the unsigned value of the actual intensity measurement. It can be treated like background noise in an acoustic measurement and much the same rules apply in terms of how much lower than the measurement it must be (preferably 10 dB but meaningful measurements can be made if the residual intensity is between 3 and 10 dB below the measurement). The residual intensity is determined by taking measurements along a closed line (such as a circle) that does not enclose a source or sink. The mean intensity measured along this line is the residual intensity.

Craik et al. (1995) showed that the residual intensity index for the bi-axial method of intensity measurement was slightly higher than for the two accelerometer array method.

Errors in structural power transmission measurement also arise because of the presence of wave types other than the one that is to be measured, for two reasons. First, a large portion of the total vibratory power may be missed and it may turn up in the other parts of the structure as the wave type being measured (due to energy conversion at structural discontinuities). Second, the presence of other wave types results in phase and amplitude errors in the data provided by the measurement transducers, the former becoming particularly critical in cases involving a significant reactive field.

It is interesting to note that bending wave intensity measurements made in the far field using two accelerometers appear to be more accurate than measurements made using four accelerometers, probably because the four accelerometer technique is more sensitive to relative phase errors and the finite difference approximation implicit in the finite separation distance between the accelerometers.

4.4.5   Power Transmission into Structures via a Machine Support Point or a Shaker

The vibrational power transmission through a machine support point into a supporting structure can be measured by measuring the force input, $F(t)$, and acceleration, $a(t)$, (or velocity, $u(t)$), at either the machine side of the mount or the structure side of it. In this case, the instantaneous power transmission is:

$$\Pi(t) = F(t)u(t) = F(t) \int a(t) \tag{4.252}$$

For sinusoidal excitation, the active power transmission is:

$$\Pi_a = \frac{1}{2}\mathrm{Re}\left[\hat{F}\hat{u}^*\right] \tag{4.253}$$

where:

$$u(t) = \hat{u}\mathrm{e}^{\mathrm{j}\omega t} \tag{4.254}$$

$$F(t) = \hat{F}\mathrm{e}^{\mathrm{j}\omega t} \tag{4.255}$$

The reactive component is:

$$\Pi_r = \frac{1}{2}\mathrm{Im}\left[\hat{F}\hat{u}^*\right] \tag{4.256}$$

The cross-spectral formulation for harmonic excitation is:

$$\Pi_a(\omega) = \text{Re}\,[G_{Fu}(\omega)] = \frac{1}{\omega}\text{Im}\,[G_{Fa}(\omega)] \tag{4.257}$$

For random signals, the power spectrum, $G_{Fu}(\omega)$, is replaced by the power spectral density, $G_{DFu}(\omega)$, $G_{Fa}(\omega)$ is replaced by $G_{DFa}(\omega)$ and the power becomes the power per hertz. For single frequency signals or harmonics, the power associated with each harmonic is found by multiplying Equation (4.257) by the bandwidth of each FFT filter (or the frequency resolution).

The reactive power (corresponding to stored energy) is:

$$\Pi_r(\omega) = \text{Im}\,[G_{F,u}(\omega)] = \frac{1}{\omega}\text{Re}\,[G_{F,a}(\omega)] \tag{4.258}$$

In Equations (4.257) and (4.258), a is the measured acceleration, F is the measured force and u is the measured velocity.

The active power transmission can also be measured using an ordinary sound intensity analyser, placing the force transducer signal in one channel and the acceleration transducer in the other. Equation (4.252) can be used to express the active power transmission as:

$$\Pi_a = \left\langle F(t) \int a(t)\mathrm{d}t \right\rangle \tag{4.259}$$

which for a stationary signal can be shown to be equal to:

$$\Pi_a = \left\langle \frac{F+a}{2} \int\limits_{-\infty}^{t} (F-a)\mathrm{d}\tau \right\rangle \tag{4.260}$$

which is directly proportional to the result obtained with a sound intensity analyser with the force signal input to one channel and the acceleration input to the other. The intensity analyser result would have to be multiplied by $\rho\Delta$ to give the correct absolute value for Π_a. Here, ρ is the density of air and Δ is the microphone spacing assumed by the sound intensity analyser. Equation (4.260) can be shown to be equal to Equation (4.259) by multiplying terms in Equation (4.260) and noting that $F \int a\mathrm{d}t = -a \int F\mathrm{d}t$ for stationary signals and $\langle F \int F\mathrm{d}t \rangle = \langle a \int a\mathrm{d}t \rangle = 0$.

The force and acceleration at the structure side or machine side of the mount may be measured using a piezoelectric force transducer and a piezoelectric accelerometer. However, it is rarely practical to be able to position a force transducer between a machine and its support structure. In most cases it is far easier to measure the structural input impedance at the mounting point using an electrodynamic shaker and then to calculate the power transmission from the machine to the support structure through the mounting point using:

$$\Pi_a = \frac{1}{2}\text{Re}\{Z\}|\hat{u}|^2 \tag{4.261}$$

Using a shaker attached to a structure, the structural input impedance as well as the power input by the shaker can be measured using force and acceleration transducers. Procedures for measuring the input power have already been outlined and the structural input impedance can be derived from the structural velocity and input force measurements, when the structure is excited by a shaker, using:

$$Z_{\text{in}} = F_{\text{in}}/u \tag{4.262}$$

In some cases, successful impedance measurements can be taken using an impedance head transducer which has the force and accelerometer transducers mounted together in a single unit. However, care must be taken to ensure that the accelerometer crystal is not between the force crystal and the mount (for measurements of power transmission into the mount) and not between the force crystal and the structure (for measurements of power transmission out of the mount). Otherwise the true force into the mount or into the structure will not be measured, due to the lack of stiffness of the accelerometer crystal causing a small amplitude and phase shift (see Figure 4.12). In many cases, however, it is preferable not to use an impedance head to measure the input power to a structure, but rather, to use a separate force transducer and acceleration transducer. Two suitable arrangements are shown in Figure 4.13.

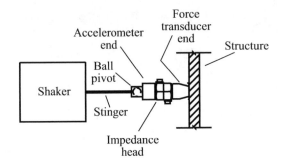

FIGURE 4.12 Correct arrangement to measure power flow into a structure using an impedance head.

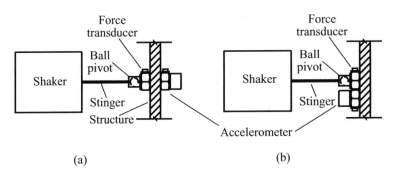

FIGURE 4.13 Arrangements for input power measurements from a shaker to a structure using separate force and acceleration transducers.

These arrangements are only suitable provided that the structure thickness in Figure 4.13(a) is much less than a structural wavelength and the distance between the centre of the force transducer and the centre of the acceleration transducer in Figure 4.13(b) is much less than a structure wavelength.

4.4.6 Power Transmission into Structures via Vibration Isolators

Prior to discussing the measurement of power transmission through vibration isolators, it will be useful to discuss the underlying principles of vibration isolation, beginning with a single-degree-of-freedom system involving a single isolator, then extending the discussion to systems with multiple isolators.

Two types of vibration-isolating applications will be considered: (1) those where the intention is to prevent transmission of vibratory forces from a machine to its foundation, and (2) those where the intention is to reduce the transmission of motion of a foundation to a device mounted on it. Rotating equipment, such as motors, fans, turbines, etc. mounted on vibration isolators are examples of the first type. An electron microscope, mounted resiliently in the basement of a hospital, is an example of the second type.

4.4.6.1 Single-Degree-of-Freedom Systems

To understand vibration isolation, it is useful to gain familiarity with the behaviour of single-degree-of-freedom systems, such as illustrated in Figure 4.14 (Church, 1963; Tse et al., 1978; Rao, 2016; Inman, 2014). In the figure, the two cases considered here are illustrated with a spring, mass and dashpot. In the first case, the mass is driven by an externally applied force, F, while in the second case, the base is assumed to move with some specified vibration displacement, y_1 (Tse et al., 1978).

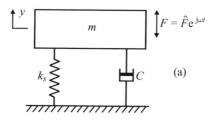

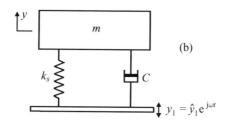

FIGURE 4.14 Single-degree-of-freedom system: (a) forced mass, rigid base; (b) vibrating base.

The equation of motion for the single-degree-of-freedom oscillator shown in Figure 4.14(a), is:

$$m\ddot{y} + C\dot{y} + k_s y = F \tag{4.263}$$

where m (kg) is the oscillator mass, C (N-s/m) is its damping constant, k_s (N/m) is its stiffness, and y (m) is the displacement of the oscillator mass, generated by the forcing function, F (N).

For sinusoidal motion, the force may be represented for convenience (see Figure 1.9 and the associated discussion) as $F = \hat{F}e^{j\omega t}$ and the resulting displacement as $y = \hat{y}e^{j\omega t}$ (m) at radian frequency, $\omega = 2\pi f$, with f the frequency in Hz. As the motion is sinusoidal, $\ddot{y} = \omega^2 y$ and $\dot{y} = j\omega y$. Equation (4.263) can then be written as:

$$-m\omega^2 y + C\omega y + k_s y = F = \hat{F}e^{j\omega t} \tag{4.264}$$

Thus:

$$\frac{y}{F} = \frac{1}{k_s - m\omega^2 + jC\omega} = \frac{(k_s - m\omega^2) - jC\omega}{(k_s - m\omega^2)^2 + (C\omega)^2} \tag{4.265}$$

where y and F are complex numbers, indicating that the displacement is not in-phase with the force, as a result of the presence of damping. Note that $y/F = \hat{y}/\hat{F}$. The modulus (or amplitude) of the complex ratio, $|y/F|$, may be written as:

$$\left|\frac{y}{F}\right| = \sqrt{\frac{(k_s - m\omega^2)^2 + (C\omega)^2}{[(k_s - m\omega^2)^2 + (C\omega)^2]^2}} = \left[(k_s - m\omega^2)^2 + (C\omega)^2\right]^{-1/2} \tag{4.266}$$

In the absence of any excitation force, F, or damping, C, the system, once disturbed, will vibrate sinusoidally at a constant amplitude (dependent on the amplitude of the original disturbance) at its undamped resonance frequency, $f_0 = \omega_0/(2\pi)$. Solution of Equation (4.263) with $F = C = 0$ gives for the undamped resonance frequency:

$$f_0 = \frac{1}{2\pi}\sqrt{\frac{k_s}{m}} \quad \text{(Hz)} \tag{4.267}$$

The static deflection, d, of the mass supported by the spring is given by $d = mg/k_s$, where g is the acceleration due to gravity, so that Equation (4.267) may be written in the following alternative form:

$$f_0 = \frac{1}{2\pi}\sqrt{\frac{g}{d}} \quad \text{(Hz)} \tag{4.268}$$

Substitution of the value of g equal to 9.81 m/s gives the following useful equation (where d is in metres):

$$f_0 = 0.5/\sqrt{d} \quad \text{(Hz)} \tag{4.269}$$

The preceding analysis is for an ideal system in which the spring has no mass, which does not reflect the actual situation. If the mass of the spring is denoted m_s, and it is uniformly distributed along its length, it is possible to get a first order approximation of its effect on the resonance frequency of the mass-spring system by using Rayleigh's method and setting the maximum kinetic energy of the mass, m, plus the spring mass, m_s, equal to the maximum potential energy of the spring. The velocity of the spring is zero at one end and a maximum of $\dot{y} = \omega y$ at the other end. Thus, the kinetic energy in the spring may be written as:

$$E_{ks} = \frac{1}{2}\int_0^L u_m^2 \mathrm{d}m_s \tag{4.270}$$

where u_m is the velocity of the segment of spring of mass, $\mathrm{d}m_s$, and L is the length of the spring. The quantities, u_m and $\mathrm{d}m_s$, may be written as:

$$u_m = \frac{x\dot{y}}{L} \text{ and } \mathrm{d}m_s = \frac{m_s}{L}\mathrm{d}x \tag{4.271}$$

where x is the distance from the spring support to segment, $\mathrm{d}m_s$. Thus, the KE in the spring may be written as:

$$E_{ks} = \frac{1}{2}\int_0^L \left(\frac{x\dot{y}}{L}\right)^2 \frac{m_s}{L}\mathrm{d}x = \frac{m_s\dot{y}^2}{2L^3}\int_0^L x^2\mathrm{d}x = \frac{1}{2}\frac{m_s\dot{y}^2}{3} \tag{4.272}$$

Equating the maximum KE in the mass, m, and spring, with the maximum PE in the spring gives:

$$\frac{1}{2}\frac{m_s}{3}\dot{y}^2 + \frac{1}{2}m\dot{y}^2 = \frac{1}{2}k_s y^2 \tag{4.273}$$

Substituting $\dot{y} = \omega y$ in the above equation gives the resonance frequency as:

$$f_0 = \frac{1}{2\pi}\sqrt{\frac{k_s}{m + (m_s/3)}} \tag{4.274}$$

Thus, according to Equation (4.274), more accurate results will be obtained if the suspended mass is increased by one-third of the spring mass.

The mass, m_s, of the spring is the mass of the active coils, which, for a spring with flattened ends, is two less than the total number of coils. Alternatively, the number of active coils is equal to the number that are free to move plus 0.5. For a coil spring of overall diameter, D, wire diameter, d, and with n_C active coils of material density ρ_m, the mass is:

$$m_s = n_C \frac{\pi d^2}{4} \pi D \rho_m \qquad (4.275)$$

For a coil spring with a helix angle, α (usually just a few degrees), the stiffness (N/m) or the number of Newtons required to stretch it by 1 metre is:

$$k_s = \frac{d^4 \cos \alpha}{8 n_C D^3} \left[\frac{\cos^2 \alpha}{G} + \frac{2 \sin^2 \alpha}{E} \right]^{-1} \qquad (4.276)$$

where E is the modulus of elasticity (Young's modulus) of the spring material, $G = E/[2(1+\nu)]$ is the modulus of rigidity (or shear modulus) and ν is Poisson's ratio (=0.3 for spring steel).

Of critical importance to the response of the systems shown in Figure 4.14 is the damping ratio, $\zeta = C/C_c$, where C_c is the critical damping coefficient defined as:

$$C_c = 2\sqrt{k_s m} \qquad \text{(kg/s)} \qquad (4.277)$$

When the damping ratio is less than unity, the transient response is cyclic, but when the damping ratio is unity or greater, the system transient response ceases to be cyclic.

In the absence of any excitation force, F, but including damping, $C < 1$, the system of Figure 4.14, once disturbed, will oscillate approximately sinusoidally at its damped resonance frequency, f_d. Solution of Equation (4.263) with $F = 0$ and $C \neq 0$ gives for the damped resonance frequency:

$$f_d = f_0 \sqrt{1 - \zeta^2} \qquad \text{(Hz)} \qquad (4.278)$$

With a periodic excitation force, $F = \hat{F} e^{j\omega t}$, the system of Figure 4.14 will respond sinusoidally at the driving frequency $\omega = 2\pi f$. Let $f/f_0 = X$, then Equation (4.266) can be rewritten as:

$$\frac{|y|}{|F|} = \left| \frac{y}{F} \right| = \frac{1}{k_s} \left[\left(1 - X^2\right)^2 + 4\zeta^2 X^2 \right]^{-1/2} \qquad (4.279)$$

The frequency of maximum displacement, which is obtained by differentiation of Equation (4.263), is:

$$f_{\text{maxdis}} = f_0 \sqrt{1 - 2\zeta^2} \qquad (4.280)$$

The amplitude of velocity $|\dot{y}| = 2\pi f |y|$ is obtained by differentiation of Equation (4.279), and is written as:

$$\frac{|\dot{y}|}{|F|} = \left| \frac{\dot{y}}{F} \right| = \frac{1}{\sqrt{k_s m}} \left[\left(\frac{1}{X} - X \right)^2 + 4\zeta^2 \right]^{-1/2} \qquad (4.281)$$

Inspection of Equation (4.281) shows that the frequency of maximum velocity amplitude is the undamped resonance frequency:

$$f_{\text{maxvel}} = f_0 \qquad (4.282)$$

Similarly, it may be shown that the frequency of maximum acceleration amplitude is:

$$f_{\text{maxacc}} = \frac{f_0}{\sqrt{1 - 2\zeta^2}} \qquad (4.283)$$

Alternatively, if the structure represented by Figure 4.14 is hysteretically damped, which in practice is the more usual case, then the viscous damping model is inappropriate. This case may be investigated by setting $C = 0$ and replacing k_s in Equation (4.263) with complex $k_s(1 + j\eta)$, where η is the structural loss factor. Solution of Equation (4.263) with these modifications gives for the displacement amplitude, $|y'|$, of the hysteretically damped system:

$$\frac{|y'|}{|F|} = \frac{1}{k_s} \left[\left(1 - X^2\right)^2 + \eta^2 \right]^{-1/2} \qquad (4.284)$$

For the case of hysteretic (or structural) damping the frequency of maximum displacement occurs at the undamped resonance frequency of the system, as shown by inspection of Equation (4.284):

$$f'_{\text{maxdis}} = f_0 \qquad (4.285)$$

Similarly the frequencies of maximum velocity and maximum acceleration for the case of hysteretic damping may be determined.

The preceding analysis shows clearly that maximum response depends on what is measured and the nature of the damping in the system under investigation. Where the nature of the damping is known, the undamped resonance frequency and the damping constant may be determined using appropriate equations; however, in general where damping is significant, resonance frequencies can only be determined by curve fitting frequency response data (see Section 5.4.4). Alternatively, for small damping, the various frequencies of maximum response are essentially all equal to the undamped frequency of resonance.

Referring to Figure 4.14(a), the fraction of the exciting force, F, acting on the mass, m, which is transmitted through the spring to the support is of interest. Alternatively, referring to Figure 4.14(b), the fraction of the displacement of the base, which is transmitted to the mass, is often of greater interest. Either may be expressed in terms of the transmissibility, T_F, which in Figure 4.14(a) is the ratio of the amplitude of the force transmitted to the foundation ($|F_f| = |k_s y + j\omega C y|$) to the amplitude of the exciting force, $|F|$, acting on the machine, and in Figure 4.14(b) it is the ratio of the displacement of the machine to the displacement of the foundation. Using Equations (4.265) and (4.279), we can write:

$$T_F = \frac{|F_f|}{|F|} = \frac{|k_s y + j\omega C y|}{|y(k_s - m\omega^2 + jC\omega|)} = \sqrt{\frac{1 + (2\zeta X)^2}{(1 - X^2)^2 + (2\zeta X)^2}} \qquad (4.286)$$

Figure 4.15 shows the fraction, expressed in terms of the transmissibility, T_F, of the exciting force (system (a) of Figure 4.14) transmitted from the vibrating body through the isolating spring to the support structure. The transmissibility is shown for various values of the damping ratio, ζ, as a function of the ratio of the frequency of the vibratory force to the resonance frequency of the system.

When the transmissibility is identified with the *displacement* of the mass, m, of the system illustrated in Figure 4.14(b), then Figure 4.15 shows the fraction, T_D, of the exciting displacement amplitude transmitted from the base through the isolating spring to the supported mass, m. The figure allows determination of the effectiveness of the isolation system for a single-degree-of-freedom system.

The vibration amplitude of a single-degree-of-freedom system is dependent on its mass, stiffness and damping characteristics as well as the amplitude of the exciting force. This conclusion can be extended to apply to multi-degree-of-freedom systems, such as machines and structures. Consideration of Equation (4.286) shows that as X tends to zero, the force transmissibility, T_F, tends to one; the response is controlled by the stiffness, k_s. When X is approximately one, the force transmissibility is approximately inversely proportional to the damping ratio; the response is controlled by the damping, C. As X tends to large values, the force transmissibility tends to zero as the square of X; the response is controlled by the mass, m.

The energy transmissibility, T_E, is related to the force transmissibility, T_F, and displacement transmissibility, T_D, by $T_E = T_F T_D$. As $T_F = T_D$, then $T_E = T_F^2$. The energy transmissibility,

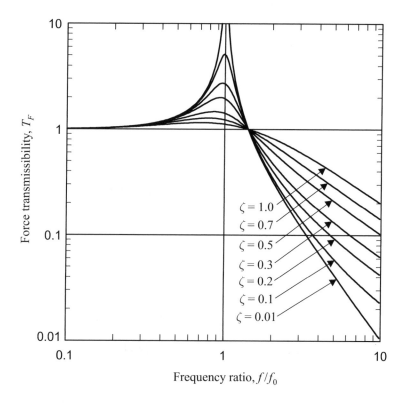

FIGURE 4.15 Force or displacement transmissibility of a viscously damped mass-spring system. The quantities, f and f_0, are the excitation and undamped mass-spring resonance frequencies, respectively, ζ is the system damping ratio and T_F is the fraction of excitation force transmitted by the spring to the foundation. Note that for values of frequency ratio greater than 2, the force transmissibility increases with increasing damping ratio.

T_E, can be related to the expected increase or decrease, ΔL_W, in sound power radiated by the supported structure over that radiated when the vibrating mass is rigidly attached to the support structure as:

$$\Delta L_W = 10 \log_{10} T_E = 20 \log_{10} T_F \qquad (4.287)$$

Differentiation of Equation (4.286) or use of Equation (4.280) gives for the frequency of maximum force transmissibility for a viscously damped system:

$$f_F = f_0 \sqrt{1 - 2\zeta^2} \qquad \text{(Hz)} \qquad (4.288)$$

The preceding equations and figures refer to viscous damping (where the damping force is proportional to the vibration velocity), as opposed to hysteretic or structural damping (where the damping force is proportional to the vibration displacement). Generally, the effects of hysteretic damping are similar to those of viscous damping up to frequencies of $f = 10 f_0$. Above this frequency, hysteretic damping results in larger transmission factors than shown in Figure 4.15.

Referring to Figure 4.15, it can be seen that below resonance (ratio of unity on the horizontal axis) the force transmission is greater than unity and no isolation is achieved. In practice, the amplification obtained below a frequency ratio of 0.5 is rarely of significance so that, although no benefit is obtained from the isolation at these low frequencies, no significant detrimental effect is experienced either. However, in the frequency ratio range 0.5–1.4, the presence of isolators significantly increases the transmitted force and the amplitude of motion of the mounted body.

In operation, this range is to be avoided. Above a frequency ratio of 1.4 the force transmitted by the isolators is less than that transmitted with no isolators, resulting in the isolation of vibration; the higher the frequency the greater the isolation. Thus, for an isolator to be successful, its stiffness must be such that the mounted resonance frequency is less than 0.7 times the minimum forcing frequency.

All practical isolators have some damping, and Figure 4.15 shows the effect of damping; increasing the damping decreases the isolation achieved. For best isolation, no damping would be desirable. On the other hand, damping is necessary for installations involving rotating equipment because the equipment rotational speed (and hence forcing frequency) will pass through the mounted resonance frequency on shutdown and start-up. In these cases, the amplitude of the transmitted force will exceed the exciting force and indeed could build up to an alarming level.

Sometimes the rotational speed of the equipment can be accelerated or decelerated rapidly enough to pass through the region of resonance so quickly that the amplitude of the transmitted force does not have time to build up to the steady-state levels indicated by Figure 4.15. However, in some cases, the rotational speed can only be accelerated slowly through the resonant range, resulting in a potentially disastrous situation if the isolator damping is inadequate. In this case, the required amount of damping could be large; a damping ratio $\zeta = 0.5$ could be required.

An external damper can be installed to accomplish the necessary damping, but always at the expense of reduced isolation at higher frequencies. An alternative to using highly damped isolators is to use rubber snubbers to limit excessive motion of the machine at resonance. Snubbers can also be used to limit excessive motion. These have the advantage of not limiting high-frequency isolation.

4.4.6.2 Surging in Coil Springs

Surging in coil springs is a phenomenon where high-frequency transmission occurs at frequencies corresponding to the resonance frequencies of wave motion in the coils. This limits the high-frequency performance of such springs and in practical applications rubber inserts above or below the spring are used to minimise the effect.

4.4.6.3 Four-Isolator Systems

In most practical situations, more than one isolator is used to isolate a particular machine. This immediately introduces the problem of more than one system resonance frequency at which the force transmission will be large. If possible, it is desirable to design the isolators so that none of the resonance frequencies of the isolated system correspond to any of the forcing frequencies.

The most common example of a multi-degree-of-freedom system is a machine mounted symmetrically on four isolators (Crede, 1965). In general, a machine, or body mounted on springs, has six degrees of freedom. There will be one vertical translational mode of resonance frequency, f_0, one rotational mode about the vertical axis and two rocking modes in each vertical plane, as illustrated in Figure 4.16. The calculation of resonance frequencies for such a system in terms of the resonance frequency, f_0, will now be considered. The latter frequency may be calculated using either Equation (4.267) or (4.268), as for a single-degree-of-freedom system, with one spring having the combined stiffness of the four shown in Figure 4.16. Note that stiffnesses add linearly when springs are in parallel; that is, $k_s = k_{s1} + k_{s2}$, etc.

Rocking and horizontal mode resonance frequencies may be determined by reference to Figure 4.17. The resonance frequencies, f_a and f_b, for roll and horizontal motion are given, respectively, in parts (a) and (b) of the figure. The parameters in these figures are defined as: $W = (\delta/b)\sqrt{(k_{sx}/k_{sy})}$, $M = a/\delta$ and $\Omega = (\delta/b)(f_i/f_0)$, where the subscript $i = a$ in Figure 4.17(a) and $i = b$ in Figure 4.17(b). k_{sx} and k_{sy} are the isolator stiffnesses in the x- and y-directions, respectively, and δ is the radius of gyration for rotation about the horizontal z-axis through the centre of gravity (see Figure 4.16, where the dimensions, a and b, are also defined).

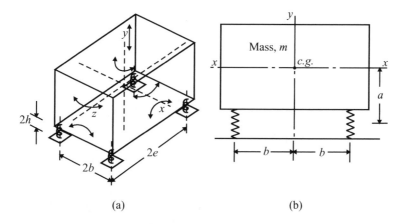

(a)   (b)

FIGURE 4.16 Vibration modes for a machine mounted on four isolators. The origin of the co-ordinates is coincident with the assumed centre of gravity (c.g.) at height $a+h$ above the mounting plane.

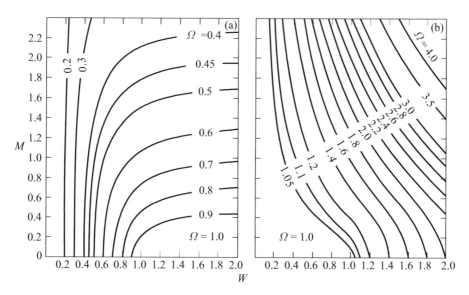

FIGURE 4.17 Charts for determining roots, Ω, of the characteristic equation: $\Omega^4 - \Omega^2(1+W^2 + M^2W^2) + W^2 = 0$.

For motion in the orthogonal vertical plane, the same figures (4.17(a) and (b)) are used, with the quantities, x and b, replaced with z and e, respectively (see Figure 4.16), and with δ now the radius of gyration for rotation about the x-axis.

The resonance frequency of the rotational vibration mode about the vertical y-axis is:

$$f_y = \frac{1}{\pi}\sqrt{(b^2 k_{sz} + e^2 k_{sx})/I_y} \quad \text{(Hz)} \qquad (4.289)$$

The quantities, $2b$ and $2e$, are the distances between centrelines of the support springs, k_{sz} is the isolator stiffness in the z-direction, usually equal to k_{sx}, and I_y is the moment of inertia of the body about the y-axis.

Values for the stiffnesses, k_{sx}, k_{sy} and k_{sz}, are usually available from the isolator manufacturer. Note that for rubber products, static and dynamic stiffnesses are often different. It is the dynamic stiffnesses that are required here. For a rectangular cross section of dimensions $2d \times 2q$, the radius of gyration, δ, about an axis through the centre and perpendicular to the plane of the section is:

$$\delta = \sqrt{(d^2 + q^2)/3} \qquad (4.290)$$

When placing vibration isolators beneath a machine, it is good practice to use identical isolators and to place them symmetrically with respect to the centre of gravity of the machine. This results in equal loading and deflection of the isolators.

The calculation of the force transmission for a multi-degree-of-freedom system is complex and not usually contemplated in conventional isolator design. However, the analysis of various multi-degree-of-freedom systems has been discussed in the literature (Mustin, 1968; Smollen, 1966). Generally, for a multi-degree-of-freedom system, good isolation is achieved if the frequencies of all the resonant modes are less than about two-fifths of the frequency of the exciting force. However, a force or torque may not excite all the normal modes, and then the natural frequencies of the modes that are not excited do not need to be considered, except to ensure that they do not actually coincide with the forcing frequency.

4.4.6.4   Two-Stage Vibration Isolation

Two-stage vibration isolation is used when the performance of single-stage isolation is inadequate and it is not practical to use a single-stage system with a lower resonance frequency. As an example, two-stage isolation systems have been used to isolate diesel engines from the hull of large submarines.

A two-stage isolator is illustrated in Figure 4.18, where the machine to be isolated is represented as mass, m_2, and the intermediate mass is represented as mass, m_1. The intermediate mass should be as large as possible, but should be at least 70% of the machine mass being supported.

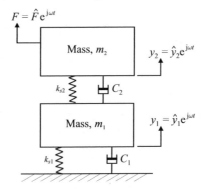

FIGURE 4.18 Two-stage vibration isolator.

The equations of motion of the masses, m_1 and m_2, in Figure 4.18 may be written as:

$$m_1 \ddot{y}_1 + C_1 \dot{y}_1 + k_{s1} y_1 - C_2 (\dot{y}_2 - \dot{y}_1) - k_{s2} (y_2 - y_1) = 0 \qquad (4.291)$$

and:

$$m_2 \ddot{y}_2 + C_2 (\dot{y}_2 - \dot{y}_1) + k_{s2} (y_2 - y_1) = F \qquad (4.292)$$

These equations can be solved to give the complex displacements of each mass as:

$$\frac{y_1}{F} = \frac{k_{s2} + j\omega C_2}{(k_{s1} + k_{s2} - \omega^2 m_1 + j\omega C_1 + j\omega C_2)(k_{s2} - \omega^2 m_2 + j\omega C_2) - (k_{s2} + j\omega C_2)^2} \qquad (4.293)$$

and:

$$\frac{y_2}{F} = \frac{k_{s1} + k_{s2} - \omega^2 m_1 + j\omega C_1 + j\omega C_2}{(k_{s1} + k_{s2} - \omega^2 m_1 + j\omega C_1 + j\omega C_2)(k_{s2} - \omega^2 m_2 + j\omega C_2) - (k_{s2} + j\omega C_2)^2} \quad (4.294)$$

The complex force transmitted to the foundation is:

$$F_T = y_1(k_{s1} + j\omega C_1) \quad (4.295)$$

and thus the transmissibility, $T_F = |F_T/F|$, is:

$$T_F = \left| \frac{(k_{s1} + j\omega C_1)(k_{s2} + j\omega C_2)}{(k_{s1} + k_{s2} - \omega^2 m_1 + j\omega C_1 + j\omega C_2)(k_{s2} - j\omega^2 m_2 + j\omega C_2) - (k_{s2} + j\omega C_2)^2} \right| \quad (4.296)$$

The damping constants, C_1 and C_2, are found by multiplying the damping ratios, ζ_1 and ζ_2, by the critical damping, C_{c1} and C_{c2}, given by Equation (4.277), using stiffnesses, k_{s1} and k_{s2}, and masses, m_1 and m_2, respectively.

As a two-stage isolation system has two degrees of freedom, it will have two resonance frequencies corresponding to high force transmissibility. The undamped resonance frequencies of the two-stage isolator may be calculated using (Muster and Plunkett, 1988; Ungar and Zapfe, 2006):

$$\left(\frac{f_a}{f_0}\right)^2 = Q - \sqrt{Q^2 - B^2} \quad \text{and} \quad \left(\frac{f_b}{f_0}\right)^2 = Q + \sqrt{Q^2 - B^2} \quad (4.297)$$

where:

$$Q = 0.5\left(B^2 + 1 + \frac{k_{s1}}{k_{s2}}\right) \quad (4.298)$$

and:

$$B = \frac{f_1}{f_0} \quad (4.299)$$

$$f_1 = \frac{1}{2\pi}\sqrt{\frac{k_{s1} + k_{s2}}{m_1}} \quad (4.300)$$

$$f_0 = \frac{1}{2\pi}\sqrt{\frac{k_{s1} k_{s2}}{m_2(k_{s1} + k_{s2})}} \quad (4.301)$$

The quantity, f_1, is the resonance frequency of mass, m_1, with mass, m_2, held fixed and f_0 is the resonance frequency of the single-degree-of-freedom system with mass, m_1, removed. The upper resonance frequency, f_b, of the combined system is always greater than either f_1 or f_0, while the lower frequency is less than either f_1 or f_0.

At frequencies above twice the second resonance frequency, f_b, the force transmissibility for an undamped system will be approximately equal to $(f^2/(f_1 f_0))^2$, proportional to the fourth power of the excitation frequency, compared to a single-stage isolator, for which it is approximately $(f/f_0)^2$ above twice the resonance frequency, f_0.

In Figure 4.19, the force transmissibility for a two-stage isolator for a range of ratios of masses and stiffnesses is plotted for the special case where $\zeta_1 = \zeta_2$.

4.4.6.5 Measurement of Power Transmission through a Vibration Isolator

For a spring isolator separating a machine from a structure, the axial force transmitted through the isolator to the structure is given by the product of the difference in displacement between the two ends of the isolator and the spring constant. Thus, the real or active power transmitted through the isolator to the support structure will be:

$$\Pi_a = k \langle u_2(w_1 - w_2) \rangle \quad (4.302)$$

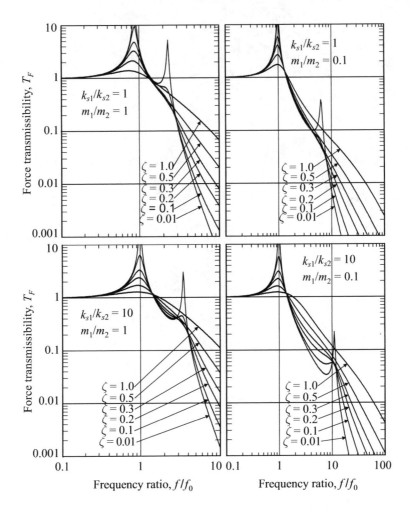

FIGURE 4.19 Force transmissibility for a two-stage vibration isolation system for various values of stiffness and mass ratio. In all figures, $\zeta_1 = \zeta_2 = \zeta$.

where k is the isolator axial stiffness, u_2 is the velocity of the base of the isolator and w_1 and w_2 are, respectively, the displacements of the top and base of the isolator.

If the supporting structure is much more rigid than the isolator, then the difference, $(w_1 - w_2)$, can be approximated simply with the displacement, w_2, of the equipment side of the isolator (Pavic, 1977). Thus:

$$\Pi_a = k \langle u_2 w_1 \rangle_t \tag{4.303}$$

Equation (4.303) is only valid for frequencies below the first resonance frequency of the isolation system. At frequencies around and above the first resonance frequency, frequency dependent corrections are necessary. If the vibration field can be considered stationary, Equation (4.303) is equivalent to Equation (4.80) where acoustic pressure has been replaced by structural displacement. Thus, below the first resonance frequency, where the power transmission into the isolator equals the power transmission out, and the measurements are made using accelerometers, Equation (4.303) can be written as:

$$\Pi_a(\omega) = \frac{k}{2\omega^2} \left\langle (a_2 + a_1) \int\limits_{-\infty}^{t} (a_1 - a_2) \mathrm{d}\tau \right\rangle_t \tag{4.304}$$

which is directly proportional to the result obtained by placing the two accelerometer signals into the two channels of a sound intensity analyser. The calibration (or proportionality) constant is $\dfrac{\rho \Delta k}{\omega^2}$, where Δ is the microphone spacing assumed by the intensity analyser.

Alternatively, if a spectrum analyser rather than an intensity analyser is available, the power transmission can be written as:

$$\Pi_a = \frac{k}{\omega^3} \operatorname{Im} \left[G_{a_1,a_2}(\omega) \right] \tag{4.305}$$

5

Modal Analysis

5.1 Introduction

In simple terms, modal analysis is the process (analytical or experimental or both) that determines the properties of the normal modes that describe the dynamic response of a structure. Each normal mode of vibration is characterised in terms of a resonance frequency, mode shape and damping, much like the simple single-degree-of-freedom oscillator (see Section 5.2.1). The overall response of a structure to a specified excitation force over a specified frequency range can be determined by summing the contributions of each mode at each specific frequency of interest in the frequency range under consideration. When summing the contributions at specific frequencies, relative phases must be taken into account and it should be remembered that a vibration mode can contribute to the structural response at frequencies other than its resonance frequency. It also should be noted that in order to obtain accurate results for the overall structural or acoustic response, it is necessary to include in the sum, modes with resonance frequencies up to twice the upper frequency of interest in the analysis and down to 50% of the lowest frequency of interest in the analysis.

More specifically, modal analysis is the process of analytically or experimentally determining the dynamic properties (resonance frequencies, damping and mode shapes) of a system made up of a number of particles (or point masses) connected together in some way by a number of massless stiffnesses. Such a system may also be a continuous elastic structure that has been idealised as a number of point masses by use of some discretisation procedure such as finite element analysis. Modal properties can also be determined by measuring the structural response to a known excitation. Once estimated, the modal model of a structure can be used, together with an estimate or measure of the system damping, to calculate the response of the system to any other applied forces, which may be point forces, distributed forces, point moments or distributed moments. When applied to an elastic structure, modal analysis results in complete representation of the dynamic properties of a structure in terms of its modes of vibration. Modes of vibration are solutions to an eigenvalue problem, as formulated from the differential equations that describe the motion of the structure or system of particles. The modal parameters (mode shapes and resonance frequencies) can be found either analytically or experimentally, while the modal damping parameter can only be determined experimentally. It should be remembered that both analytical and experimental modal analyses are generally only practical for identifying the lowest order (first 20 or 30) vibration modes of a structure.

Very simple continuous structures such as beams, plates and cylinders are also amenable to analysis from first principles, by using the appropriate wave equation, together with suitable boundary conditions, to determine resonance frequencies and mode shapes. Examples of this type of analysis were discussed in Chapter 2.

In this chapter, the principles of analysis of discretised systems will be developed, and the results obtained for the system response at a particular location (or locations) to a particular force or forces as a function of system resonance frequencies, modal damping and mode shapes will also be applicable to the simple systems analysed using the wave equation, and indeed should give identical results, provided that sufficient modes are used in the determination of the structural response. However, the modal or discretised approach discussed here is practical to use for any structure, whereas the exact approach is only suitable for simple structures and even for many of those, the resulting equations are extremely complex and difficult to evaluate.

5.2   Modal Analysis: Analytical

5.2.1   Single-Degree-of-Freedom System with Viscous Damping

The purpose of this section is to develop the ideas and principles that will be of use in the analysis of multi-degree-of-freedom discrete systems.

The nomenclature used here has been adjusted to be consistent with the structural mechanics nomenclature used in Chapter 2. Thus, displacement is denoted w, and the corresponding vector location on a structure is generally $\boldsymbol{x}$ and for a specific location, it is $\boldsymbol{x}_i$.

An idealised single-degree-of-freedom system may be described as one that consists of a mass, not free to rotate and free to move in only one direction. The mass is attached to a rigid boundary through a massless spring and a massless viscous damper, as shown in Figure 5.1. In terms of dynamics, Figure 5.1 is similar to Figure 4.14, except that the motion is in the horizontal w-direction instead of the vertical y-direction.

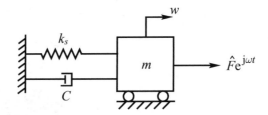

FIGURE 5.1 Single-degree-of-freedom system.

Using Newton's laws, the equation of motion for this system may be written as:

$$m\ddot{w} + C\dot{w} + k_s w = F = \hat{F}e^{j\omega t} \tag{5.1}$$

For harmonic motion, $w = \hat{w}e^{j\omega t}$ and Equation (5.1) may be written as:

$$-m\omega^2\hat{w} + j\omega C\hat{w} + k_s\hat{w} = F = \hat{F}e^{j\omega t} \tag{5.2}$$

where ω is the frequency of oscillation in rad/s.

If $C = F = 0$, Equation (5.2) may be solved to give the undamped natural frequency of the system as:

$$\omega_n = \sqrt{\frac{k_s}{m}} \quad (\text{rad/s}) \tag{5.3}$$

A single-degree-of-freedom system is critically damped if, after removal of an exciting force, its motion decays rapidly to zero, with no oscillation about a mean value. This critical damping value, C_c, for a SDOF system is:

$$C_c = 2\sqrt{km} \tag{5.4}$$

The critical damping ratio, ζ, of a SDOF system is defined as the ratio of the damping, C, to the critical damping, C_c, and is:

$$\zeta = \frac{C}{2\sqrt{k_s m}} \tag{5.5}$$

Substituting $s = j\omega$ into Equation (5.2), using Equations (5.3) and (5.5) and rearranging gives:

$$s^2 + 2\zeta\omega_n s + \omega_n^2 = 0 \tag{5.6}$$

When $\zeta < 1$ (as in most practical systems) the solutions to Equation (5.6) are two complex conjugates:

$$s = -\zeta\omega_n \pm j\omega_n\sqrt{1 - \zeta^2} \tag{5.7}$$

Thus, the natural frequency of vibration is complex, with an imaginary part, $\omega_n\sqrt{1 - \zeta^2}$, representing the oscillating response, and a real part, $\zeta\omega_n$, representing the decaying response of the system. As discussed in Section 4.4.6.1, $\omega_n\sqrt{1 - \zeta^2}$, is often referred to as ω_d, the damped natural frequency, although the frequency at which the system displacement will be a maximum (for an input force which is constant with frequency) is $\omega_n\sqrt{1 - 2\zeta^2}$. It is interesting to note that the frequency at which the system velocity will be a maximum is ω_n and the frequency of maximum acceleration is $\omega_n[1 - 2\zeta^2]^{-1/2}$.

The solution for the motion, w, is $w = \hat{w}e^{st}$, or:

$$w = \hat{w}e^{-\zeta\omega_n t}e^{j(\omega_n\sqrt{1-\zeta^2})t} \tag{5.8}$$

which represents a cyclic decaying motion, as shown in Figure 5.2.

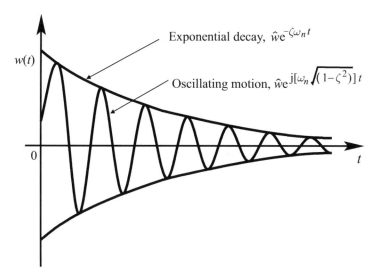

FIGURE 5.2 Free vibration characteristics of a damped single-degree-of-freedom system.

If the forcing function, F, is now non-zero and equal to $\hat{F}e^{j\omega t}$, it can be shown that the response of the damped system is:

$$w(t) = \frac{\hat{F}e^{j\omega t}}{(k_s - \omega^2) + j\omega C} \tag{5.9}$$

The receptance function, defined as w/F, is then:

$$\frac{w}{F} = \alpha(\omega) = \frac{1}{(k_s - \omega^2 m) + j\omega C} = \frac{1}{m\omega_n^2\left[1 - (\omega/\omega_n)^2 + 2j(\omega/\omega_n)\zeta\right]} \tag{5.10}$$

The quantity, w/F, is dependent upon the frequency, ω, of the excitation and is known as the receptance frequency response function, because w represents displacement amplitude of the point mass. If the velocity, u $(= w\omega)$, were used instead of the displacement, then u/F would be the mobility frequency response function.

5.2.2 Single-Degree-of-Freedom System with Hysteretic Damping

Up to now, the type of damping that has been considered has been viscous; that is, the damping force is proportional to the velocity of the point mass. Although this type of damping is representative of acoustic spaces, it is not very representative of structures, where it is often observed that damping varies at a rate approximately proportional to frequency. Structural damping is more accurately represented as hysteretic damping, which is frequency dependent and of the form, $h = C\omega$, for which the equation of motion for a single-degree-of-freedom (SDOF) system may be written as:

$$m\ddot{w} + k(1 + \mathrm{j}\eta)w = F = \hat{F}\mathrm{e}^{\mathrm{j}\omega t} \tag{5.11}$$

If the excitation force, F, is equal to 0, and we assume a solution:

$$w = \hat{w}\mathrm{e}^{\mathrm{j}\omega'_n t} \tag{5.12}$$

then Equation (5.11) can be rewritten as:

$$-(\omega'_n)^2 + \omega_n^2(1 + \mathrm{j}\eta) = 0 \tag{5.13}$$

and the damped resonance frequency, ω'_n, is then:

$$\omega'_n = \omega_n\sqrt{(1 + \mathrm{j}\eta)} \tag{5.14}$$

When the excitation force, F, is non-zero, the receptance function is:

$$\frac{w}{F} = \alpha(\omega) = \frac{1}{(k_s - \omega^2 m) + \mathrm{j}h} = \frac{1}{m\omega_n^2[1 - (\omega/\omega_n)^2 + \mathrm{j}\eta]} \tag{5.15}$$

where $\eta = h/k_s$ is the structural loss factor.

From Equation (5.15), it may be concluded that the maximum displacement occurs at a frequency of $\omega = \omega_n$, and the maximum velocity occurs at a frequency of $\omega = \omega_n\sqrt{(1 + \eta^2)}$ (assuming a forcing function that is constant with frequency).

It can be seen from Equations (5.10) and (5.15) that the phase between the excitation force and displacement changes by 180° as the frequency of the excitation passes from below resonance to above resonance. For an undamped system, the phase changes instantaneously at the resonance frequency and as the damping becomes larger, the phase change becomes less sudden.

Various damping measures and methods for measuring them are discussed in detail in Section 6.6.

5.2.3 Multi-Degree-of-Freedom Systems

This section and following parts of this text are concerned with structures having many degrees of freedom (or modes of vibration), and their analysis is dependent upon the use and manipulation of matrices and vectors in a general way. To help visualise these analysis processes, a two-degree-of-freedom system will be used as an example, although the general equations and solutions will apply to systems with any number of degrees of freedom.

Consider the damped two-degree-of-freedom system shown in Figure 5.3.

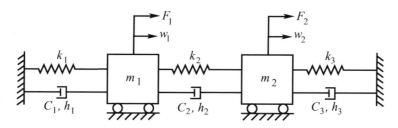

FIGURE 5.3 Damped two-degree-of-freedom system.

The type of damping for the system illustrated in Figure 5.3 has not been specified but may be viscous or structural (hysteretic). For the case of viscous damping, the two equations of motion are:

$$\left.\begin{array}{l} m_1\ddot{w}_1 + (C_1 + C_2)\dot{w}_1 + (k_1 + k_2)w_1 - C_2\dot{w}_2 - k_2w_2 = F_1(t) \\[2mm] -C_2\dot{w}_1 - k_2w_1 + m_2\ddot{w}_2 + (C_2 + C_3)\dot{w}_2 + (k_2 + k_3)w_2 = F_2(t) \end{array}\right\} \quad (5.16)$$

where $F_1(t) = \hat{F}_1 e^{j\omega t}$ and $F_2(t) = \hat{F}_2 e^{j\omega t}$. As the forcing function is harmonic at frequency, ω (rad/s), all variables in the equations to follow are functions of ω.

In matrix form, Equation (5.16) may be written as:

$$\begin{bmatrix} m_1 & 0 \\ 0 & m_2 \end{bmatrix}\begin{bmatrix} \ddot{w}_1 \\ \ddot{w}_2 \end{bmatrix} + \begin{bmatrix} C_1 + C_2 & -C_2 \\ -C_2 & C_1 + C_2 \end{bmatrix}\begin{bmatrix} \dot{w}_1 \\ \dot{w}_2 \end{bmatrix}$$

$$+ \begin{bmatrix} k_1 + k_2 & -k_2 \\ -k_2 & k_1 + k_2 \end{bmatrix}\begin{bmatrix} w_1 \\ w_2 \end{bmatrix} = \begin{bmatrix} F_1(t) \\ F_2(t) \end{bmatrix} \quad (5.17)$$

In the absence of damping or any external excitation force, Equation (5.17) becomes:

$$\begin{bmatrix} m_1 & 0 \\ 0 & m_2 \end{bmatrix}\begin{bmatrix} \ddot{w}_1 \\ \ddot{w}_2 \end{bmatrix} = \begin{bmatrix} k_1 + k_2 & -k_2 \\ -k_2 & k_1 + k_2 \end{bmatrix}\begin{bmatrix} w_1 \\ w_2 \end{bmatrix} = 0 \quad (5.18)$$

Assuming harmonic solutions of the form, $w_1 = \hat{w}_1 e^{j\omega t}$ and $w_2 = \hat{w}_2 e^{j\omega t}$, Equation (5.18) may be written in matrix form as:

$$-\omega^2 \boldsymbol{M}\hat{\boldsymbol{w}} + \boldsymbol{K}\hat{\boldsymbol{w}} = 0 \quad (5.19)$$

for which the only non-trivial solution is:

$$\det|\boldsymbol{K} - \omega^2\boldsymbol{M}| = 0 \quad (5.20)$$

That is:

$$\det\begin{vmatrix} k_1 + k_2 - m_1\omega^2 & -k_2 \\ -k_2 & k_2 + k_3 - m_2\omega^2 \end{vmatrix} = 0 \quad (5.21)$$

or:

$$\omega^4 - \left(\frac{k_1 + k_2}{m_1} + \frac{k_2 + k_3}{m_2}\right)\omega^2 + \frac{k_1k_3 + k_1k_2 + k_2k_3}{m_1m_2} = 0 \quad (5.22)$$

This equation has two positive solutions, ω_1 and ω_2. Substituting these solutions into Equation (5.19) and rearranging gives:

$$\left(\frac{\hat{w}_1}{\hat{w}_2}\right)_1 = \frac{k_2}{k_1 + k_2 - \omega_1^2 m_1} \quad (5.23)$$

$$\left(\frac{\hat{w}_1}{\hat{w}_2}\right)_2 = \frac{k_2}{k_1 + k_2 - \omega_2^2 m_1} \quad (5.24)$$

Equations (5.23) and (5.24) represent the relative displacements of the masses m_1 and m_2 (or the mode shape) for modes 1 and 2, respectively. The modal matrix, $\mathbf{\Psi}$, is a 2×2 matrix and consists of one column for each mode. Thus:

$$\mathbf{\Psi} = \left[\begin{array}{cc} a_{11} & a_{12} \\ a_{21} & a_{22} \end{array} \right] = \left[\begin{array}{cc} 1 & 1 \\ (\hat{w}_2/\hat{w}_1)_1 & (\hat{w}_2/\hat{w}_1)_2 \end{array} \right] \tag{5.25}$$

Note that the coordinates x_1 and x_2 are not unique generalised coordinates. We could just as easily have selected $q_1 = w_1$ and $q_2 = w_2 - w_1$, which results in the following equation for the undamped system:

$$\left[\begin{array}{cc} m_1 & 0 \\ m_2 & m_2 \end{array} \right] \left[\begin{array}{c} \ddot{q}_1 \\ \ddot{q}_2 \end{array} \right] + \left[\begin{array}{cc} k_1 & -k_2 \\ k_3 & k_2 + k_3 \end{array} \right] \left[\begin{array}{c} q_1 \\ q_2 \end{array} \right] = 0 \tag{5.26}$$

Thus, the most general expression for Equation (5.17) is:

$$\left[\begin{array}{cc} m_{11} & m_{12} \\ m_{21} & m_{22} \end{array} \right] \left[\begin{array}{c} \ddot{q}_1 \\ \ddot{q}_2 \end{array} \right] + \left[\begin{array}{cc} C_{11} & C_{12} \\ C_{21} & C_{22} \end{array} \right] \left[\begin{array}{c} \dot{q}_1 \\ \dot{q}_2 \end{array} \right] + \left[\begin{array}{cc} k_{11} & k_{12} \\ k_{21} & k_{22} \end{array} \right] \left[\begin{array}{c} q_1 \\ q_2 \end{array} \right] = \left[\begin{array}{c} Q_1(t) \\ Q_2(t) \end{array} \right] \tag{5.27}$$

or:

$$\boldsymbol{M\ddot{q}} + \boldsymbol{C\dot{q}} + \boldsymbol{Kq} = \boldsymbol{Q}(t) \tag{5.28}$$

which also applies to a system with n degrees of freedom, where n is any integer number. The matrices $\boldsymbol{M}$, $\boldsymbol{C}$ and $\boldsymbol{K}$ are $n \times n$ matrices and $\ddot{\boldsymbol{q}}$, $\dot{\boldsymbol{q}}$, $\boldsymbol{q}$ and $\boldsymbol{Q}(t)$ are $n \times 1$ vectors.

If the system is undamped and not subject to any external forces, the complete solution may be expressed in two $n \times n$ matrices as:

$$\left[\begin{array}{ccc} \diagdown & & \\ & \omega_i^2 & \\ & & \diagdown \end{array} \right] \quad \text{and} \quad \mathbf{\Psi} \tag{5.29}$$

where ω_i is the resonance frequency for the ith mode, and the ith column in $\mathbf{\Psi}$ is a vector that describes the ith mode shape. The matrix, $\mathbf{\Psi}$, is not unique; any column may be multiplied by a scaling factor, which will be a function of the normalisation used in the calculation procedure.

Vibration modes of a system are characterised by resonance frequencies and mode shapes, and possess orthogonal properties that may be stated mathematically as:

$$\mathbf{\Psi}^{\mathrm{T}} \boldsymbol{M} \mathbf{\Psi} = \left[\begin{array}{ccc} \diagdown & & \\ & m_i & \\ & & \diagdown \end{array} \right] \tag{5.30}$$

and:

$$\mathbf{\Psi}^{\mathrm{T}} \boldsymbol{K} \mathbf{\Psi} = \left[\begin{array}{ccc} \diagdown & & \\ & k_i & \\ & & \diagdown \end{array} \right] \tag{5.31}$$

where the resonance frequency for the ith mode is:

$$\omega_i^2 = k_i/m_i \tag{5.32}$$

Although m_i and k_i are both affected by the scaling used for the mode shapes, the ratio k_i/m_i is unique for a given mode, i. The quantities, m_i and k_i, are often referred to as the modal (or generalised) mass and modal (or generalised) stiffness, respectively.

A result of modal orthogonality is that the scalar product of any two column vectors in the mode shape matrix must be equal to zero. In other words, the eigenvectors (or mode shapes) obey the relation:

$$\iint\limits_{S} \rho_m(x,y)h(x,y)\psi_m(x,y)\psi_n(x,y)\mathrm{d}x\mathrm{d}y = 0 \quad \text{for } m \neq n \tag{5.33}$$

where the integration is over the surface of area, $S = x_{\max}y_{\max}$, $\rho_m(x,y)$ is the structural density at (x,y) and $h(x,y)$ is the thickness of the surface at (x,y). The structural location (x,y) will be denoted by the vector, $\boldsymbol{x}$.

For a surface of uniform density and thickness, Equation (5.33) implies that the space averaged product of any two mode shapes is zero. Equation (5.33) is derived by Cremer et al. (1973) who also show that the eigenvectors of systems that are not closed need not satisfy the orthogonality relation. That is, if energy can be removed from the system, orthogonality of modes is generally violated. A system is closed if its edges are free or completely restrained; added masses or non-conducting impedances also extract no energy. A system is "open" if it is connected to other systems or if its edges can dissipate energy into its supports. The equations of motion can be uncoupled by the correct choice, p, of the generalised coordinates so that for mode, i, we can write the following independent equation of motion:

$$m_i\ddot{p}_i + k_ip_i = 0 \tag{5.34}$$

where p_i is referred to as the ith principal coordinate. The principal coordinates are related to the generalised coordinates, $\boldsymbol{x}$, by:

$$\boldsymbol{p} = \boldsymbol{\Psi}^{-1}\boldsymbol{x} \tag{5.35}$$

For a discrete system with the mass defined at N_p locations (m_k; $k = 1, \ldots, N_p$), the modal mass, m_i, for the ith mode may be calculated from the mode shape vector for the ith mode using:

$$m_i = \sum_{k=1}^{N_p} m_k\psi_k^{\,2} \tag{5.36}$$

where m_k is the mass of the structure associated with location, k.

For a continuous, non-discrete surface for which the mode shape for mode, i, is known, the modal mass for mode, i, is:

$$m_i = \iint\limits_{S} m_s(\boldsymbol{x})\psi^2(\boldsymbol{x})\mathrm{d}\boldsymbol{x} \tag{5.37}$$

where $m_s(\boldsymbol{x})$ is the mass per unit area at location, $\boldsymbol{x}$, on the surface, of total area, S.

5.2.3.1 Forced Response of Undamped Systems

The general equation for a multi-degree-of-freedom system with no damping, but with external forces acting at each point mass is:

$$\boldsymbol{M}\ddot{\boldsymbol{w}}(t) + \boldsymbol{K}\boldsymbol{w}(t) = \boldsymbol{F}(t) \tag{5.38}$$

The components of the force vector, $\boldsymbol{F}(t)$, may have any amplitudes and phases, but it is assumed here, that all forces are of the same frequency, such that $F(t) = \hat{F}\mathrm{e}^{\mathrm{j}\omega t}$. Forces at different frequencies may be treated separately and the results combined using superposition. Thus, the force vector is defined as:

$$\boldsymbol{F}(t) = \hat{\boldsymbol{F}}\mathrm{e}^{\mathrm{j}\omega t} \tag{5.39}$$

and the solution vector is:

$$\boldsymbol{w}(t) = \hat{\boldsymbol{w}} \mathrm{e}^{\mathrm{j}\omega t} \tag{5.40}$$

Note that $\hat{\boldsymbol{F}}$ and $\hat{\boldsymbol{w}}$ are complex vectors; that is, each element of each vector is characterised by an amplitude and a phase. Equation (5.38) may now be rewritten using Equations (5.39) and (5.40) to give:

$$(\boldsymbol{K} - \omega^2 \boldsymbol{M}) \hat{\boldsymbol{w}} \mathrm{e}^{\mathrm{j}\omega t} = \hat{\boldsymbol{F}} \mathrm{e}^{\mathrm{j}\omega t} \tag{5.41}$$

Equation (5.41) may be rearranged to solve for the unknown responses, $\hat{\boldsymbol{w}}$:

$$\hat{\boldsymbol{w}} = (\boldsymbol{K} - \omega^2 \boldsymbol{M})^{-1} \hat{\boldsymbol{F}} \tag{5.42}$$

which can be written as:

$$\hat{\boldsymbol{w}} = \boldsymbol{\alpha} \hat{\boldsymbol{F}} \tag{5.43}$$

where $\boldsymbol{\alpha}(\omega)$ is the $n \times n$ receptance matrix for the system, and is a function of frequency, ω. The general element, $\alpha_{jk}(\omega)$, in the receptance matrix is defined as the ratio of the displacement at location, j, to the force applied at location, k, which is causing the displacement at location, j, at frequency, ω. That is:

$$\alpha_{jk}(\omega) = \hat{w}(\boldsymbol{x}_j) / \hat{F}(\boldsymbol{x}_k) \tag{5.44}$$

Equation (5.44) represents the individual receptance function, which is similar to that defined in Equation (5.15) for the SDOF system.

Values for the elements of $\boldsymbol{\alpha}$ can be determined by substituting appropriate values for the mass, stiffness and force matrices into Equation (5.42). However, this involves inversion of the system matrix, which is impractical for systems with a large number of modes of vibration. Also, no insight is gained into the form of the various properties of the frequency response function.

Returning to Equation (5.42), the following can be written:

$$(\boldsymbol{K} - \omega^2 \boldsymbol{M}) = \boldsymbol{\alpha}^{-1} \tag{5.45}$$

Pre-multiplying both sides by $\boldsymbol{\Psi}^{\mathrm{T}}$ and post-multiplying both sides by $\boldsymbol{\Psi}$ gives:

$$\boldsymbol{\Psi}^{\mathrm{T}}(\boldsymbol{K} - \omega^2 \boldsymbol{M})\boldsymbol{\Psi} = \boldsymbol{\Psi}^{\mathrm{T}} \boldsymbol{\alpha}^{-1} \boldsymbol{\Psi} \tag{5.46}$$

or:

$$\begin{bmatrix} \ddots & & \\ & m_i & \\ & & \ddots \end{bmatrix} \begin{bmatrix} \ddots & & \\ & \omega_i^2 - \omega^2 & \\ & & \ddots \end{bmatrix} = \boldsymbol{\Psi}^{\mathrm{T}} \boldsymbol{\alpha}^{-1} \boldsymbol{\Psi} \tag{5.47}$$

which gives:

$$\boldsymbol{\alpha} = \boldsymbol{\Psi} \left\{ \begin{bmatrix} \ddots & & \\ & m_i & \\ & & \ddots \end{bmatrix} \begin{bmatrix} \ddots & & \\ & \omega_i^2 - \omega^2 & \\ & & \ddots \end{bmatrix} \right\}^{-1} \boldsymbol{\Psi}^{\mathrm{T}} \tag{5.48}$$

It is clear that the receptance matrix defined in Equation (5.45) is symmetric and this is recognised as a principle of reciprocity which applies to many characteristics of practical systems. In this case, the implication is that:

$$\alpha_{jk}(\omega) = \hat{w}(\boldsymbol{x}_j) / \hat{F}(\boldsymbol{x}_k) = \alpha_{kj}(\omega) = \hat{w}(\boldsymbol{x}_k) / \hat{F}(\boldsymbol{x}_j) \tag{5.49}$$

which demonstrates the principle of reciprocity.

Any individual frequency response function can be expressed using Equation (5.48), and summing the contributions from N_m modes, as:

$$\alpha_{jk}(\omega) = \sum_{i=1}^{N_m} \frac{\psi_i(\boldsymbol{x}_j)\psi_i(\boldsymbol{x}_k)}{m_i(\omega_i^2 - \omega^2)} \tag{5.50}$$

In Equation (5.50) the quantity $\psi_i(\boldsymbol{x}_j)$, is the mode shape function for mode, i, evaluated at location, $\boldsymbol{x}_j$.

Any arbitrary normalisation of the mode shape matrix, $\boldsymbol{\Psi}$, will be reflected in the modal

mass matrix, $\begin{bmatrix} \ddots & & \\ & m_i & \\ & & \ddots \end{bmatrix}$, so that $\alpha_{jk}(\omega)$ is independent of any normalisation process.

Note that up to now, as damping has been excluded from the analysis, the mode shapes, modal masses and modal stiffnesses are all real quantities.

5.2.3.2 Damped MDOF Systems: Proportional Damping

Proportional damping is a special type of damping that simplifies system analysis. The damping may be viscous or hysteretic but the damping matrix must be proportional to either or both of the mass and stiffness matrices. Proportional viscous damping is the type usually assumed in the analysis of acoustic waves in fluid media.

The advantage in using proportional damping is that the mode shapes for both the damped and undamped cases are the same and the modal resonance frequencies are also very similar. Thus, the properties of a proportionally damped system may be determined by analysing in full the undamped system and then making a small correction for the damping. Although this is done in many commercial software packages, it is only valid for this very special type of damping.

Proportional Viscous Damping

As derived earlier, the general equation describing the motion of a viscously damped multi-degree-of-freedom system subject to external forces is:

$$\boldsymbol{M}\ddot{\boldsymbol{w}} + \boldsymbol{C}\dot{\boldsymbol{w}} + \boldsymbol{K}\boldsymbol{w} = \boldsymbol{F}(t) \tag{5.51}$$

If the damping matrix, $\boldsymbol{C}$, is proportional to the mass matrix, then:

$$\boldsymbol{C} = \beta\boldsymbol{M} \tag{5.52}$$

If the damping matrix is pre- and post-multiplied by the eigenvector (or mode shape) matrix for the undamped system, as was done previously for the mass and stiffness matrices, then:

$$\boldsymbol{\Psi}^{\mathrm{T}}\boldsymbol{C}\boldsymbol{\Psi} = \beta \begin{bmatrix} \ddots & & \\ & m_i & \\ & & \ddots \end{bmatrix} = \begin{bmatrix} \ddots & & \\ & C_i & \\ & & \ddots \end{bmatrix} \tag{5.53}$$

where the diagonal elements, C_i, represent the generalised (or modal) damping values.

Because this matrix is diagonal, the mode shapes of the damped system are identical to those of the undamped system. This can be shown by taking the general equation of motion, Equation (5.51), with no excitation forces, $\boldsymbol{f}(t)$, and transforming it to principal coordinates using the undamped system mode shape matrix, $\boldsymbol{\Psi}$, to obtain:

$$\begin{bmatrix} \ddots & & \\ & m_i & \\ & & \ddots \end{bmatrix}\ddot{\boldsymbol{P}} + \begin{bmatrix} \ddots & & \\ & C_i & \\ & & \ddots \end{bmatrix}\dot{\boldsymbol{P}} + \begin{bmatrix} \ddots & & \\ & k_i & \\ & & \ddots \end{bmatrix}\boldsymbol{P} = 0 \tag{5.54}$$

where $\boldsymbol{P} = \boldsymbol{\Psi}^{-1}\boldsymbol{w}$. The ith individual equation is then:

$$m_i\ddot{p}_i + C_i\dot{p}_i + k_ip_i = 0 \tag{5.55}$$

which is clearly the equation of motion for a single-degree-of-freedom system or for a single mode of a multi-degree-of-freedom system. This mode has a complex resonance frequency of:

$$\omega_i' = \zeta_i\omega_i + \mathrm{j}\omega_i\sqrt{1 - \zeta_i^2} \tag{5.56}$$

where the first term represents the oscillating part of the response and the second term represents the decaying part of the response (see Figure 5.2) and where $\omega_i^2 = k_i/m_i$ and:

$$\zeta_i = \frac{C_i}{2\sqrt{k_i m_i}} = \frac{\beta}{2\omega_i} \tag{5.57}$$

A more general form of viscous proportional damping is where the damping matrix is related to the mass and stiffness matrices as follows:

$$C = \beta M + \gamma K \tag{5.58}$$

In this case, the damped system will have the same (real) mode shape vectors as the undamped system and the resonance frequencies will be:

$$\omega_i' = \omega_i \sqrt{1 - \zeta_i^2} \tag{5.59}$$

where:

$$\zeta_i = \frac{\beta}{2\omega_i} + \frac{\gamma\omega_i}{2} \tag{5.60}$$

Forced Response Analysis - Proportional Viscous Damping

For viscous proportional damping, the receptance matrix for forced excitation is:

$$\boldsymbol{\alpha} = \left[K + \mathrm{j}\omega C - \omega^2 M \right]^{-1} \tag{5.61}$$

The receptance, $\alpha_{jk}(\omega)$, which is the ratio of the displacement, $\hat{w}$, at location, $\boldsymbol{x}_i$, to a force, F, of frequency, ω, at location, $\boldsymbol{x}_k$, is:

$$\alpha_{jk}(\omega) = \sum_{i=1}^{N_m} \frac{\psi_i(\boldsymbol{x}_j)\psi_i(\boldsymbol{x}_k)}{m_i(\omega_i^2 - \omega^2) + \mathrm{j}\omega C_i} \tag{5.62}$$

or:

$$\alpha_{jk}(\omega) = \sum_{i=1}^{N_m} \frac{\psi_i(\boldsymbol{x}_j)\psi_i(\boldsymbol{x}_k)}{\omega_i^2 m_i \left[1 - (\omega/\omega_i)^2 + 2\mathrm{j}\,(\omega/\omega_i)\,\zeta_i \right]} = \sum_{i=1}^{N_m} \frac{\psi_i(\boldsymbol{x}_j)\psi_i(\boldsymbol{x}_k)}{m_i \left[\omega_i^2 - \omega^2 + 2\mathrm{j}\omega\,\omega_i\,\zeta_i \right]} \tag{5.63}$$

where: $\zeta_i = \dfrac{C_i}{2\sqrt{k_i m_i}}$.

Proportional Hysteretic Damping

The same procedure as outlined for proportional viscous damping can be followed for proportional hysteretic damping. The equations of motion are written as:

$$M\ddot{w} + (K + \mathrm{j}H)\,w = F(t) \tag{5.64}$$

and the hysteretic damping matrix, H, is proportional to the mass and stiffness matrices as follows:

$$H = \beta M + \gamma K \tag{5.65}$$

Again, the mode shapes for the damped system are identical to those for the undamped system and the damped resonance frequencies are:

$$\omega_i' = \omega_i \sqrt{1 + \mathrm{j}\eta}; \quad \eta_i = \gamma + \beta/\omega_i^2 \;;\; \omega_i^2 = k_i/m_i \tag{5.66}$$

Forced Response Analysis - Proportional Hysteretic Damping

The expression for a receptance element (response at $\boldsymbol{x}_j$ due to a force at $\boldsymbol{x}_k$) of the general frequency response function matrix is written as:

$$\alpha_{jk}(\omega) = \sum_{i=1}^{N_m} \frac{\psi_i(\boldsymbol{x}_j)\psi_i(\boldsymbol{x}_k)}{m_i(\omega_i^2 - \omega^2) + \mathrm{j}\eta_i m_i \omega_i^2} \tag{5.67}$$

where N_m is sufficiently large to obtain an accurate estimate of $\alpha_{jk}(\omega)$. To achieve this, it is usually necessary to include modes with resonance frequencies up to twice the highest frequency of interest.

In many practical structures and acoustic spaces, even though the damping may not be strictly proportional, it is often sufficiently small that for the purposes of estimating the resonance frequencies, mode shapes and frequency response function, it is sufficiently accurate to assume proportional damping; this assumption is commonly used in the analysis of acoustic spaces and space structures. With this assumption it is possible to calculate the undamped resonance frequencies and mode shapes and then calculate the actual resonance frequencies using Equation (5.66) and the actual frequency response function using Equation (5.67).

Damping only significantly affects the structural response at frequencies near resonance frequencies, so it is a reasonable approximation to replace $\omega\omega_i$ with ω_i^2 in Equation (5.63), which, for small damping, such that $\eta = 2\zeta$, results in Equation (5.63) being the same as Equation (5.67).

The output from most finite element software packages are the resonance frequencies, mode shapes, and modal masses for the undamped system. The forced response at any frequency, ω, can then be calculated using additional software, which uses Equation (5.67) and estimates of structural (or hysteretic) damping, η.

5.2.3.3 Damped MDOF Systems: General Viscous Damping

Although hysteretic damping is representative of the damping found in structures, it is not representative of the damping in acoustic spaces. In this latter case the damping is closer to viscous; so the analysis of a multi-degree-of-freedom system with general viscous damping is important.

The analysis of the general viscous damping case is a complex problem which will only be briefly considered here. The general equation for forced excitation with viscous damping may be written as:

$$\boldsymbol{M}\ddot{\boldsymbol{w}} + \boldsymbol{C}\dot{\boldsymbol{w}} + \boldsymbol{K}\boldsymbol{w} = \boldsymbol{F}(t) \tag{5.68}$$

As before, the case of zero excitation force is considered with an assumed a solution of the form:

$$\boldsymbol{w} = \hat{\boldsymbol{w}}\mathrm{e}^{st} \tag{5.69}$$

Substituting this solution into Equation (5.68) with zero excitation force, gives:

$$(s^2\boldsymbol{M} + s\boldsymbol{C} + \boldsymbol{K})\hat{\boldsymbol{w}} = 0 \tag{5.70}$$

This equation has $2M$ eigenvalue solutions (resonance frequencies) that occur in complex conjugate pairs of the following form, which is similar to Equation (5.7) for a SDOF system:

$$s_i = \omega_i \left(-\zeta_i \pm \mathrm{j}\sqrt{1 - \zeta_i^2} \right) \tag{5.71}$$

where $\omega_i\sqrt{1 - \zeta^2}$ is the damped natural frequency and ζ_i is the critical damping ratio of mode i. The corresponding eigenvectors (or mode shapes) $\boldsymbol{\psi}_i^*$ and $\boldsymbol{\psi}_i$, which result from substitution of the solutions of Equation (5.71) into Equation (5.70), are also complex conjugates.

It can be shown that:

$$\omega_i^2 = k_i/m_i \tag{5.72}$$

and:

$$2\omega_i\zeta_i = C_i/m_i \tag{5.73}$$

where:

$$\left. \begin{array}{l} C_i = \boldsymbol{\psi}_i^* \boldsymbol{C} \boldsymbol{\psi}_i \\ m_i = \boldsymbol{\psi}_i^* \boldsymbol{M} \boldsymbol{\psi}_i \\ k_i = \boldsymbol{\psi}_i^* \boldsymbol{K} \boldsymbol{\psi}_i \end{array} \right\} \tag{5.74}$$

When the system being analysed is an acoustic system, the above analysis may be used with the following substitutions in Equation (5.68):

- displacement, w, is replaced with sound pressure, p, at a location in the enclosed space;
- force, $\boldsymbol{F}$, is replaced with the volume velocities, $\boldsymbol{v} = \boldsymbol{u}S$, of the sources; and
- the mass matrix, $\boldsymbol{M}$, damping matrix, $\boldsymbol{C}$, and the stiffness matrix, $\boldsymbol{K}$, remain the same but do not represent physical realisable quantities.

This equivalence means that the modal analysis approach for modal parameter estimation used for structures can also be used for acoustic spaces.

Forced Response Analysis: General Viscous Damping

To derive expressions for the frequency response function, it is necessary to take into account the coupling of the equations of motion, which was not an issue for the SDOF system discussed in Section 5.2.1. De-coupling is achieved using a new coordinate vector defined as:

$$\boldsymbol{y} = \left[\begin{array}{c} \boldsymbol{w} \\ \dot{\boldsymbol{w}} \end{array} \right] \tag{5.75}$$

Substituting this in the equation of motion (with no force) gives:

$$[\boldsymbol{C} : \boldsymbol{M}]\dot{\boldsymbol{y}} + [\boldsymbol{K} : \boldsymbol{0}]\boldsymbol{y} = \boldsymbol{0} \tag{5.76}$$

which represents n equations in $2n$ unknowns. Thus, we need an identity equation of the type:

$$[\boldsymbol{M} : \boldsymbol{0}]\dot{\boldsymbol{y}} + [\boldsymbol{0} : -\boldsymbol{M}]\boldsymbol{y} = \boldsymbol{0} \tag{5.77}$$

Combining Equations (5.76) and (5.78) gives:

$$\left[\begin{array}{cc} \boldsymbol{C} & \boldsymbol{M} \\ \boldsymbol{M} & \boldsymbol{0} \end{array} \right] \dot{\boldsymbol{y}} + \left[\begin{array}{cc} \boldsymbol{K} & \boldsymbol{0} \\ \boldsymbol{0} & -\boldsymbol{M} \end{array} \right] \boldsymbol{y} = \boldsymbol{0} \tag{5.78}$$

which can be written as:

$$\boldsymbol{A}\dot{\boldsymbol{y}} + \boldsymbol{B}\boldsymbol{y} = \boldsymbol{0} \tag{5.79}$$

These $2n$ equations are now in the standard eigenvalue form and by assuming a solution of the form:

$$y = \hat{y}\mathrm{e}^{st} \tag{5.80}$$

the $2N$ eigenvalues s_i and eigenvectors $\boldsymbol{\Psi}_i$ of the system can be obtained, which together satisfy the general equation:

$$(s_i\boldsymbol{A} + \boldsymbol{B})\boldsymbol{\Psi}_i = \boldsymbol{0}; \qquad i = 1, \ldots, 2N \tag{5.81}$$

These eigenvalues will exist as complex conjugate pairs with orthogonality properties, which may be expressed as:

$$\boldsymbol{\Psi}^{\mathrm{T}} \boldsymbol{A} \boldsymbol{\Psi} = \begin{bmatrix} \ddots & & \\ & a_i & \\ & & \ddots \end{bmatrix} \tag{5.82}$$

$$\boldsymbol{\Psi}^{\mathrm{T}} \boldsymbol{B} \boldsymbol{\Psi} = \begin{bmatrix} \ddots & & \\ & b_i & \\ & & \ddots \end{bmatrix} \tag{5.83}$$

and which have the characteristic that:

$$s_i = -b_i/a_i \tag{5.84}$$

The forcing vector may now be expressed in terms of the new coordinate as:

$$\boldsymbol{P} = \begin{bmatrix} \boldsymbol{F} \\ \boldsymbol{0} \end{bmatrix} \tag{5.85}$$

Assuming a harmonic forcing function and response, and following a similar development as outlined in Equations (5.41) to (5.50) the following is obtained:

$$\begin{bmatrix} \hat{\boldsymbol{w}} \\ \cdots \\ \mathrm{j}\omega\hat{\boldsymbol{w}} \end{bmatrix} = \sum_{i=1}^{2M} \frac{\boldsymbol{\psi}_i^{\mathrm{T}} \hat{\boldsymbol{P}} \boldsymbol{\psi}_i}{a_i(\mathrm{j}\omega - s_i)} \tag{5.86}$$

As the eigenvalues and eigenvectors appear in complex conjugate pairs, Equation (5.88) may be written as:

$$\begin{bmatrix} \hat{\boldsymbol{w}} \\ \cdots \\ \mathrm{j}\omega\hat{\boldsymbol{w}} \end{bmatrix} = \sum_{i=1}^{M} \left[\frac{\boldsymbol{\psi}_i^{\mathrm{T}} \hat{\boldsymbol{P}} \boldsymbol{\psi}_i}{a_i(\mathrm{j}\omega - s_i)} + \frac{\boldsymbol{\psi}_i^{*\mathrm{T}} \hat{\boldsymbol{P}} \boldsymbol{\psi}_i^*}{a_i^*(\mathrm{j}\omega - s_i^*)} \right] \tag{5.87}$$

Extracting a single displacement element, as a result of a force, $\hat{F}(\boldsymbol{x}_k, \omega)$, at location, k, and frequency, ω, the following is obtained:

$$\alpha_{jk}(\omega) = \frac{\hat{w}(\boldsymbol{x}_j, \omega)}{\hat{F}(\boldsymbol{x}_k, \omega)} = \sum_{i=1}^{M} \frac{\psi_i(\boldsymbol{x}_j)\psi_i(\boldsymbol{x}_k)}{a_i[\omega_i\zeta_i + \mathrm{j}(\omega - \omega_i\sqrt{1 - \zeta_i^2})]} + \frac{\psi_i^*(\boldsymbol{x}_j)\psi_i^*(\boldsymbol{x}_k)}{a_i^*[\omega_i\zeta_i + \mathrm{j}(\omega + \omega_i\sqrt{1 - \zeta_i^2})]} \tag{5.88}$$

This expression can be further reduced to:

$$\hat{w}(\boldsymbol{x}_j, \omega) = \hat{F}(\boldsymbol{x}_k, \omega) \sum_{i=1}^{M} \frac{\theta_{ijk}(\boldsymbol{x}_j) + \mathrm{j}(\omega/\omega_i)\phi_{ijk}(\boldsymbol{x}_k)}{\omega_i^2 - \omega^2 + 2\mathrm{j}\omega\omega_i} \tag{5.89}$$

where:

$$\theta_{ijk} = 2\left(\zeta_i \mathrm{Re}\{G_{ijk}\} - \mathrm{Im}\{G_{ijk}\}\sqrt{1 - \zeta_i^2} \right) \tag{5.90}$$

$$\phi_{ijk} = 2\mathrm{Re}\{G_{ijk}\} \tag{5.91}$$

and:

$$G_{ijk} = \frac{1}{a_i}\psi_{ik}\psi_{ij} \tag{5.92}$$

For frequencies close to a resonance frequency, Equation (5.89) is very similar in form to Equation (5.100), which, together with the fact that damping only significantly affects the response near a resonance, is why viscous damping (proportional to vibration velocity) is often represented in calculations in practice as hysteretic damping (proportional to vibration displacement), or as proportional damping with the loss factor, $\eta_i = 2\zeta_i$ (see Section 6.6).

Note that damping must be measured or determined on the basis of experience with similar structures or acoustic spaces. It is also usually dependent on excitation frequency, so it must be estimated as a function of frequency. To obtain a structural or acoustic response prediction with an accuracy of 1 dB, damping must be estimated with a 20% accuracy (Ewins, 2000).

5.2.3.4   Damped MDOF Systems: General Hysteretic Damping

Although viscous damping is representative of that found in acoustic spaces, hysteretic damping is more representative of the damping found in structures. In the practical analysis of structural vibrations, there are some cases where the assumption of proportional damping cannot be made. Thus, here we will consider the properties of a system with general hysteretic (or structural) damping.

The equation of motion may be written as before as:

$$M\ddot{w} + (K + jH)\,w = F(t) \tag{5.93}$$

Considering first the case where $F(t) = 0$ a solution of the following form is assumed:

$$w = \hat{w}e^{j\omega_i' t} \tag{5.94}$$

Substituting this solution into Equation (5.93) yields a solution consisting of complex eigen frequencies and mode shapes. The complex mode shape matrix consists of elements defined by a relative amplitude and phase so that the relative phase of the motion of the point masses can vary between 0 and 180° instead of being confined to one or the other as was the case for undamped or proportionally damped systems.

The ith eigenvalue may be written as:

$$(\omega_i')^2 = \frac{k_i + jk_i\eta_i}{m_i} = \omega_i^2(1 + j\eta_i) \tag{5.95}$$

where $\omega_i = \sqrt{k_i/m_i}$ is a natural frequency and η_i is the loss factor for the ith mode. Note that ω_i is not exactly equal to the natural frequency for the undamped system, although it is close (Ewins, 2000).

The eigenvectors possess the same type of orthogonality properties as the undamped system and these may be defined by the following equations:

$$\mathbf{\Psi}^{\mathrm{T}}M\mathbf{\Psi} = \begin{bmatrix} \ddots & & \\ & m_i & \\ & & \ddots \end{bmatrix} \tag{5.96}$$

$$\mathbf{\Psi}^{\mathrm{T}}(K + jH)\mathbf{\Psi} = \begin{bmatrix} \ddots & & \\ & k_i + jk_i\eta_i & \\ & & \ddots \end{bmatrix} \tag{5.97}$$

The generalised (or modal) stiffness parameters, $(k_i + jk_i\eta_i)$, are now complex and the eigen solution for mode i is:

$$\lambda_i^2 = \frac{k_i + jk_i\eta_i}{m_i} \tag{5.98}$$

Forced Response Analysis: General Hysteretic Damping

For the forced response analysis, a direct solution to Equation (5.93) for a single frequency (harmonic) exciting force vector is:

$$\hat{w} = \left(K + jH - \omega^2 M\right)^{-1}\hat{F} = \alpha\hat{F} \tag{5.99}$$

where $F(t) = \hat{F}e^{j\omega t}$.

Following the same procedure as for the proportionally damped system, it can be shown that for frequency, ω, a force input at location, $\boldsymbol{x}_k$, and a response at location, $\boldsymbol{x}_j$:

$$\alpha_{jk}(\omega) = \frac{\hat{w}(\boldsymbol{x}_j, \omega)}{\hat{F}(\boldsymbol{x}_k, \omega)} = \sum_{i=1}^{N_m} \frac{\psi_i(\boldsymbol{x}_j)\psi_i(\boldsymbol{x}_k)}{m_i(\omega_i^2 - \omega^2 + \mathrm{j}\eta_i\omega_i^2)} \tag{5.100}$$

The only difference between Equation (5.100) and the equivalent Equation (5.67) is that in Equation (5.100) the mode shapes are complex numbers.

If it is desired to know the response of the system at frequency, ω, due to a number of forces acting simultaneously, the following expression may be used:

$$\hat{\boldsymbol{w}}(\boldsymbol{x}, \omega) = \sum_{i=1}^{N_m} \frac{\boldsymbol{\psi}_i^{\mathrm{T}}(\boldsymbol{x})\hat{\boldsymbol{F}}(\boldsymbol{x}, \omega)\boldsymbol{\psi}_i(\boldsymbol{x})}{m_i(\omega_i^2 - \omega^2 + \mathrm{j}\eta_i\omega_i^2)} \tag{5.101}$$

where $\hat{\boldsymbol{F}}(\boldsymbol{x}, \omega)$ is the force amplitude vector at frequency, ω, in which elements are set equal to zero for locations where there is no external force applied and $\boldsymbol{\psi}_i(\boldsymbol{x})$ is a vector that represents the mode shape function values at all locations, $\boldsymbol{x}$, for mode, i.

5.2.4 Summary

From the preceding sections it can be seen that the resonance frequencies and mode shapes for a vibroacoustic system may be determined by discretising it; that is, dividing it into a number of point masses connected by massless springs and dampers.

The equations of motion for such a system may be derived using Newton's laws or Lagrange's equations as outlined in Section 2.2.2.5. Once the equations of motion have been derived they may be expressed in matrix form in terms of mass, stiffness and damping matrices and a force vector, which can be solved for resonance frequencies and mode shapes as well as for the response of the structure at any specified location and for any excitation frequency. In practice, the use of commercial finite element software packages makes the derivation of the equations of motion, the corresponding mass, stiffness and damping matrices and the solution of the equations of motion transparent to the user. However, it is still necessary for the user to have a fundamental understanding of the physical principles involved (as outlined briefly in this section) so that the limitations of any analysis may be properly evaluated.

Of particular concern in the use of finite element analysis, or indeed in the use of any theoretical analysis for the determination of the response of a structure or acoustic space to one or more excitation forces, is the accurate estimation of the structural or acoustic damping quantity. No analysis can provide a value for this; it is a required input to the analysis and the results of the response analysis are crucially dependent on its value. Damping values are usually determined from measurements on other similar structures or acoustic spaces, and sometimes the analyst just uses values derived from past experience. Thus, in many cases, the results from a response analysis will be approximate only; their accuracy almost solely depends on the accuracy in the estimation of the structural (hysteretic) and/or acoustic damping.

5.3 Modal Analysis: Numerical

Chapters 1 and 2 discuss means of obtaining resonance frequencies and mode shapes for simple enclosures and structures. However, when the enclosure or structure shape becomes more complex, it is necessary to use numerical methods, known as the boundary element method (BEM) or finite element analysis (FEA), to obtain mode shapes and resonance frequencies. FEA can be used to solve coupled problems such as calculation of the sound pressure level in an enclosure as a result of vibration of the enclosing structure. However, this usually requires substantial computational resources; thus it is more efficient to use BEM or FEA to calculate mode shapes and resonance frequencies of the enclosure surrounded by a rigid structure, and also for the structure acting in a vacuum. The two sets of results are then coupled together using modal coupling analysis (see Section 5.3.1) so that the sound pressure field in the enclosure as a result of vibration of the enclosing structure can be calculated. Details of the BEM and FEA modelling procedures, as well as examples and some practical suggestions, are provided in Chapter 11 of Bies et al. (2017).

The procedures for calculating resonance frequencies and mode shapes of an acoustic or structural system are implemented in most commercially available finite element and boundary element software, and the underlying theory has been discussed in detail in a number of text books (Wu, 2000; von Estorff, O., 2000; Marburg and Nolte, 2008). Kirkup (2007) provides software on a CD-ROM that accompanies his textbook, which can be used for modal analysis of a volume using boundary element analysis.

5.3.1   Modal Coupling Analysis

Modal coupling analysis is used to calculate the sound pressure level in an enclosed space as a result of vibration of all or part of the enclosing structure. It involves calculating the displacement mode shapes and resonance frequencies of the non-rigid part of the enclosing structure, assuming that the structure is in a vacuum. Then the pressure mode shapes and resonance frequencies of the enclosed fluid are calculated, assuming that the entire enclosing structure is rigid. Modal coupling analysis is valid when the fluid in the enclosure has a low density (such as air), but when the fluid is relatively dense (such as water), it will generate erroneous results because it does not account for the mass loading of the fluid on the structure, nor does it account for coupling between fluid modes. In this case, FEA, involving two-way fluid–structure interaction, is an appropriate analysis technique, and it is described by Howard and Cazzolato (2015) and Bies et al. (2017).

One of the main advantages of using the modal coupling method is that the computational time taken to solve the system of equations is significantly less than conducting a full fluid–structure interaction analysis using FEA. This is very important if optimisation studies are to be conducted that involve many FEA evaluations while converging to an optimum solution.

Modal coupling analysis begins with expressions for the displacement response of the enclosing structure in terms of the amplitudes of the structural modes of vibration and the the acoustic pressure response of the enclosed space in terms of acoustic pressure modes. Equations are then developed that describe how the structural modes couple with the enclosure fluid modes. Finally, the total vibration displacement at any point on the enclosing structure and the acoustic pressure response at any point within the enclosure, as a result of any external forces acting on the structure or any acoustic sources within the enclosure, are calculated.

The in vacuo displacement amplitude response, $w(\boldsymbol{x}, \omega)$, at location, $\boldsymbol{x} = (x, y)$, on the enclosing structure at angular frequency, ω, is given as the sum of the modal responses of each contributing mode at frequency, ω, as:

$$\hat{w}(\boldsymbol{r}_S, \omega) = \sum_{\ell=1}^{N_s} w_\ell(\omega) \varphi_\ell(\boldsymbol{r}_S) \qquad (5.102)$$

where $w = \hat{w} e^{j\omega t}$, $\varphi_\ell(\boldsymbol{x}_S)$, is the displacement mode shape of the ℓth structural mode at arbitrary location, $\boldsymbol{x}$, on the surface of the structure, and $w_\ell(\omega)$ is the modal participation factor (or displacement contribution, expressed as a fraction of the total contributions to the response from all modes) of the ℓth mode at frequency, ω. Although the value of N_s should theoretically be infinity, it is sufficiently accurate to choose a value of N_s such that the highest order mode considered has a resonance frequency at least twice that of the highest frequency of interest in the analysis. The N_s structural mode shapes and resonance frequencies can be evaluated using FEA or BEM software, and the nodal displacements for a mode, ℓ, are described as a vector, φ_ℓ, and then collated into a matrix, $[\varphi_1, \varphi_2, ...\varphi_{Ns}]$, for all the modes from 1 to N_s.

The acoustic pressure response of the enclosed space with rigid boundaries at angular frequency, ω, is:

$$\hat{p}(\boldsymbol{r}, \omega) = \sum_{n=1}^{N_a} p_n(\omega)\psi_n(\boldsymbol{r}) \tag{5.103}$$

where $p = \hat{p}e^{j\omega t}$, $\psi_n(\boldsymbol{r})$ is the acoustic pressure mode shape of the nth mode at arbitrary location, $\boldsymbol{r}$, within the enclosure, and p_n is the modal participation factor (or acoustic pressure contribution, expressed as a fraction of the total contributions to the response from all modes) of the nth mode at frequency, ω. Similarly to the structural mode case, N_a is chosen such that the highest order mode considered has a resonance frequency of at least twice that of the highest frequency of interest in the analysis. As for the structural case, the resonance frequencies and mode shapes can be evaluated using FEA or BEM software, where the nodal pressures for a mode, n, are described as a vector ψ_n and then collated into a matrix $[\psi_1, \psi_2, ... \psi_{Na}]$ for all the modes from 1 to N_a.

The response of the enclosing structure for structural mode, ℓ, with coupling to the interior space included is (Fahy, 1985; Fahy and Gardonio, 2007):

$$-\omega^2 w_\ell + \omega_\ell^2(1 + j\eta_\ell)w_\ell = \frac{S}{m_\ell} \sum_{n=1}^{N_a} p_n C_{n\ell} + \frac{F_\ell}{m_\ell} \tag{5.104}$$

where pressures, forces and displacements are all a function of frequency, ω. The quantities, ω_ℓ and η_ℓ, are the structural resonance frequency and structural (or hysteretic) loss factor, respectively, for the ℓth mode, m_ℓ is the modal mass (see Equation (5.37)) and F_ℓ is the modal force applied to the structure for the ℓth mode, S is the surface area of the structure and $C_{n\ell}$ is the dimensionless coupling coefficient between structural mode, ℓ, and acoustic mode, n, given by the integral of the product of the structural, φ_ℓ, and acoustic pressure, ψ_n, mode shape functions over the surface of the structure, as:

$$C_{n\ell} = \frac{1}{S} \iint_S \psi_n(\boldsymbol{r}_s)\varphi_\ell(\boldsymbol{r}_s) \, \mathrm{d}S \tag{5.105}$$

The right-hand side of Equation (5.104) describes the forces that are applied to the structure in terms of modal forces. The first term describes the modal force exerted on the structure due to the acoustic pressure in the enclosure acting on the enclosing structure. The second term describes the forces that act directly on the structure. As an example, consider a point force, $F_s(x_s, y_s, z_s)$, acting normal to the structure at nodal location, (x_s, y_s, z_s), for which the displacement mode shapes and resonance frequencies have been evaluated using FEA. As the force acts on the structure at a point, the modal force, F_ℓ, at frequency, ω, for mode, ℓ is:

$$F_\ell(\omega) = \varphi_\ell(x_s, y_s, z_s)F_s(x_s, y_s, z_s, \omega) \tag{5.106}$$

where $\varphi_\ell(x_s, y_s, z_s)$ is the ℓth structural displacement mode shape at the nodal location, (x_s, y_s, z_s).

The dimensionless coupling coefficient, $C_{n\ell}$, is calculated from FEA or BEM results as:

$$C_{n\ell} = \frac{1}{S} \sum_{i=1}^{J_s} \psi_n(\boldsymbol{x}_i)\varphi_\ell(\boldsymbol{x}_i)S_i \tag{5.107}$$

where S is the total surface area of the structure in contact with the acoustic fluid, S_i is the nodal area of the ith node on the surface (and hence $S = \sum_{i=1}^{J_s} S_i$), J_s is the total number of nodes on the surface, $\psi_n(\boldsymbol{x}_i)$ is the acoustic pressure mode shape for the nth acoustic mode at node location, $\boldsymbol{x}_i$, on the surface of the enclosing structure, and $\varphi_\ell(\boldsymbol{x}_i)$ is the displacement mode shape of the ℓth structural mode at node location, $\boldsymbol{x}_i$, on the surface of the enclosing structure. The area associated with each node of a structural finite element can usually be extracted using FEA or BEM software. The nodal areas can also be calculated by using the nodal coordinates that form the elements.

The acoustic pressure response of the fluid in the enclosure with coupling to the vibrating structure, which acts as the enclosure boundary for mode, n, at frequency, ω, is given by (Fahy, 1985; Fahy and Gardonio, 2007):

$$-\omega^2 p_n + 2\zeta_n \omega_n \mathrm{j}\omega p_n + \omega_n^2 p_n = -\left(\frac{\rho c^2 S}{\Lambda_n}\right) \sum_{\ell=1}^{N_s} -\omega^2 w_\ell C_{n\ell} + \left(\frac{\rho c^2}{\Lambda_n}\right) \mathrm{j}\omega Q_n \qquad (5.108)$$

where ω_n, is the resonance frequency of cavity mode, n, ζ_n is the viscous critical damping ratio for mode n, ρ is the density of the fluid in the enclosure, c is the speed of sound in the fluid, and Λ_n is the modal volume defined as the integration over the enclosure volume of the square of the acoustic pressure mode shape function:

$$\Lambda_n = \iiint_V \psi_n^2(\boldsymbol{r}) \, \mathrm{d}V \qquad (5.109)$$

and Q_n is a modal volume velocity excitation applied to the volume for the nth mode. For a single source of volume velocity, Q_b, acting at nodal location, (x_b, y_b, z_b), the modal volume velocity is defined as:

$$Q_n(\omega) = \psi_n(x_b, y_b, z_b) Q_b(\omega) \qquad (5.110)$$

where $\psi_n(x_b, y_b, z_b)$ is the mode shape at the nodal location (x_b, y_b, z_b).

For N_s acoustic sources located within the enclosed volume, $Q_n(\omega)$ is defined as:

$$Q_n(\omega) = \sum_{i=1}^{N_s} \psi_n(x_i, y_i, z_i) Q(x_i, y_i, z_i, \omega) \qquad (5.111)$$

where $Q(x_i, y_i, z_i, \omega)$ is the complex amplitude of the volume velocity of interior acoustic source, i, at frequency, ω, and nodal location (x_i, y_i, z_i); and $\psi_n(x_i, y_i, z_i)$ is the nth mode shape at the nodal location (x_i, y_i, z_i). The volume velocity of the source is the acoustic particle velocity (equal to the source velocity) multiplied by the area assigned to the node at which the source acts.

An important point to note is that because the acoustic mode shapes used in the structural-acoustic modal coupling method are for a rigid-walled enclosure, corresponding to a normal acoustic particle velocity at the wall surface equal to zero, the acoustic velocity at the surface resulting from the modal coupling method is incorrect (Jayachandran et al., 1998). However, the acoustic pressure at the surface is correct (albeit with a finite error), and this is all that is required for correctly coupling the structural vibration and acoustic pressure modal equations of motion.

The modal volume can be determined directly from a finite element analysis of the enclosed volume, by normalising the mode shapes to the mass matrix, $[\boldsymbol{M}]$ (Cazzolato, 1999), such that:

$$\boldsymbol{\Psi}_n^{\mathrm{T}}[\boldsymbol{M}]\boldsymbol{\Psi}_n = 1 \qquad (5.112)$$

where $\boldsymbol{\Psi}_n$ is the mass normalised mode shape function vector. The modal volume, Λ, can then be written in terms of the maximum value of $\boldsymbol{\Psi}_n$ as:

$$\Lambda_n = \frac{c^2}{\max(\boldsymbol{\Psi}_n^2)} \quad (5.113)$$

The unity normalised mode shapes can be calculated as:

$$\hat{\boldsymbol{\Psi}}_n = \frac{\boldsymbol{\Psi}_n}{\max(\boldsymbol{\Psi}_n)} \quad (5.114)$$

Equations (5.104) and (5.108) can form a matrix equation as:

$$\begin{bmatrix} \Lambda_\ell((1+\mathrm{j}\eta_\ell)\omega_\ell^2 - \omega^2) & -S[\boldsymbol{C}_{n\ell}] \\ S\omega^2[\boldsymbol{C}_{n\ell}]^{\mathrm{T}} & \dfrac{\Lambda_n}{\rho c^2}(\omega_n^2 + \mathrm{j}2\zeta_n\omega_n\omega - \omega^2) \end{bmatrix} \begin{bmatrix} \boldsymbol{w}_\ell \\ \boldsymbol{p}_n \end{bmatrix} = \begin{bmatrix} \boldsymbol{F}_\ell \\ \dot{\boldsymbol{Q}}_n \end{bmatrix} \quad (5.115)$$

where all the ℓ structural and n acoustic modes are included in the matrices, so that the square matrix on the left-hand side of Equation (5.115) has dimensions $(\ell + n) \times (\ell + n)$. The left-hand matrix in Equation (5.115) can be made symmetric by dividing all terms in the lower equation by $-\omega^2$. The structural modal participation factor, w_ℓ, for structural mode, ℓ, is an element of the vector, $\boldsymbol{w}_\ell$, and the acoustic modal participation factor, p_n, is an element of $\boldsymbol{p}_n$. $\boldsymbol{w}_\ell$ and $\boldsymbol{p}_n$ are frequency dependent and can be calculated by pre-multiplying each side of Equation (5.115) by the inverse of the square matrix on the left-hand side. Once these factors are calculated, the vibration displacement of the structure can be calculated from Equation (5.102) and the acoustic pressure inside the enclosure can be calculated using Equation (5.103).

5.4   Modal Analysis: Experimental

It is not the intention of this section to cover the subject of experimental modal analysis in an exhaustive manner. The reader is referred to the excellent book by Ewins (2000) for this treatment. However, it is intended here to introduce the concepts necessary for the understanding of the use of experimental modal analysis as an aid for the design, optimisation and performance evaluation of structural systems. In this text, we will use experimental modal analysis in two ways; the first is the traditional use, involving the determination of the resonance frequencies and mode shapes for a complex structure that cannot be analysed from first principles. The second, less common use, involves determining the contributions of each vibration mode to the total vibration response of a structure at any particular excitation frequency for one or more excitation sources acting on a structure for which the theoretical mode shapes are known a priori. This latter technique is a useful aid for both structural design and performance evaluation.

Four assumptions are basic to the traditional experimental determination of modal resonance frequencies, mode shapes and damping of an elastic structure. They are listed as follows:

1. The structure is assumed to be linear. Associated with any displacement from equilibrium there will arise a restoring force of opposite sign proportional (to a first approximation) to the displacement. For example, a restraint that is of much greater stiffness in one direction than in the opposite direction is excluded from this analysis. Linearity has the consequence that the sum of the effects of two forces is the same as the effect of applying the sum of the two forces.

2. The structure's behaviour is time invariant. This is important, as repeated testing can be used to obtain statistical accuracy. Alternatively, if the system response varies with time, then the statistical approach implicit in the methods of modal analysis are obviously inappropriate.

3. The structural response is observable. That is, enough vibration modes or degrees of freedom can be measured to obtain an adequate behavioural model of the structure.

4. Maxwell's law of reciprocity is assumed to hold. This law, which follows from system linearity, may be stated as follows: 'the displacement at position A due to a unit force applied at position B is equal to the displacement at B due to a unit force applied at A.'

The same four assumptions, represented by the first sentence in each of the four preceding items, are applicable to the modal analysis of acoustic spaces.

Traditional experimental modal analysis is often referred to as modal testing. Modal testing is used to verify theoretical models of structures and acoustic spaces and is also used to create accurate dynamic models of existing structures or acoustic spaces (modal resonance frequencies, damping and mode shapes) so that the effect of proposed structural or acoustic enclosure modifications can be evaluated.

Experimental modal analysis (or modal testing) began in the early days of the space programme in the USA, in the 1950s. There were no FFT analysers or laboratory based minicomputers then, so modal testing was done mostly with analogue instrumentation such as oscillators, amplifiers and oscilloscopes. With this type of testing, commonly referred to as sine testing or normal mode testing, a structure is excited one mode at a time with sinusoidal excitation. This excitation is provided by attaching one or more electrodynamic shakers to the structure and driving them with a sinusoidal signal. The frequency of the signal is adjusted to coincide with the natural frequency of one of the structure's modes. When this is done, the structure will readily absorb the energy and its predominant motion will be the mode shape of the mode being excited. Modal damping is measured by shutting off the shaker(s) and measuring the decay rate of the sinusoidal motion in the structure, as the mode is naturally damped out (see Equation (6.38)). The same testing procedure can be used for the modal analysis of acoustic spaces by replacing shakers with loudspeakers and structural modes with acoustic modes. However, when measuring modal damping by shutting off the excitation source, it is important to account for or eliminate the effects of the internal damping of the shaker when lightly damped structures are being tested, and to account for or eliminate the effects of loudspeaker internal damping when reverberant acoustic enclosures are being tested. Shaker damping effects can be eliminated by using a fuse arrangement to disconnect the shaker from the structure when it is switched off. Loudspeaker damping elimination is more difficult, but requires a mechanism to isolate the loudspeaker from the acoustic space when it is switched off. Alternatively the effects of the loudspeaker damping can be determined using a reverberant enclosure and measuring the decay rate with it excited by the test loudspeaker and then by a horn driver source which has very low damping associated with it. This latter test is only useful in the low-frequency range, which is where the loudspeaker damping is greater than the reverberant room damping.

With the discovery of the fast Fourier transform in 1965 and the advent of FFT analysers soon afterwards, a whole new approach to modal testing was born. This has become known as the transfer function method. This method is faster and easier to perform, and is much cheaper to implement than the normal mode method. Another real advantage is that a variety of different excitation signals (transient, random and swept sine) can be used to excite the structure or acoustic space. The structure or acoustic space is excited over a broad band of frequencies with these signals, and consequently many modes are excited at once. To identify modal parameters, a special type of FFT analyser must be used, which can measure the frequency response function (FRF) between two points (A and B) on the structure or in the acoustic space (see Section 7.3.14). An FRF measurement contains all of the information necessary to describe the dynamic response of the structure or acoustic space at point B due to an excitation force at point A.

To identify modal parameters for a structure, a set of these FRFs must be measured, typically between the force at a single excitation point and the displacement (for assumed hysteretic-type damping) or velocity (for assumed viscous-type damping) at many response points. (Alternatively, a set of measurements between many excitation points and a single response point can also be used. For the case of structural modal analysis, this latter set is typically measured when a small portable hammer is used to excite the structure.) A real advantage of this method is that these measurements can be made one at a time, thus requiring fewer transducers and less signal conditioning equipment than the normal mode method.

For acoustic spaces, the transfer function must be measured between a source volume velocity and the sound pressure at many spatial locations. For a loudspeaker with a small, airtight, backing enclosure, the volume velocity may be determined using a microphone in the enclosed volume to sample the acoustic pressure. This is valid for frequencies up to the frequency corresponding to a backing enclosure dimension greater than about one-tenth of a wavelength. In most cases, it is necessary to repeat the transfer function measurements with a few different source locations. Source locations should be chosen with the idea of the source being able to excite as many modes as possible. For example, in a rectangular-shaped room, all modes have antinodes in the room corners so the ideal location for a sound source would be near one of the room corners.

Once a set of FRF measurements is obtained with the FFT analyser and stored in a digital mass storage memory, they are then put through a 'curve fitting' process to identify the modal parameters of the structure or acoustic space.

5.4.1   Structural Modal Analysis

A complete modal analysis test covers three phases: test preparation and set-up; transfer function (frequency response) measurements; and modal parameter identification. Best results are obtained if the input force as well as the structural response is measured, allowing the transfer function (response spectrum divided by the input force spectrum) to be determined. If the input force is assumed to be of uniform amplitude over the frequency range of interest, and is thus not measured, then by measuring the power spectrum of the structural or acoustic response, it is still possible to obtain reasonable results for resonance frequencies and mode shapes, although damping results will have large errors.

The equipment required for experimental modal analysis of a structure consists of:

1. swept frequency oscillator (or random noise generator) and power amplifier connected to a shaker, which, in turn, is attached to a structure via a thin rod called a stinger; or instrumented hammer (with piezoelectric load cell and charge amplifier);
2. accelerometer and amplifier;
3. two channel FFT spectrum analyser and some form of interface to a personal computer with modal analysis software installed, or a computer with a data acquisition system, FFT analysis software and modal analysis software.

5.4.1.1   Test Set-Up

The first step in a modal analysis test is to support the structure so that it is unconstrained and is not affected significantly by its environment. To simulate this condition, the structure can be suspended from flexible cords, preferably attached to structural vibration nodes. Alternatively, a resilient support such as a foam mat may be used, on which the structure is placed. The support should be designed so that the frequency of the highest frequency rigid body mode is less than 10% of the frequency of the lowest frequency bending mode, to minimise bending wave modal distortion. If a constrained support is not possible (for example, if the structure is very heavy), then it is sometimes acceptable to test the structure by resting it on a hard concrete slab. This is referred to as the 'grounded' condition, but it is less than ideal. Rigid body modes are those associated with the body considered as rigid and mounted on a flexible suspension system. Essentially, these are the modes of the suspension system. The bending wave modes, on the other hand, are associated with flexure of the body, treated as a system with distributed mass and stiffness. In the former case, energy absorption would be confined to losses in the suspension system, whereas in the latter case, energy absorption may occur throughout the suspended body as well.

Prior to beginning a modal analysis, it is necessary to decide the locations that will be used to excite the structure and locations that will be used to measure its response, so that the overall structural response can be determined. Convenient local coordinates, for example, Cartesian, cylindrical or spherical coordinate systems, may be used for locating identification points. A computer can be used to put these into one global coordinate system. The emphasis should be on convenience for subsequent interpretation. If results from a finite element analysis are available, they can be used as a guide to determine the most suitable points. Alternatively, one or two preliminary frequency sweeps and response measurements at a few locations may be used to obtain a rough idea of resonance frequencies and mode shapes prior to a full modal analysis.

The next step in a structural modal analysis test is to decide which type of excitation to use and how to apply it to the structure. There are three types of excitation that are in common use: step relaxation, shaker and impact. Each of these types will be discussed in detail in the following paragraphs. In addition to the three standard excitation types discussed below, it is also possible to excite the structure by using an electromagnetic driver (such as the coil and magnet from a loudspeaker) as discussed by Ewins (2000). However, this technique is only used to excite rotating objects due to the complexity of measuring the excitation force and so is not discussed here.

5.4.1.2   Excitation by Step Relaxation

With step relaxation, the structure is preloaded (often in the shape of the mode of interest if a single mode is being studied) with a measured force and then released. The transient response thus generated propagates through and energises the structure. This method is often used in modal analysis of piping systems.

5.4.1.3   Excitation by Electrodynamic Shaker

An electromagnetic shaker converts an electrical signal into a mechanical force applied to the structure. Shakers can be attached at different locations on the structure to ensure even distribution of the exciting force, but often only one shaker is used at one or two different positions. For each position of the shaker, accelerometer measurements are taken at many points on the structure, and the number of accelerometer locations required is generally a little more than twice the order of the highest mode to be investigated. The structure is usually excited at one or two locations and the vibration pickup (usually an accelerometer) is moved about to take measurements at a large number of locations (whose geometrical positions have been previously entered into the computer). The force input to the structure is measured using a piezoelectric force transducer attached to the structure at the point of application of the force.

The shaker may be driven by a single sine wave, the frequency of which is incremented after each test, or it may be driven by a swept sine signal, by pseudo-random noise (see Section 7.3.17) or by random noise. The main advantage of the sine or swept sine excitation methods is the higher signal to noise ratio and low crest factor (ratio of peak amplitude to RMS), but this is at the expense of a longer measuring time. The sweep rate of the frequency sweep must be sufficiently slow not to distort the frequency response of the structure. This is especially important for lightly damped structures. One way of checking whether the sweep rate is acceptable is to compare the frequency response obtained by sweeping upwards in frequency with that obtained by sweeping downwards at the same rate. If the two results are not identical, the sweep rate is too fast.

The advantages of pseudo-random noise (random amplitude at a fixed set of frequencies) over random noise (random frequency and amplitude) include better signal to noise ratio and elimination of leakage. Leakage is defined here as movement of energy at one frequency in the spectrum to other frequencies. This leakage problem occurs with random noise because all of the excitation frequencies do not coincide with the finite set represented by the spectrum analyser. The FFT analysis of a finite sample length means that only certain frequencies are represented in the resulting spectrum whereas for random noise, energy exists at all frequencies. This results in a leakage of all the signal energy into the set of discrete frequencies which are represented. On the other hand, if pseudo-random noise is used, which is represented by energy at the same discrete frequencies as those which characterise the analyser, no leakage occurs and the structural response signal will be much less 'noisy'.

5.4.1.4 Excitation by Impact Hammer

Impact excitation is generally implemented using an instrumented hammer, so that the impact force applied to the structure can be measured using a force transducer embedded in the head of the hammer. For this method to be successful, the force applied with each impact should not vary too much between impacts. With impact excitation, either one of two techniques may be employed. The first involves exciting the structure at one or two locations and measuring the response with accelerometers at many other locations. Alternatively, to avoid shifting accelerometers around, equally good results are obtained by fixing one or two accelerometers to the structure and exciting it with the impact hammer at many other locations. This latter technique is generally faster to implement.

The impact hammer (see Figure 5.4) comprises four parts:

1. handle;
2. head;
3. force transducer; and
4. tip.

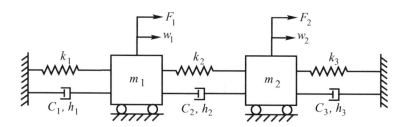

FIGURE 5.4 Impact hammer and response: (a) hammer force as a function of time for tips of varying hardness; (b) impact hammer.

The impact hammer excites the structure over a broad frequency range. Both the magnitude and the frequency content of the excitation applied to the structure can be controlled by the mass of the hammer and the velocity of the hammer impact.

The frequency range that is excited by the impact hammer is dependent on the duration of the transient spike generated in the time domain, and this is controlled by the stiffness of the contact surfaces and the mass of the impact hammer. Shorter transient durations result in higher frequency excitation.

There are various ways of controlling the magnitude of the input force for impact hammer excitation:

1. The input force to the structure is dependent upon the velocity of the impact hammer at the instant of impact between the tip and the structure. If consistency can be established between the contact velocity of each hit, the impact force magnitude can be accurately controlled. However, this is very hard to accomplish manually.

2. The magnitude of the input force can also be varied by changing the mass of the hammer head. By adding mass to the hammer (increasing the weight of the head), the input force magnitude will be increased. Changing the mass of the head is easily accomplished by attaching a different impact head.

The frequency content of the input force can be controlled by controlling the duration of the transient time domain signal. This can be accomplished in two ways:

1. Adding mass to the hammer will increase the time domain pulse width of the impact. This in turn will decrease the effective frequency range of excitation.

2. Decreasing the stiffness of the contact surface between the hammer and the test structure will also decrease the effective frequency range of excitation. Without altering the structure, the easiest means of altering the interface stiffness is by changing the tip of the impact hammer. Typically, three tips are supplied with a hammer kit. This gives the user the ability to excite different frequency ranges, depending on the requirements of the test. The softer an impact tip, the lower the interface stiffness, the longer the time domain duration of the impact, and the lower the effective frequency range of excitation.

An impact hammer is best calibrated by using it to measure the dynamic behaviour of a simple structure with known dynamic behaviour. The calibration is usually conducted using a suspended mass (at least 100 times the accelerometer mass), as shown in Figure 5.5. The accelerometer is attached to one side of the mass and the other side is impacted with the hammer. The spectrum analyser is used to measure the dimensionless inertance frequency response function, E_a/E_f (acceleration/force). Since the inertance frequency response of the pendulum mass is a constant $1/m$, any dynamic characteristics detected are those of the instrumentation. The setup in Figure 5.5 can also be used to calibrate a system for which an electrodynamic shaker, attached to the force transducer, is used as the force input.

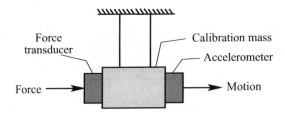

FIGURE 5.5 Calibrating the measuring system.

From Newton's law, $(E_f/S_f)(1/m) = (E_a/S_a)$. A quantity, R, may be defined which can be used in subsequent measurements to convert voltages to $\mathrm{m}/(\mathrm{s}^2\mathrm{N})$:

$$R = S_a/S_f = mE_a/E_f \qquad (5.116)$$

where m is the calibration mass, E_a and E_f are voltages measured at calibration and S_a and S_f are the sensitivities of the accelerometer and force transducer, respectively. Any subsequent frequency response measurements (inertances) made using this measurement system will be calibrated in both amplitude and phase when divided by R.

The main advantage of the impact hammer excitation of the structure is its short measurement time, but this is generally at the expense of a poor signal to noise ratio, and a relatively high crest factor, which can exacerbate any non-linearity effects in the structural response. As with random noise excitation, leakage can be a problem. However, this problem may be minimised by correct windowing of the input force and response signals, as discussed a little later in this section. When doing FRF measurements with an impact hammer, it is necessary to average measurements for a particular excitation point and response point by taking between 5 and 20 measurements, so that the averaged response curve does not change significantly when another measurement is averaged. There are two aspects to the measurements that must be taken into account if large errors in the FRF are to be avoided. The first is that there must be a significant signal to noise ratio (a minimum of 5 dB but preferably more than 10 dB) over the frequency range of interest. This requires an examination of the auto power spectra of both the accelerometer signal and the force signal for the two conditions of 'excitation' and 'no excitation' and the difference in dB between the two conditions is the signal to noise ratio. The second source of large errors in the FRF measurement is the 'double hit' phenomenon. Sometimes this is difficult to avoid, as it is often caused by the structure vibrating from the first hit and hitting the hammer before it is removed. All spectra corresponding to double hits should be removed from the averaged spectra. Minimising the number of double hits sometimes takes a bit of practice with one's technique and less dexterous operators may never be able to achieve single hits. The presence of double hits is easily seen in the time domain impact hammer signal.

5.4.1.5   Structural Response Transducers

The third step in setting up a modal test is to select and mount the structural response transducers. The transducers most commonly used to measure the structural or acoustic response are accelerometers connected to charge amplifiers, so that a voltage output proportional to the structural acceleration is generated. For acoustic spaces, microphones are used.

In placing structural response sensors, for example accelerometers, care should be taken not to mass load the structure. Alternatively, if a finite element analysis is available, it may conveniently be altered to account for the accelerometer masses. Subsequently, their effects can be investigated analytically by letting masses tend to zero. If a finite element analysis is not available, tests may be repeated with a lighter (or heavier) accelerometer to experimentally investigate any loading effect. If mass loading is a problem, then the structure's vibration response should be measured using a non-contacting sensor such as a close mounted microphone or laser doppler velocimeter. Care must also be taken to ensure that accelerometers are securely fastened to the structure under test. However, this problem is generally only acute at high frequencies. Care should also be taken in the use of force transducers to measure the force input to the structure. The force transducer should be mounted as close as possible to the point of force input to the structure. Any mass or damping (such as a linear bearing in a shaker) between the transducer and the structure will lead to errors. In addition, piezoelectric force transducers are sensitive to bending moments, so care should be taken to avoid bending of the transducer.

A more convenient (although relatively expensive) way of measuring the structural response is to use a laser doppler velocimeter (LDV) (or vibrometer), which is capable of mapping the entire structural response in three dimensions, to any excitation, in a very short time. Most LDV systems include modal analysis software that provides mode shapes, resonance frequencies and damping for modes resonant in a specified frequency range.

5.4.2 Acoustic Modal Analysis

The test setup for experimental modal analysis of an acoustic space consists of:

1. swept frequency oscillator (or random noise generator) and power amplifier attached to a loudspeaker with an airtight backing enclosure, in which a microphone is located to provide a signal proportional to the volume velocity of the speaker diaphragm;

2. microphone and amplifier;

3. two channel FFT spectrum analyser and some form of interface to a personal computer with modal analysis software installed, or a computer with a data acquisition system, FFT analysis software and modal analysis software.

The acoustic space is usually excited at several locations by moving a loudspeaker with an airtight backing enclosure. The acoustic field is usually sampled with a microphone at a large number of locations (whose geometrical positions have been previously entered into the computer). The number of microphone locations required is generally a little more than twice the order of the highest mode to be investigated. If results from a finite element analysis are available, they can be used as a guide to determine the most suitable locations for the microphones. Alternatively, one or two preliminary frequency sweeps and response measurements at a few locations may be used to obtain a rough idea of resonance frequencies and mode shapes prior to a full modal analysis.

For very reverberant spaces excited at low frequencies, the loudspeaker could add significantly to the damping in the space. In these cases, it may be preferable to use a special source such as the one described in Peeters et al. (2014), although it may be difficult to determine the source volume velocity. However, the volume velocity will be proportional to the voltage driving the source and in this case it is still possible to determine modal resonance frequencies, mode shapes and damping.

The transfer function required is between the source volume velocity and the sound pressure (that is, p/v, or acoustic impedance, Z_A) at a number of locations in the enclosed space. When a small loudspeaker with an airtight backing enclosure is used, the volume velocity, v is related to the sound pressure, p_{in}, inside the backing enclosure at frequency, ω, radians/sec, by:

$$v = j\frac{p_{in}V\omega}{\rho c^2} \qquad (5.117)$$

where V is the volume of the backing enclosure and ρ and c are the density of gas and speed of sound, respectively, in the backing enclosure. The sound pressure inside the backing enclosure can be measured with a microphone mounted in an airtight hole in the enclosure wall. In cases where an alternative acoustic source is used (such as the one used by Peeters et al. (2014)), the voltage signal driving the source can be used as it will be proportional to the source volume velocity (although the proportionality relationship generally will not be constant with frequency, resulting in errors in damping estimates).

The loudspeaker is driven by a power amplifier which, in turn, may be driven by a single sine wave, the frequency of which is incremented after each test, or it may be driven by a swept sine signal, by pseudo-random noise (see Section 7.3.17) or by random noise (see Section 5.4.1.3).

5.4.3 Measuring the Transfer Function (or Frequency Response)

Until now, we have considered harmonic excitation forces of frequency, ω (rad/s). However, if a structure or acoustic space is excited by a band of frequencies simultaneously, it is desirable to obtain the frequency response function phase and amplitude for the band of frequencies simultaneously. The most convenient way to achieve this is by Fourier analysis, which is discussed in Chapter 7.

When taking transfer function measurements using either random noise or impact excitation (structure only), it is necessary to take many averages to ensure that the system 'noise' is minimised. A way of checking that this is so is to look at the coherence function γ^2 (see Chapter 7) which is automatically calculated by most spectrum analysers at the same time as the transfer function. The coherence function gives an indication of how much of the output signal is caused by the input signal and varies between 0 and 1. A value of 1 represents a valid measurement, while a value of less than about 0.7 represents an invalid measurement.

However, great care should be exercised when interpreting coherence data, as its meaning is dependent on the number of averages taken. For example, a single average will give a coherence of unity which certainly does not imply an error free frequency response measurement. As a guide, the expected random error in a frequency response measurement as a function of the coherence value and number of averages taken is given in Figure 5.6 which is adapted from the standards, ISO 7196-1 (2011); ISO 7196-2 (2015); ISO 7196-5 (1995).

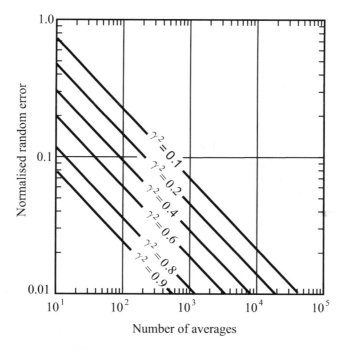

FIGURE 5.6 Accuracy of FRF estimates vs. number of averages for various values of the coherence, γ^2, between the input force and system response.

In general, low coherence values can indicate one or more of five problems:

1. insufficient signal level, as illustrated in Figure 5.7 (turn up gain on the analyser);
2. poor signal to noise ratio;
3. presence of other extraneous forcing functions;
4. insufficient averages taken;
5. leakage.

Insufficient signal level (case 1 above) is characterised by a rough plot of coherence vs. frequency, even though the average may be close to one (see Figure 5.7). To overcome this, the gains of the transducer preamplifiers should be turned up or the attenuator setting on the spectrum analyser adjusted, or both.

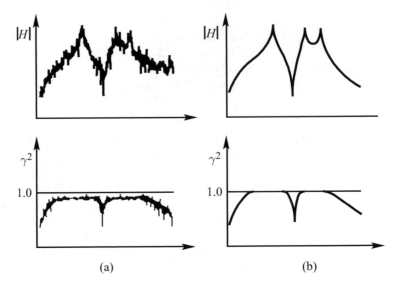

FIGURE 5.7 Effect of insufficient signal level on the frequency response function, H, and the coherence, γ^2: (a) measurement with noise due to incorrect attenuator setting (insufficient signal level); (b) same measurement with optimum attenuator setting.

For excitation by random noise via a shaker (for a structure) or loudspeaker (for an acoustic space), poor coherence between the excitation and response signals is often measured at system resonances. This is partly due to the leakage problem discussed previously and partly because the input force (for a structure) or input sound (for an acoustic space) is very small (close to the instrumentation noise floor) at these frequencies.

Poor signal to noise ratio (case 2 above) can have two causes in the case of impact excitation of a structure. The first is due to the bandwidth of the input force being less than the frequency range of interest or the frequency range set on the spectrum analyser. This results in force zeros at higher frequencies, giving the false indication of many high frequency vibration modes. Conversely, it is not desirable for the bandwidth of the force to extend beyond the frequency range of interest to avoid the problem of exciting vibration modes above the frequency range of interest, thus contaminating the measurements with extraneous signals. A good compromise is for the input force auto power spectrum to be between 10 and 20 dB down from the peak value at the highest frequency of interest.

The second reason for poor signal to noise ratio during impact testing is the short duration of the force pulse in relation to the duration of the time domain data block. In many cases, the pulse may be defined by only a few sample points comprising only a small fraction of the total time window, the rest being noise. Thus, when averaged into the measurement, the noise becomes significant. This S/N problem can be minimised by using special force and response data windows that are specifically designed for impact testing. Such windows allow the force pulse to pass unattenuated and then selectively attenuate the following noise. An ideal window for the impact force signal is illustrated in Figure 5.8.

The first portion is represented by a multiplier of one while the second portion is a half Hanning window. The results of using the window are shown in Figure 5.9.

In practice, it is sufficient to use an adjustable-width, rectangular window for the impact force signal and an adjustable length single-sided exponential window for the response signal. This ensures that the force signal and the structural response caused by it are weighted strongly compared with any extraneous noise that may be present. These windows, which are available in many commercial spectrum analysers, are illustrated in Figure 5.10.

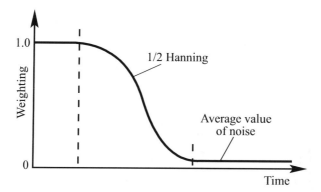

FIGURE 5.8 Ideal window for the input force signal for impact testing.

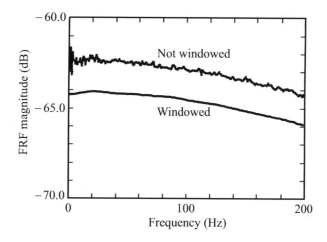

FIGURE 5.9 Effect of correct windowing of the impact force input signal.

Another way (if the above mentioned alternative is not possible) to overcome the problem of short force duration is to use a number of randomly spaced impacts over the duration of the sample record and use a standard Hanning window. If it is possible to use only one impact, and the spectrum analyser does not have the optimum impact window mentioned above, then a rectangular window must be used, otherwise the force pulse will be effectively missed and the transfer function of noise will be measured.

Items 3 and 4 above have obvious solutions and will not be discussed further. The leakage problem (item 5 above) can be caused by incorrect windowing of the data in the time domain. Leakage results in a broadening of the resonance peaks and always occurs when a signal is truncated by the measurement time window. Leakage is minimised for random noise excitation by using either a Hanning or Kaiser–Bessel time window function. For impact excitation, leakage will not be a problem if the windows suggested above to minimise the S/N problem are used. The only possible problem may occur when a rectangular window is used and the force pulse or response pulse duration is sufficiently long that it had not decayed sufficiently by the end of the time window.

In some cases where the system is lightly damped, the resonance peaks may be so sharp that the frequency resolution of the analyser in the chosen baseband is insufficient. In this case, the response peak will follow the shape of the window function used and the calculated damping (using modal analysis software) will be incorrect. In this case, it is necessary to use zoom analysis (not simply expanding the displayed frequency response spectrum, but a new analysis with finer

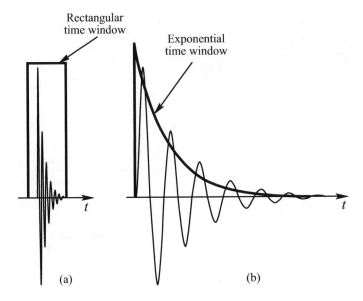

FIGURE 5.10 Typical impact force and response windows available in many commercial FFT analysers: (a) rectangular force window; (b) exponential response window.

frequency resolution) to increase the frequency resolution. This can be done on individual parts of the spectrum by some spectrum analysers or with software on a personal computer, and the results combined to cover the entire frequency range.

In some cases, the resonance peak may appear inverted (see Figure 5.11) due to the extremely small value of the cross spectrum $G_{xy}(j\omega)$ being divided by an extremely small value of the power spectrum of the applied force $G_{xx}(j\omega)$. Note that in this latter case, the value of the coherence function between the applied force and the structural response is also small, indicating unreliable data.

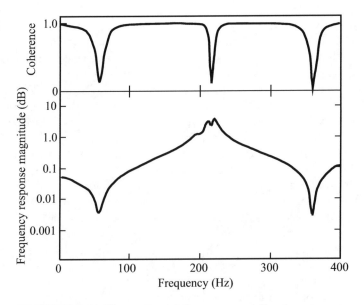

FIGURE 5.11 Illustration of the resonance dip phenomenon.

The transfer function measurement is used almost exclusively in modal testing, as it allows the structural or acoustic response to be normalised to the input force or input sound at each frequency. Sometimes if swept sine excitation is used, a feedback system keeps the input force or sound constant and then only the response power spectrum is recorded, using the 'peak' averaging facility on the spectrum analyser (or computer with data acquisition system and spectrum analysis software), which allows recording of the maximum response at each frequency during the sweep. Swept sine excitation is not in common use today, given the shorter testing times and higher accuracy available with random excitation. However, for cases where structures or acoustic spaces are heavily damped, a slow sine sweep will often provide better results.

Overloading of transducer pre-amplifiers and the spectrum analyser amplifiers (or the data acquisition system input) can also lead to serious errors in the transfer function measurements. Where spectrum analysers are used, this is best avoided by setting the spectrum analyser input attenuators low enough so that the spectrum analyser will overload before the transducer pre-amplifiers will. Some spectrum analysers can be set up to discard a data record if it contains any overload measurements, otherwise the overload light should be carefully monitored during testing. Where a data acquisition system and computer are used with spectrum analysis software, there must be a procedure for identifying and discarding records that contain overload data.

5.4.4 Modal Parameter Identification

The transfer function data corresponding to the ratio of the response at many locations to the input force at another location or vice versa is used to produce modal resonance frequencies, modal damping and mode shape data for each vibration mode or acoustic mode that is identified. This is generally done by downloading all the transfer function data and associated coordinate locations to a personal computer which contains one of the commercially available modal analysis software packages.

For structures that have several modes, most software has the option of employing specialised curve fitting procedures that determine modal resonance frequencies and damping for all modes at once (multimodal curve fitting). If modes are closely spaced in frequency and/or well damped, then treating each one individually and curve fitting each one individually (single mode curve fitting) will lead to large errors; thus, the reason for treating all modes together. However, the multimodal curve fitting procedures can be very time consuming and memory hungry, and thus suited to environments where accuracy is more important than speed.

5.4.4.1 Mode Shapes

Evaluation of the mode shapes involves calculation of the relative vibration amplitude at each test point on the structure; however, almost all modal analysis software packages use only linear interpolation between adjacent geometric points on the structure, resulting in a distorted view of the true mode shape if an insufficient number of points is used. Most software packages also allow an animated view of the mode shape for a particular mode.

Once the transfer functions and mode shapes have been calculated for a particular structure, most modal analysis software allows calculation of the response of the structure (see Equations (5.100) and (5.101)) at each node to a defined excitation force, which may be in the form of a time history, random noise or a number of sinusoids. Some software also presents an animated total response shape as a result of a single frequency sinusoidal force applied at any number of nodes. However, a limitation of most modal analysis software is its inability to provide information regarding the contributions of each mode to the overall response when the structure is excited by a known force.

A capability offered by some modal analysis software is the calculation of the effect of adding or subtracting mass at various nodes, the effect of changing the stiffness between nodes or between the ground and particular nodes, and the effect of adding damping between nodes or between the ground and various nodes. This capability is invaluable when it is desired to modify the structural response to suit particular operating requirements.

5.4.4.2   SDOF Curve Fit of FRF Data - Peak Amplitude Method

The peak amplitude method for extracting modal resonance frequencies and damping assumes that:

1. the system resonance frequencies correspond to the peaks in the FRF; and
2. the damping associated with each peak may be calculated from the bandwidth corresponding to the two data points either side of the peak, which are less than the peak amplitude by a factor of $\sqrt{2}$, or 3 dB (see Section 6.6).

The structural (or hysteretic) damping is given by (see Figure 5.12):

$$\eta_i = \frac{(\omega_a^2 - \omega_b^2)}{2\omega_i^2} = \frac{(\omega_a + \omega_b)(\omega_a - \omega_b)}{2\omega_i^2} \approx \frac{(\omega_a - \omega_b)(2\omega_i)}{2\omega_i^2} = \frac{(\omega_a - \omega_b)}{\omega_i} = \frac{\Delta\omega}{\omega_i} \qquad (5.118)$$

If viscous damping rather than structural (hysteretic) damping is present, η_i in Equation (5.118) may be replaced with $2\zeta_i$ and the mobility FRF used in place of the receptance FRF. For acoustic systems, the structural velocity in the mobility FRF is replaced with the acoustic pressure.

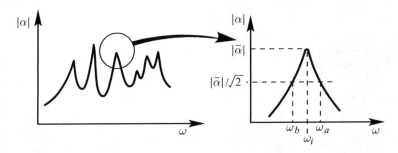

FIGURE 5.12 Modal parameter estimation using the amplitude method (after Ewins (2000)).

5.4.4.3   SDOF Curve Fit of FRF Data - Circle Fit Method, Structural Damping

The most useful single-degree-of-freedom curve fitting technique is known as the circle fit. A curve fitting operation is performed on FRF data in the vicinity of a resonance peak and this allows extraction of resonance frequency and damping information for a single mode. For structural (hysteretic) damping, the receptance FRF must be used, as other FRF's do not result in data points forming the arc of a circle; rather, they will form part of an ellipse.

If the FRF data for a single-degree-of-freedom system are presented in the form of a Nyquist plot (real vs. imaginary), then the data should fall on the circumference of a circle. In MDOF systems, the data around a particular resonance peak will fall on the arc of a circle, provided the peaks are reasonably well separated. This arc of data points is then fitted to a circle.

The circle fit method is a refinement of the very simple amplitude method; however, the circle fit method includes the effects of other modes, provided they are not too close in frequency to the mode under consideration. Thus, the circle fit method only assumes that in the vicinity of a resonance peak, the system response is dominated by a single mode and that in the small frequency range about resonance, the effects of all other modes are essentially constant and independent of frequency. This assumption can be expressed as follows for mode, i. From Section 5.2:

$$\alpha_{jk}(\omega) = \sum_{i=1}^{M} \frac{\psi_i(\boldsymbol{x}_j)\psi_i(\boldsymbol{x}_k)}{m_i(\omega_i^2 - \omega^2 + \mathrm{j}\eta_i\omega_i^2)} \tag{5.119}$$

which can be written as:

$$\alpha_{jk}(\omega) = \frac{\psi_i(\boldsymbol{x}_j)\psi_i(\boldsymbol{x}_k)}{m_i(\omega_i^2 - \omega^2 + \mathrm{j}\eta_i\omega_i^2)} + \sum_{\substack{r=1 \\ r \neq i}}^{M} \frac{\psi_r(\boldsymbol{x}_j)\psi_r(\boldsymbol{x}_k)}{m_r(\omega_r^2 - \omega^2 + \mathrm{j}\eta_r\omega_r^2)} \tag{5.120}$$

It is assumed that the second term in Equation (5.120) is independent of frequency (over the small frequency range under consideration) and Equation (5.120) may then be written as:

$$\alpha_{jk}(\omega) = \frac{\psi_i(\boldsymbol{x}_j)\psi_i(\boldsymbol{x}_k)}{m_i(\omega_i^2 - \omega^2 + \mathrm{j}\eta_i\omega_i^2)} + B_{ijk} = \frac{A_{ijk}}{\omega_i^2 - \omega^2 + \mathrm{j}\eta_i\omega_i^2} + B_{ijk} \tag{5.121}$$

where A_{ijk} is referred to as the modal constant. This can be demonstrated using a four DOF system, as shown in Figure 5.13. For this system, the receptance properties were computed with Equation (5.121) over a small frequency range around the second mode resonance frequency, between ω_1 and ω_2. Each of the two terms in Equation (5.121) has been plotted separately, in Figures 5.13(a) and 5.13(b), respectively, with the total receptance over the same frequency range shown in Figure 5.13(c). The first term, which only includes the effect of the mode of interest and is plotted in Figure 5.13(a), varies substantially over the narrow frequency range that encompasses the resonance region, with values falling on the expected circular arc, while the second term in Equation (5.121), which includes the combined effects of all the other modes, varies by only a small amount over the narrow frequency range between ω_1 and ω_2. As can be seen from Figures 5.13(a) and 5.13(c), the circle representing the receptance for the single mode under consideration has had its centre displaced down and across to obtain the total receptance. This means that the combined effect of the other modes on the FRF can be represented as a constant term for frequencies around the resonance frequency of the mode under consideration.

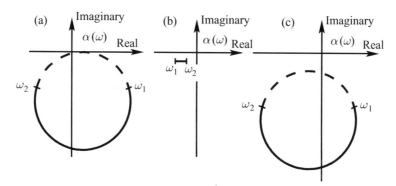

FIGURE 5.13 Nyquist plot of four-degree-of-freedom receptance data: (a) first term; (b) second term; (c) total (after Ewins (2000)).

When undertaking a circle fit, the basic function of interest is:

$$\alpha(\omega) = \frac{1}{\omega_i^2 - \omega^2 + \mathrm{j}\eta_i\omega_i^2} \tag{5.122}$$

since the effect of including the product, $(\psi_i(\boldsymbol{x}_j)\psi_i(\boldsymbol{x}_k)/m_i)$ (referred to as the modal constant, A_{ijk}), is to scale the size of the circle by $|\psi_i(\boldsymbol{x}_j)\psi_i(\boldsymbol{x}_k)/m_i|$ and to rotate it by the phase of

$\psi_i(\boldsymbol{x}_j)\psi_i(\boldsymbol{x}_k)$. Equation (5.122) is plotted in Figure 5.14(a) and the result after including the A_{ijk} term is shown in Figure 5.14(b). Note that Figure 5.14 represents a single mode only — the effects of other non-resonant modes on the response of the mode under consideration have not been included. The effect of these other modes is discussed below.

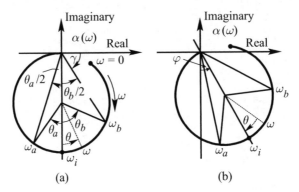

FIGURE 5.14 Properties of the modal circle: (a) unscaled modal circle; (b) scaled modal circle (after Ewins (2000)).

It can be seen from Figure 5.14(a) that for any frequency ω:

$$\tan\gamma = \eta_i/(1 - (\omega/\omega_i)^2) \tag{5.123}$$

$$\tan(90° - \gamma) = \tan(\theta/2) = \frac{(\omega_i^2 - \omega^2)}{\omega_i^2 \eta_i} = \frac{1 - (\omega/\omega_i)^2}{\eta_i} \tag{5.124}$$

from which:

$$\omega^2 = \omega_i^2[1 - \eta_i \tan(\theta/2)] \tag{5.125}$$

If Equation (5.125) is differentiated with respect to θ:

$$\frac{\mathrm{d}\omega^2}{\mathrm{d}\theta} = -\left(\frac{\omega_i^2 \eta_i}{2}\right)\left\{1 + \frac{[1 - (\omega/\omega_i)^2]^2}{\eta_i^2}\right\} \tag{5.126}$$

The reciprocal of Equation (5.126) is a measure of the rate at which the locus sweeps around the circular arc and reaches a maximum value (maximum sweep rate) when $\omega = \omega_i$, the natural frequency of the mode under consideration. This is demonstrated by differentiating Equation (5.126) with respect to frequency, which shows that:

$$\frac{\mathrm{d}}{\mathrm{d}\omega}\left(\frac{\mathrm{d}\omega^2}{\mathrm{d}\theta}\right) = 0 \quad \text{when} \quad (\omega_i^2 - \omega^2) = 0 \tag{5.127}$$

This result is useful for analysing MDOF system data, since examination of the relative spacing of the measured data points around the circular arc near each resonance allows the resonance frequency to be determined.

A value for the modal loss factor, η_i, can also be obtained from this basic modal circle. Consider two specific points on the circle, one corresponding to a frequency, ω_b, below the natural frequency, and the other to one, ω_a, above the natural frequency. Referring to Figure 5.13:

$$\tan(\theta_b/2) = (1 - (\omega_b/\omega_i)^2)/\eta_i \tag{5.128}$$

$$\tan(\theta_a/2) = ((\omega_a/\omega_i)^2 - 1)/\eta_i \tag{5.129}$$

Using Equations (5.128) and (5.129), an expression for the damping of the mode can be obtained as:

$$\eta_i = \frac{\omega_a^2 - \omega_b^2}{\omega_i^2[\tan(\theta_a/2) + \tan(\theta_b/2)]} \qquad (5.130)$$

If the damping is light ($\eta_i < 0.03$). then Equation (5.130) can be simplified to:

$$\eta_i = \frac{2(\omega_a - \omega_b)}{\omega_i[\tan(\theta_a/2) + \tan(\theta_b/2)]} \qquad (5.131)$$

and using the two points for which $\theta_a = \theta_b = 90°$ (the half-power points), the familiar relationship is obtained for light damping:

$$\eta_i = (\omega_a - \omega_b)/\omega_i \qquad (5.132)$$

If $\eta_i > 0.03$, then Equation (5.130) must be used and for $\theta_a = \theta_b = 90°$:

$$\eta_i = (\omega_a^2 - \omega_b^2)/\omega_i^2 \qquad (5.133)$$

The circle diameter corresponding to Equation (5.122) is given by $D_{ijk} = 1/(\omega_i^2 \eta_i)$. When scaled by the product $[\psi_i(\boldsymbol{x}_j)\psi_i(\boldsymbol{x}_k)]/m_i$, the circle diameter becomes:

$$D_{ijk} = \frac{|\psi_i(\boldsymbol{x}_j)\psi_i(\boldsymbol{x}_k)|}{\omega_i^2 \eta_i m_i} = \frac{|A_{ijk}|}{\omega_i^2 \eta_i} \qquad (5.134)$$

The whole circle is rotated so that the principal diameter, which passes through ω_i, is orientated at an angle, φ, to the imaginary axis and is equal to the phase angle of A_{ijk}. At this point, no allowance has been made for the effect on the response of modes other than the one under consideration. This is discussed in Section 5.4.4.6.

5.4.4.4   SDOF Curve Fitting of FRF Data - Circle Fit Method, Viscous Damping

If viscous damping rather than structural (hysteretic) damping is present, then the mobility FRF function must be used for structural modal analysis and the acoustic impedance function must be used for acoustic modal analysis, as other FRF's do not result in data points forming the arc of a circle; rather, they will form part of an ellipse. This means that for acoustic systems, the structural velocity in the structural mobility FRF is replaced with the acoustic pressure and the structural force is replaced with the acoustic source volume velocity, thus producing an acoustic impedance function in place of the structural mobility function.

As the structural mobility function, Y (or its acoustical equivalent, Z_A), must be used, Equation (5.121) becomes:

$$Y_{jk}(\omega) = \frac{j\omega\left[\theta_{ijk}(\boldsymbol{x}_j) + j(\omega/\omega_i)\phi_{ijk}(\boldsymbol{x}_k)\right]}{m_i(\omega_i^2 - \omega^2 + 2j\zeta_i\omega\omega_i)} + B_{ijk} = j\omega\frac{A_{ijk}}{\omega_i^2 - \omega^2 + 2j\zeta_i\omega\omega_i} + B_{ijk} \qquad (5.135)$$

where A_{ijk} and B_{ijk} are represented by different expressions to those for hysteretic (or structural) damping.

Equation (5.122) becomes:

$$Y(\omega) = \frac{j\omega}{\omega_i^2 - \omega^2 + 2j\zeta_i\omega\omega_i} \qquad (5.136)$$

The angle, θ, defining points on the circle is:

$$\tan(\theta/2) = \frac{\text{Im}\{Y(\omega)\}}{\text{Re}\{Y(\omega)\}} = \frac{\omega_i^2 - \omega^2}{2\zeta_i\omega\omega_i} \qquad (5.137)$$

and following the same procedure as for structural (hysteretic) damping,

$$\tan(\theta_b/2) = \frac{\omega_i^2 - \omega_b^2}{2\zeta_i\omega_b\omega_i} \tag{5.138}$$

$$\tan(\theta_a/2) = \frac{\omega_a^2 - \omega_i^2}{2\zeta_i\omega_a\omega_i} \tag{5.139}$$

Thus:

$$\zeta_i = \frac{\omega_a^2 - \omega_b^2}{2\omega_i[\omega_a \tan(\theta_a/2) + \omega_b \tan(\theta_b/2)]} \tag{5.140}$$

If the half power points are selected so that $\theta_a = \theta_b = 90°$:

$$\zeta_i = (\omega_a - \omega_b)/2\omega_i \tag{5.141}$$

which is valid for all values of damping.

The circle diameter is the same as for structural (hysteretic) damping (Equation (5.134)), except that η_i is replaced with $2\zeta_i$ and A_{ijk} is replaced with $j\omega A_{ijk}$. The circle fit for a mobility analysis is therefore rotated by 90° relative to an equivalent circle for a receptance function. The effect of other modes on the frequency response function is discussed in Section 5.4.4.6.

5.4.4.5   Circle Fit Analysis Procedure

The steps involved in a circle fit are:

1. select the points to be used;
2. fit the circle and estimate the quality of fit;
3. determine the natural frequency by locating it on the circle;
4. calculate multiple damping estimates (using various choices for θ_a and θ_b), and the standard deviation of the results;
5. calculate the modal constant, $A_{ijk} = \psi_i(\boldsymbol{x}_j)\psi_i(\boldsymbol{x}_k)/m_i$.

Step 1 can be either manual or automatic. If automatic, a fixed number of points on either side of any identified maximum in the response modulus can be selected. However, if manual, the operator has the option of rejecting suspect data points that are possibly influenced significantly by neighbouring modes. If possible, the data points should encompass about 270° of the circle, although this is not often achieved in practice. However, if the range is limited to less than 180°, the result becomes more and more sensitive to the accuracy of the data. At least six points should be used.

Step 2 relies on the use of curve fitting software to find a circle that minimises the squares of the deviations of the data points from the circle circumference. 'Errors' of the order of 1–2% should be expected. An example circle fit to measured data is shown in Figure 5.15(a). The quality of fit is related to the mean square error between the data points and the fitted circle.

Step 3 can be implemented by numerically constructing radial lines from the circle centre to a succession of points equally spaced in frequency around the resonance frequency and by noting the angles subtended from each line to its neighbours (see Figure 5.15(b)). The frequency at which the frequency is changing maximally between the points on the circle can be found and the location refined using a finite difference method, which usually results in an estimate of the natural frequency with a precision of about 10% of the frequency increments between the points.

Step 4 involves computing a set of damping estimates using every possible combination from the selected data points of one point below resonance with one above resonance using Equation (5.130). The mean value and standard deviation of all the estimates are then computed and values are examined individually to see whether there are any particular trends. If the

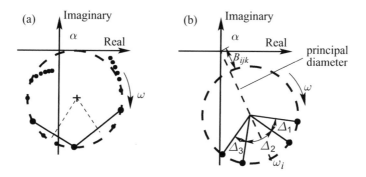

FIGURE 5.15 Circle fit to FRF data: (a) circle fit; (b) natural frequency location (after Ewins (2000)), where B_{ijk} is defined in Equation (5.121).

standard deviation is less than 4–5%, then the analysis may be considered to be good (Ewins, 2000). However, a scatter of 20 or 30%, would indicate a problem. If the large variations in damping estimate are random, then the scatter is probably due to measurement errors but if it is systematic, then it could be caused by poor experimental setup, interference from neighbouring modes or non-linear behaviour, none of which should be averaged out.

The final step 5 involves determining the magnitude and phase of the modal constant, A_{ijk}, (using Equation (5.134)) from the circle diameter passing through the point ω_i, and from its orientation relative to the real and imaginary axes. This calculation can proceed once the natural frequency and damping estimates have been determined.

5.4.4.6 Reconstructing Frequency Response Curves

If it is desired to construct a theoretically regenerated FRF plot against which to compare the original measured data, it is necessary to determine the contribution to this resonant response of the other modes. Although these other modes are not analysed directly, they do affect the FRF data for the modes that are analysed directly. The effect of other modes on the FRF function can be determined by evaluation of the quantity, B_{ijk}, of Equations (5.121) or (5.135). The value of B_{ijk} is the distance from the top of the principal diameter (the one passing through the point ω_i) to the origin, as shown in Figure 5.15. The effect of other modes on the reconstructed FRF function can also be characterised in terms of a pair of residual functions, one of which accounts for modes with lower resonance frequencies and one that accounts for modes with higher resonance frequencies.

To regenerate a receptance FRF curve from the modal parameters extracted from measured data for the case of hysteretic damping, the following equation may be used:

$$\alpha_{jk} = \sum_{i=M_1}^{M_2} \frac{A_{ijk}}{\omega_i^2 - \omega^2 + j\eta_i\omega_i^2} \tag{5.142}$$

However, the equation that would fit the measured FRF more closely is:

$$\alpha_{jk} = \sum_{i=1}^{M} \frac{A_{ijk}}{\omega_i^2 - \omega^2 + j\eta_i\omega_i^2} \tag{5.143}$$

which can be written as:

$$\alpha_{jk} = \left(\underbrace{\sum_{i=1}^{M_1-1}}_{\substack{\text{(low frequency} \\ \text{modes)}}} + \sum_{i=M_1}^{M_2} + \underbrace{\sum_{i=M_2+1}^{M}}_{\substack{\text{(high frequency} \\ \text{modes)}}} \right) \frac{A_{ijk}}{\omega_i^2 - \omega^2 + \mathrm{j}\eta_i \omega_i^2} \qquad (5.144)$$

Figure 5.16 shows typical values of each of the three terms separately, and the middle one is all that is computed using modal data extracted from the modal analysis. To make the model accurate within the frequency range of the tests, it is necessary to correct the regenerated plot within the central frequency range to take account of the low frequency and high frequency modes.

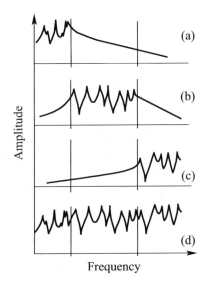

FIGURE 5.16 Contribution of various terms to the total FRF: (a) low frequency modes; (b) identified modes; (c) high frequency modes; (d) all modes.

From Figure 5.16, it can be seen that in the frequency range of interest, the first term of Equation (5.144) approximates to a mass-like behaviour while the third term approximates to a stiffness-like behaviour. On this basis the residual terms may be quantified and Equation (5.144) may be rewritten as:

$$\alpha_{jk} = -\frac{1}{\omega^2 M_{jk}^R} + \sum_{i=M_1}^{M_2} \left(\frac{A_{ijk}}{\omega_i^2 - \omega^2 + \mathrm{j}\eta_i \omega_i^2} \right) + \frac{1}{K_{jk}^R} \qquad (5.145)$$

where M_{jk}^R and K_{jk}^R are the residual mass and stiffness for that particular FRF. The residual terms are calculated using the difference between the actual FRF and the FRF constructed from the modal parameters at a few points at each end of the FRF curve in the frequency range of interest. The points at the low frequency end allow calculation of the required value of the constant, M_{jk}^R, that will make the regenerated FRF values similar to the measured values. Similarly the points at the high frequency end allow the value of K_{jk}^R to be calculated. As the mass residual constant may also affect the FRF at higher frequencies in the range of interest and the stiffness residual constant may affect the FRF at lower frequencies in the range of interest, the above-mentioned process may need to be conducted iteratively in some cases.

5.4.4.7 Multi-Degree-of-Freedom Curve Fitting FRF Data

As mentioned previously, most commercially available modal analysis software packages utilise some form of MDOF curve fitting, where a section of the FRF, containing several modes, is curve fitted to extract the modal parameters for all modes present in the frequency range selected (see Ewins (2000)). The user is usually asked how many modes he/she would like included in the fit. A sufficient number of modes will produce an almost perfect fit, but will invariably produce computational modes, which are included expressly to satisfy the curve-fitting requirement and which do not necessarily represent any physical behaviour of the structure at all. Taken to extremes, computational modes will 'fit' small irregularities or errors in the measured data. Clearly, it is an important requirement that such computational modes can be distinguished from the genuine modes, and eliminated from the analysis.

5.4.4.8 Computational Mode Elimination

One way of distinguishing computational modes from real modes is to specify more modes than really exist in the fitting procedure. Then 10 different fits to the FRF data can be made, each fit using every 10th data point of the original FRF data and each fit using a different subset of the original data. Thus, eventually all data points are used by the 10 curve fits, but each curve fit uses a different 10% of the total data. The result will be 10 estimates of frequency and damping for each of the modes. The real modes will correspond to estimates of resonance frequency and damping that do not vary much from run to run; however, the computational modes will be characterised by large variances (greater than 5%) and can be rejected on that basis.

If the system being excited is non-linear, then different values of resonance frequency and damping will be obtained for different levels of force input. Other non-linearities such as stiffness being a non-linear function of frequency will produce distortions in the FRF curves that can lead to inconsistencies in damping estimates, which will vary depending upon the calculation method.

5.4.4.9 Global Fitting of FRF Data

This is the process of analysing the curve fitted results of a number of different FRF curves, corresponding to force inputs and response measurements at various different parts of the structure, to yield the full modal model from which animated mode shapes are derived. As all FRF curves do not provide exactly the same resonance frequencies and damping values, these quantities are generally averaged.

Up to now, means to obtain resonance frequency and damping data from a single frequency response function have been discussed. The resonance frequency and damping values so obtained differ slightly depending upon the location of the excitation force and response measurement. Thus, to obtain structural mode shape data, the barest minimum number of measurements required is one force measurement at one force impact location and acceleration measurements at this and all other measurement points on the structure. Thus, it is necessary to measure a set of FRF curves, one curve for each point on the structure, with each curve sharing the same excitation point. For structural modal analysis, it is often prudent to repeat the measurements with a second force impact location to improve the likelihood of exciting modes that may have been excited poorly at the first force location and to replace poor FRF data which is often obtained when the force input and response measurement are a long way apart. The resonance frequencies, damping values and mode shapes are determined by averaging the overdetermined data. Alternatively, for impact excitation, it may be more convenient to use just a few accelerometer locations and a much larger number of impact locations.

To obtain acoustic mode shape data, the loudspeaker sound source should be placed in part of the room where most modes have an antinode. For a rectangular room, this is in a room corner. Sound pressure measurements are then made using a microphone at many different locations in the room. In most cases, it is necessary to use several locations of the loudspeaker. In reverberant environments at low frequencies, it is likely that the loudspeaker will contribute to the modal damping. To quantify the likely problem, the loudspeaker can be tested in a reverberation chamber with hard walls and the reverberation time measured with the loudspeaker present and then with the room driven by a horn driver that would have a very low damping effect. The loss factor for both situations is then calculated using Equation (6.38) and the difference is the effect of the loudspeaker.

5.4.4.10 Response Models

These are needed to enable a prediction to be made of the response of the analysed structure (or acoustic space) to a given excitation force (or volume velocity source) at any given location. Thus, the response model is an FRF matrix whose order $(n \times n)$ is dictated by the number of coordinates, n, used in the test. The FRFs are regenerated using the resonance frequencies, damping values and modal coefficients calculated by curve fitting. The FRFs that were not measured are synthesised using the coefficients calculated from the measured FRFs. In principle, this should present no problem, as it is possible to calculate the full response model from a modal model using:

$$\boldsymbol{\alpha}_{n \times n} = \boldsymbol{\psi}_{n \times N} \left[\quad m_i(\omega_i^2 - \omega^2 + \mathrm{j}\eta_i\omega_i^2) \quad \right]^{-1} \boldsymbol{\psi}_{N \times n}^{\mathrm{T}} \tag{5.146}$$

where n is the number of coordinates and N is the number of modes considered. However, this latter process is only successful if the effect of modes outside of the analysis range is taken into account; otherwise large errors can result. The effect mentioned above is taken into account by using a residue matrix, $\boldsymbol{R}$, which must somehow be estimated. Thus, the correct response model is:

$$\boldsymbol{\alpha} = \boldsymbol{\psi} \left[\quad m_i(\omega_i^2 - \omega^2 + \mathrm{j}\eta_i\omega_i^2) \quad \right]^{-1} \boldsymbol{\psi}^{\mathrm{T}} + \boldsymbol{R} \tag{5.147}$$

where $\boldsymbol{\psi}$, m_i, ω_i and η_i are derived from measured FRFs.

The most accurate means of estimating the individual elements of $\boldsymbol{R}$ is to measure all (or more than half at least) of the elements of the FRF matrix, but this represents a large increase in effort. A second, reasonably practical, approach is to extend the frequency range of the modal test beyond the frequency range of interest. As residual terms only significantly affect the modal responses at either end of the frequency range, their effect in the frequency range of interest can be reduced to an unimportant level, provided the analysis range is sufficiently extended. The main problem with this approach is that it is difficult to interpret the word 'sufficient'. However, examination of the regenerated curves for all modes and then with the highest (and lowest if relevant) ones removed will help define 'sufficient'.

A third, less practical, approach is to assess which of the elements in the FRF matrix are likely to be associated with large residual terms and to make sure that these are measured and analysed. In practice, point receptances are expected to have the highest residual values and the remote transfer receptances the smallest. This means that FRFs with significant terms in the $\boldsymbol{R}$ matrix will generally be grouped close to the diagonal of the FRF matrix, so measurements should always include point FRFs and transfer FRFs where the point of excitation is close to the point of measurement.

Attempts to resolve the residual problem by measuring all elements or half of the elements (as the FRF matrix is symmetric) of the FRF matrix raises additional problems due to inconsistencies in estimated resonance frequencies and damping values. At the very least, natural frequencies and damping values should be averaged throughout the model and the mode shapes recalculated using the average damping values. In any case, it is usually necessary to derive the response model by regenerating the FRF functions from the estimated average coefficients.

5.4.4.11   Structural or Acoustic Response Prediction

Another reason for deriving an accurate mathematical model for the dynamics of a structure is to provide the means to predict the response of that structure to more complicated and numerous excitations than can readily be measured directly in laboratory tests. Hence the idea that, by performing a set of measurements under relatively simple excitation conditions and analysing these data appropriately, it is possible to predict the structural or acoustic space response to several excitations applied simultaneously.

The basis of this approach is summarised in the standard equation:

$$\hat{\boldsymbol{w}}\mathrm{e}^{\mathrm{j}\omega t} = \boldsymbol{\alpha}\hat{\boldsymbol{F}}\mathrm{e}^{\mathrm{j}\omega t} \tag{5.148}$$

where the required elements in the FRF matrix can be derived from the modal model by the familiar formula:

$$\boldsymbol{\alpha} = \boldsymbol{\psi}\left[\begin{array}{ccc} \ddots & & \\ & m_i(\omega_i^2 + \mathrm{j}\eta_i\omega_i^2 - \omega^2) & \\ & & \ddots \end{array}\right]^{-1}\boldsymbol{\psi}^{\mathrm{T}} \tag{5.149}$$

In general, this prediction method is capable of supplying good results provided sufficient modes are included in the modal model from which the FRF data are derived.

5.5   Modal Amplitude Determination from System Response Measurements

In many cases for simple structures, mode shapes and resonance frequencies can be determined theoretically. However, when such a simple structure is excited at a single frequency (especially a frequency not corresponding to a resonance frequency), it is of interest to determine the relative amplitudes of the vibration modes from accelerometer measurements at a number of locations on the structure. Similarly, for acoustic spaces of simple shape, excited at a single frequency, it may be of interest to find the relative amplitudes of the acoustic modes that contribute to the total response. The following discussion is based on the vibration modes of a structure, but the approach applies equally well to an acoustic space.

The displacement, $w(\boldsymbol{x}, t)$, of a structure at location, $\boldsymbol{x}$, and time, t, may be written in terms of the modal amplitudes, $a_i(t)$, and mode shape functions, $\varphi(t)$, at location, $\boldsymbol{x}$, and time, t, as:

$$w(\boldsymbol{x}, t) = \sum_{i=1}^{\infty} a_i(t)\varphi_i(\boldsymbol{x}) \tag{5.150}$$

where $a_i(t)$ is the amplitude of mode i at time t and $\varphi_i(x)$ is the mode shape function evaluated at location, $\boldsymbol{x}$. A similar expression may be written for the acoustic pressure, $p(\boldsymbol{x}, t)$, at location, $\boldsymbol{x}$, in an acoustic space, so that:

$$p(\boldsymbol{x}, t) = \sum_{i=1}^{\infty} a_i(t)\psi_i(\boldsymbol{x}) \tag{5.151}$$

If both sides of Equation (5.150) are multiplied by the mode shape function, $\varphi_i(x)$, the result integrated over the surface of the structure, and the orthogonality property of the mode shape functions used, the following is obtained:

$$a_i(t) = \iint\limits_{S} \varphi_i(\boldsymbol{x})w(\boldsymbol{x}, t)\mathrm{d}\boldsymbol{x} \tag{5.152}$$

For an acoustic space, the integration is over the volume of the space, so that:

$$a_i(t) = \iiint\limits_V \psi_i(\boldsymbol{x}) p(\boldsymbol{x}, t) \mathrm{d}\boldsymbol{x} \tag{5.153}$$

where ψ_i is the acoustic mode shape function for mode i.

Measurements of quantities such as displacement or pressure are usually provided by point sensors, such as accelerometers or microphones. It is possible to extract modal amplitudes from such measurements by actually implementing Equations (5.150) and (5.151); the discrete measurements can be interpolated to estimate the continuous function (such as structural displacement or acoustic pressure), and the integration performed numerically (see, for example, Meirovitch and Baruh (1982) for a vibration implementation, and Moore (1979) for an acoustic implementation). However, there is a second approach to resolving modal amplitudes that is better suited to real time implementation. Returning to Equation (5.150) and using it to express the displacement at a number of measurement locations, $\boldsymbol{x}_j$, results in the following equation for each measurement location, $\boldsymbol{x}_j$:

$$w(\boldsymbol{x}_j, t) = \sum_{i=1}^{\infty} a_i(t) \varphi_i(\boldsymbol{x}_j) \tag{5.154}$$

If the number of modes considered is limited to N_m and the number of measurement locations limited to N_p, then a set of N_p simultaneous equations is obtained, which can be written in matrix form as:

$$\boldsymbol{w}(t) = \boldsymbol{\Phi} A(t) \tag{5.155}$$

where:

$$\boldsymbol{w}(t) = [w(\boldsymbol{x}_1, t), w(\boldsymbol{x}_2, t), \ldots, \ w(\boldsymbol{x}_{Np}, t)]^{\mathrm{T}} \tag{5.156}$$

$$\boldsymbol{\Phi} = [\boldsymbol{\varphi}_1, \boldsymbol{\varphi}_2, \ldots, \ \boldsymbol{\varphi}_{Nm}] \tag{5.157}$$

$$\boldsymbol{\varphi}_i = [\varphi_i(\boldsymbol{x}_1), \varphi_i(\boldsymbol{x}_2), \ldots, \ \varphi_i(\boldsymbol{x}_{Nm})]^{\mathrm{T}} \tag{5.158}$$

$$\boldsymbol{A}(t) = [a_1(t), a_2(t), \ldots, \ a_{Nm}(t)]^{\mathrm{T}} \tag{5.159}$$

The amplitude of the ith mode is then:

$$a_i(t) = \sum_{j=1}^{N_p} b_{ij} w(\boldsymbol{x}_j, t) \tag{5.160}$$

where:

$$\boldsymbol{b}_i = \left[b_{i1}, \ b_{i2}, \ldots, \ b_{iN_p} \right]^{\mathrm{T}} \tag{5.161}$$

and:

$$\boldsymbol{B} = [\boldsymbol{b}_1, \boldsymbol{b}_2, \ldots, \ \boldsymbol{b}_{Nm}] = \boldsymbol{\Phi} \left[\boldsymbol{\Phi}^{\mathrm{T}} \boldsymbol{\Phi} \right]^{-1} \tag{5.162}$$

Equations (5.154) to (5.162) can also be used for an acoustic space with the structural displacement, w, replaced with the acoustic pressure, p, and the structural mode shape functions, φ, replaced with the acoustic mode shape functions, ψ.

Equation (5.162) is the generalised pseudo-inverse of the mode shape matrix, $\boldsymbol{\Phi}$, of Equation (5.155), which is used in place of the inverse when the matrix to be inverted is not square (see Section A.11). To obtain a well-conditioned matrix for use during the pseudo-inversion process, it is important that there be many more measurement points than modes to be resolved. For an optimally conditioned matrix, $\boldsymbol{\Phi}$, the measurement locations should not be uniformly spaced, and at least 2.5–3 times as many measurement points as the order of the highest order mode of interest are needed. Thus if the first 6 modes are of interest, the displacement must be measured at 15 locations, none of which should be on a modal node. In all cases, more measurement points will increase the accuracy of the modal amplitudes that are resolved. For cases where higher order modes exist that are not of interest, the amplitudes of these modes will be included (aliased) in the amplitudes of the lower order modes if the mode order is more than half the total number of measurement locations. This aliasing problem is very similar to the aliasing in the frequency domain when a time domain signal is digitised by sampling at a certain rate, and then the discrete Fourier transform used to convert it to the frequency domain (see Section 7.3). In this latter case, higher frequency components in the original signal with a frequency greater than half the sampling frequency will be aliased into the amplitudes of the lower frequencies.

For single frequency excitation, it is more convenient to express modal amplitudes as time independent complex quantities, defined by an amplitude and phase or by a real and imaginary part. In this case Equation (5.160) becomes:

$$\text{Re}\{\hat{a}_i\} = \sum_{j=1}^{N_p} b_{ij} \text{Re}\{\hat{w}(\boldsymbol{x}_j)\} \tag{5.163}$$

$$\text{Im}\{\hat{a}_i\} = \sum_{j=1}^{N_p} b_{ij} \text{Im}\{\hat{w}(\boldsymbol{x}_j)\} \tag{5.164}$$

As an example, this method will be applied to finding the complex modal amplitudes of a harmonically excited simply supported plate (Hansen, 1980). The complex displacement amplitude at location $\boldsymbol{x} = (x, y)$ on a simply supported plate is:

$$\hat{w}(x,y) = \sum_{m=1}^{N_{mx}} \sum_{n=1}^{N_{my}} a_{mn} \sin\frac{m\pi x}{L_x} \sin\frac{n\pi y}{L_y} \tag{5.165}$$

where a_{mn} is the complex modal amplitude for a mode having the indices m, n and the time dependence term has been omitted for convenience.

In matrix form this can be written as:

$$
\begin{bmatrix} \hat{w}_1 \\ \hat{w}_2 \\ . \\ . \\ \hat{w}_{N_p} \end{bmatrix} =
\begin{bmatrix}
\sin\dfrac{\pi x_1}{L_x} & \sin\dfrac{\pi y_1}{L_y} & \sin\dfrac{2\pi x_1}{L_x} & \sin\dfrac{\pi y_1}{L_y} & .. & \sin\dfrac{N_{mx}\pi x_1}{L_x} & \sin\dfrac{N_{my}\pi y_1}{L_y} \\
. & . & . & . & .. & . & . \\
. & . & . & . & .. & . & . \\
. & . & . & . & .. & . & . \\
\sin\dfrac{\pi x_{N_p}}{L_x} & \sin\dfrac{\pi y_{N_p}}{L_y} & \sin\dfrac{2\pi x_{N_p}}{L_x} & \sin\dfrac{2\pi y_{N_p}}{L_y} & .. & \sin\dfrac{N_{mx}\pi x_{N_p}}{L_x} & \sin\dfrac{N_{my}\pi y_{N_p}}{L_y}
\end{bmatrix} \times
$$

$$
\times \begin{bmatrix} a_{11} \\ a_{21} \\ . \\ . \\ . \\ a_{N_{mx}N_{my}} \end{bmatrix} \tag{5.166}
$$

or in short form:

$$\hat{\boldsymbol{w}} = \boldsymbol{\Phi A} \tag{5.167}$$

Thus:

$$\boldsymbol{A} = \boldsymbol{\Phi}\left[\boldsymbol{\Phi}^{\text{T}}\boldsymbol{\Phi}\right]^{-1} \boldsymbol{w} \tag{5.168}$$

In Equation (5.168), the real part of $\hat{\boldsymbol{w}}$ may be used to obtain the real part of $\boldsymbol{A}$ and similarly for the imaginary part. Of course in many cases, accelerometers or laser doppler velocimeters are used to measure acceleration or velocity, respectively, and the displacement is not necessarily the quantity of interest. In this case, the acceleration or velocity measurements may be used directly in place of $\hat{w}(x, y)$ and the resulting modal amplitudes are acceleration or velocity amplitudes. If the displacement amplitude is desired, then it is better to replace the quantity $\hat{w}(x, y)$ in Equation (5.165) with $j\omega\hat{w}(x, y)$ or $-\omega^2\hat{w}(x, y)$ (depending on whether velocity or acceleration is being measured) than it is to use an integrating circuit in the measurement instrumentation, because the latter is generally a significant source of error. However, when the structure is being excited by random noise and Equation (5.154) is used to obtain time varying modal amplitudes an integrating circuit must be used to obtain displacement from acceleration or velocity measurements. Alternatively, if modal amplitudes are required as a function of frequency (rather than time), then the measured vibration signal can be Fourier transformed to the frequency domain as described in Section 7.3, and each frequency component treated as described above for single frequency excitation. Note that if it is desired to average over several spectra, the power spectrum must be used; thus phase data are lost and only modal amplitude data will be available.

As mentioned earlier, it is possible to approximate the integral of Equation (5.152) directly for structures where the mode shape can be described in functional form, by using equally spaced displacement or acceleration measurement locations.

To illustrate how this may be done, two examples will be considered. The first will be a simply supported plate, the complex displacement amplitude of which can be written as:

$$\hat{w}(x, y) = \sum_{m=1}^{\infty} \sum_{n=1}^{\infty} a_{mn} \cos \frac{m\pi x}{L_x} \cos \frac{n\pi y}{L_y} \tag{5.169}$$

Note that the origin of the coordinates has been chosen to be the centre of the plate. Both the plate displacement amplitude, $\hat{w}(x, y)$, and the modal amplitude, a_{mn}, are complex, each characterised by an amplitude and phase or by a real and imaginary part. If the bottom left corner of the plate were chosen as the origin, then the cosine functions would be replaced with sine functions. The index, i, of Equation (5.150) is the mode order of mode (m, n).

Multiplying both sides of Equation (5.169) by $\cos(m\pi x/L_x)$, using modal orthogonality and integrating with respect to x from $x = -L_x/2$ to $L_x/2$ gives:

$$\frac{2}{L_x} \int_{-L_x/2}^{L_x/2} \hat{w}(x, y) \cos \frac{m\pi x}{L_x} \mathrm{d}x = \sum_{n=1}^{\infty} a_{mn} \cos \frac{n\pi y}{L_y} \tag{5.170}$$

Multiplying Equation (5.170) by $\cos(n\pi y/L_y)$, using modal orthogonality and integrating with respect to x gives:

$$\frac{4}{L_x L_y} \int_{-L_x/2}^{L_x/2} \int_{-L_y/2}^{L_y/2} \hat{w}(x, y) \cos \frac{m\pi x}{L_x} \cos \frac{n\pi y}{L_y} \mathrm{d}x\, \mathrm{d}y = a_{mn} \tag{5.171}$$

In practice, Equation (5.171) is implemented by replacing the integrals with finite sums using a series of measurement points uniformly distributed over the plate so that the spacing between points in both the x- and y-directions is uniform, although the spacing in the x-direction need not be the same as in the y-direction. It is interesting to note that this uniform spacing in both x- and y-directions can only be achieved in practice by using a diagonal array of accelerometers. Thus:

$$\frac{4}{L_x L_y} \sum_{j=1}^{N_{px}} \left[\sum_{k=1}^{N_{py}} \hat{w}(x_j, y_k) \cos \frac{n\pi y_k}{L_y} \Delta y \right] \cos \frac{m\pi x_j}{L_x} \Delta x = a_{mn} \tag{5.172}$$

As the mode shapes in both the x- and y-directions are separable as well as orthogonal, it is possible to obtain the modal amplitudes by using two line arrays of accelerometers with one array parallel to the y-axis and one parallel to the x-axis, so that in this case Equation (5.169) may be written as:

$$\hat{w}(x,y) = \hat{X}(x)\hat{Y}(y) = \sum_{m=1}^{\infty} a_m \cos\frac{m\pi x}{L_x} \sum_{n=1}^{\infty} a_n \cos\frac{n\pi y}{L_y} \qquad (5.173)$$

where $a_m a_n = a_{mn}$ of Equation (5.172).

For the accelerometer array parallel to the x-axis, and located at $y = y_c$, the displacement is only a function of the x-coordinate location. Thus:

$$\hat{w}(x,y_c) = \hat{X}(x) = \sum_{m=1}^{\infty} a_m \cos\frac{m\pi x}{L_x} \qquad (5.174)$$

Multiplying the far left and far right sides of (5.174) by $\cos(m\pi x/L_x)$, using modal orthogonality and integrating with respect to x from $x = -L_x/2$ to $L_x/2$ gives:

$$\frac{2}{L_x} \int_{-L_x/2}^{L_x/2} \hat{w}(x,y_c) \cos\frac{m\pi x}{L_x} dx = a_m \qquad (5.175)$$

Multiplying the far left and far right sides of (5.174) by $\cos(n\pi y/L_y)$, using modal orthogonality and integrating with respect to y from $y = -L_y/2$ to $L_y/2$ gives:

$$\frac{2}{L_y} \int_{-L_y/2}^{L_y/2} \hat{w}(x_c,y) \cos\frac{n\pi y}{L_y} dy = a_n \qquad (5.176)$$

Converting the integrals to finite sums as discussed previously and combining Equations (5.175) and (5.176) gives (see also Fuller et al. (1991)):

$$\frac{4\Delta x\Delta y}{L_x L_y} \left[\sum_{j=1}^{N_{px}} \hat{w}(x_j,y_c) \cos\frac{m\pi x_j}{L_x}\right] \left[\sum_{k=1}^{N_{py}} \hat{w}(x_c,y_k) \cos\frac{n\pi y_k}{L_y}\right] = a_{mn} \qquad (5.177)$$

where Δx is the spacing of the measurement points on the line at $y = y_c$, Δy is the spacing of the measurement points in the line at $x = x_c$, x_i are the measurement point locations on the line at $y = y_c$ and y_i are the measurement point locations on the line at $x = x_c$. Note that x_c or y_c must not correspond to a nodal line of any of the modes that are to be resolved. In practice they should not even be close to nodal lines or the error in a_{mn} will be large.

As the sums in Equation (5.177) are over a finite number of points, care must be taken to avoid spatial aliasing. For example, if 20 measurements are made in the x-direction, then the highest mode number that can be resolved is $m = 10$. If the amplitudes of modes of higher order than ten are significant, then these amplitudes will "fold back" into the lower order modes, giving an inaccurate estimate of the amplitudes of the lower order modes. Thus, an assumption underlying the use of the method described above is that the amplitude of any higher order mode is negligible. This can be tested by increasing the number of measurement points and observing that the decomposed modal amplitude reduces as m and n increase.

If the excitation is random and modal amplitudes as a function of frequency are desired, then a Fourier transform can be made of the data at each measurement point and the real and imaginary parts of the modal amplitudes obtained by using the corresponding real and imaginary parts of $\hat{w}$ in Equation (5.177).

For single frequency excitation, not necessarily at a modal resonance frequency, the amplitudes and phases of the displacement (relative to some fixed reference) at each measurement point may be determined and converted to give the real and imaginary components, which can then be used separately to find the real and imaginary components of the modal amplitude.

A second example of approximation of the integral of Equation (5.152) that will be considered is the resolution of circumferential modes from response measurements taken at points around the circumference of a circular cylinder at a constant axial location (Jones and Fuller, 1986).

The complex radial displacement amplitude of a cylinder can be represented as:

$$\hat{w}(\theta) = \sum_{n=0}^{\infty} [a_n \cos(n\theta) + b_n \sin(n\theta)] \tag{5.178}$$

When a cylinder is excited, circumferential waves propagate in both directions around the cylinder combining to create an interference pattern or standing wave. To solve for the complex modal amplitudes, a_n and b_n, Equation (5.178) is multiplied by $\cos(m\theta)$ and $\sin(m\theta)$, respectively, and integrated from 0 to 2π. Thus:

$$\int_0^{2\pi} \hat{w}(\theta)\cos(m\theta)\mathrm{d}\theta = \sum_{n=0}^{\infty}\left[\int_0^{2\pi} a_n \cos(n\theta)\cos(m\theta)\mathrm{d}\theta + \int_0^{2\pi} b_n \sin(n\theta)\cos(m\theta)\mathrm{d}\theta\right] \tag{5.179}$$

$$\int_0^{2\pi} \hat{w}(\theta)\sin(m\theta)\mathrm{d}\theta = \sum_{n=0}^{\infty}\left[\int_0^{2\pi} a_n \cos(n\theta)\sin(m\theta)\mathrm{d}\theta + \int_0^{2\pi} b_n \sin(n\theta)\sin(m\theta)\mathrm{d}\theta\right] \tag{5.180}$$

where $m = 0, 1, 2, 3, \ldots, \infty$. By utilising the orthogonality characteristics of the Fourier series, Equations (5.179) and (5.180) can be reduced and rearranged to solve explicitly for the modal amplitudes. The resulting equations are:

$$a_n = \frac{1}{\epsilon\pi}\int_0^{2\pi} \hat{w}(\theta)\cos(n\theta)\mathrm{d}\theta \tag{5.181}$$

$$b_n = \frac{1}{\epsilon\pi}\int_0^{2\pi} \hat{w}(\theta)\sin(n\theta)\mathrm{d}\theta \tag{5.182}$$

where $\varepsilon = 2$ for $n = 0$ and $\varepsilon = 1$ for $n > 0$; $n = 0, 1, 2, 3, \ldots, \infty$.

If $\hat{w}(\theta)$ is known completely as a function of θ, all of the modal amplitudes can be determined. In practice, however, $\hat{w}(\theta)$ is known only at discrete points around the cylinder. Therefore, the integrals of Equations (5.182) and (5.181) are represented as Fourier summations of the form:

$$a_n = \frac{\Delta\theta}{\epsilon\pi}\sum_{j=1}^{N_p} \hat{w}(\theta_j)\cos(n\theta_j) \tag{5.183}$$

$$b_n = \frac{\Delta\theta}{\epsilon\pi}\sum_{j=1}^{N_p} \hat{w}(\theta_j)\sin(n\theta_j) \tag{5.184}$$

where N_p is the number of circumferential positions where measurements are acquired and $\Delta\theta = 2\pi/N_p$ for equally spaced measuring points.

For both of the examples just given, the summations are really only approximations to integrals and rely on the contributions of modes of order greater than one half N_{px} (for m mode order) and one half N_{py} (for n mode order) being negligible. If this is not the case, aliasing will occur as explained previously. Also, the higher the mode order, the less accurate will be the final result, which implies that increasing the number of measurement points increases the accuracy of the results.

6

Statistical Energy Analysis

6.1 Introduction

Material presented in previous chapters in this book is theoretically applicable across the entire audio frequency range. However, in practice, most of the material is only useful in the low-frequency range where the resonant modes are well-separated in frequency. As the frequency of interest increases, and resonant modes become more closely spaced in frequency, it becomes increasingly difficult to separate the modal responses, so a statistical approach becomes necessary. Such an approach is known in the vibroacoustics discipline as Statistical Energy Analysis or SEA. Many books and papers have been written on this topic and it is not the intention for this chapter to provide a comprehensive survey of all the work that has been done in the past. Rather, the theoretical basis of SEA is provided and some practical considerations are outlined, which must be taken into account when applying SEA to the analysis of practical structures. SEA is a procedure for quantifying energy storage and energy transmission between sub-elements in a vibroacoustic system. It allows the calculation of mean square vibration velocity levels on a structure and mean square sound pressure levels in an enclosed space. It can also be used to estimate radiated sound power levels of a structure, thus allowing radiated sound pressure levels to be calculated.

Modal analysis, which is discussed in Chapter 5, is undertaken at single frequencies, and structures or acoustic spaces are characterised by resonance frequencies and associated displacement (or sound pressure) mode shapes. On the other hand, SEA is undertaken in frequency bands that are usually one octave or 1/3-octave wide, and modal characteristics are averaged over each frequency band. For SEA to give reasonably accurate results, it is necessary for there to be a minimum of 3 modes resonant in each frequency band. However, the quality of the results improves as the number of modes resonant in the measurement frequency band increases.

SEA assumes that vibratory energy in a structure or acoustic energy in an enclosure is distributed equally in the modes that are resonant in the frequency band being considered. When two subsystems are joined together (such as a beam joined to a plate), energy is transmitted from the system with the higher energy in each resonant mode, in the frequency band being considered, to the system with the lower energy in each of its modes, until there is equal energy in each mode in the two systems. Thus, the system with more modes resonant in the frequency band under consideration will have a greater total energy. This principle of equi-partition of energy among all system modes that are resonant in the frequency band of interest is one of the key pillars upon which SEA is based. During the energy transfer process, which achieves equi-partition of modal energy among two lightly coupled subsystems, energy is lost as a result of transmission across the junction between the two systems as well as by dissipation within each system and at subsystem boundaries. Energy is input to the vibratory system by an external vibration source and to any attached enclosed acoustic space by radiation from vibrating structural elements that make up the enclosure surrounding the acoustic space or by an acoustic source such as a loudspeaker located within the enclosed space.

Thus, in summary, the use of SEA allows average vibration levels in 1/3-octave or octave bands to be calculated for a structural system from a knowledge of average vibration levels on one structural component or from a knowledge of the vibratory input power from an external source. Where sound levels in an enclosed space are of interest, the same procedures can be used to calculate average sound pressure levels in each 1/3-octave or octave band. Where it is desirable to calculate the sound power radiated into free space by a structure in a 1/3-octave or octave frequency band, Equation (6.1) may be used.

$$\Pi_{\Delta S} = S\rho c \sigma_{\Delta S} \langle u^2 \rangle_{St\Delta} \quad \text{(watts)} \tag{6.1}$$

where the subscript, Δ, denotes frequency band average, and $\sigma_{\Delta S}$ is the band-averaged radiation efficiency for surface, S (see Section 4.2.4). The quantity, $\langle u^2 \rangle_{St\Delta}$, is the mean square surface velocity averaged over the frequency band, Δ, the surface, S, and time, t, and is calculated by using an energy balance between the vibratory power input into the structural system and the power dissipated by the system.

In the remainder of this chapter, means are discussed for describing a vibroacoustic system in a way that is amenable to SEA. This is followed by a description of the calculation procedures that can be used to quantify the various parameters that are needed in order to estimate structural vibration levels and sound pressure levels in enclosed spaces and in free field. More details on the measurement of SEA parameters and the implementation of SEA models are provided in Lyon and DeJong (1995); Norton and Karczub (2003).

6.2   Model Construction and Problem Formulation

The first step in a statistical energy analysis is to develop a simplified model of the system that allows the system to be analysed. Thus a vibroacoustic system needs to be broken down into subsystems that can be described in terms of modal density and damping factors, as well as in terms of coupling loss factors that influence the transfer of energy between adjacent subsystems. Basically, this involves separating the system into subsystems, each of which is represented by a beam, plate or cylinder element as well as one particular wave propagation (or modal) type (bending, shear or longitudinal). Once the system to be analysed has been divided into subsystems, the next step is to prepare a network diagram showing how each subsystem is connected to the others and how power is injected into the system and transferred between subsystems. The power transfer between subsystems may be via rotary or linear force coupling and coupling may be at a point, along a line or over an entire surface. In addition, waves of one type in a particular subsystem may transfer energy into waves of another type. For example, longitudinal waves in a plate can transfer energy into bending waves in a plate connected at right angles to it, as shown in Figure 6.1(a). These same longitudinal waves in a plate can also generate bending waves in a beam attached to an edge or in a beam used to join two plates at right angles, as illustrated in Figure 6.1(b).

When dividing a system into subsystems, care must be taken not to lose any modes that may characterise subsystems connected together but are not present in the individual subsystems. For example, if an I-beam is divided into three subsystems (one for the web and one for each flange), the bending modes of the complete assembly will not be captured, so in order to capture these modes, a separate subsystem, consisting of the entire I-beam must be introduced. Another example is a plate stiffened by ribs. The plate panels between ribs may be modelled as individual subsystems, one for each panel between sets of ribs. The ribs may also be modelled as separate subsystems. However, the modes associated with the stiffened plate acting as a single unit will

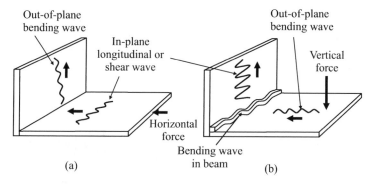

FIGURE 6.1 Coupling of different wave types at line junction between two subsystems: (a) plate–plate junction; (b) beam–plate junction.

be missed. These modes can be captured in a separate subsystem that consists of the plate and all the stiffening ribs. It is certainly not a good idea to use many small components, such as a finite element mesh, to solve an SEA problem. In such a case, elements must be combined into structural components that are sufficiently comprehensive that no modes will be lost when the structural component is considered to be a subsystem in an SEA model. In summary, if global modes exist that are not represented in any individual subsystem in the model, a new, additional subsystem must be created to account for these global modes.

The lower size limit of a subsystem is usually determined by the requirement that there be at least three resonant modes in the frequency band of interest. However, this size limit requirement can be ignored if the subsystem is well coupled to a global system with a modal overlap greater than about 0.5 to 1.0 (see Section 6.5).

The upper size limit of a subsystem is limited by the damping in the subsystem causing a significant attenuation of waves as they travel across it. Thus the maximum dimension, L, of a subsystem is given by (Lyon and DeJong, 1995):

$$L < \frac{c_g}{2\pi f \eta_d} \tag{6.2}$$

where c_g is the group speed of the energy travelling across the subsystem (see Section 1.4.10) and η_d is the loss factor, which is a measure of damping in the system (see Section 6.6). The group speed, c_g, is often approximated as the phase speed of the travelling waves at the centre frequency, f, of the frequency band under consideration and this is discussed further in Section 6.4.

6.2.1 Vibroacoustic System Analysis

In practice, vibroacoustic systems are generally broken down into many subsystems for the purpose of SEA. Subsystems are formed from groups of resonant modes that have similar properties and similar energy within the frequency band being analysed. For example, a single structural element, such as a beam, may be represented by three subsystems; one for axial modes, one for torsional modes and one for bending modes. For the purpose of establishing the basic equations, a generic vibroacoustic system with just two subsystems, as illustrated in Figure 6.2, will be analysed first. In the figure, Π_1 is the input power into system 1, Π_2 is the input power into system 2, E_1 is the energy in subsystem 1 after the two systems reach equilibrium, E_2 is the energy in subsystem 2 after the two systems reach equilibrium, η_{12} is the coupling loss factor (CLF) that is proportional to the fraction of energy that is lost when it is transmitted from system 1 to system 2, η_{21} is the CLF that is proportional to the fraction of energy that is lost when it is transmitted from system 2 to system 1, η_1 is the damping loss factor (DLF) that is proportional to the fraction of energy lost due to damping in system 1 and η_2 is the DLF that is proportional to the fraction of energy lost due to damping in system 2. The damping loss factor includes losses resulting from sound radiation as well as losses resulting from dissipation at the subsystem boundaries (see Norton and Karczub (2003)). Determination of coupling loss factors, input powers, modal densities and damping loss factors, which vary with frequency, is discussed in the remainder of this chapter.

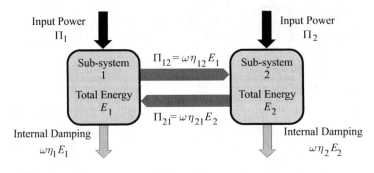

FIGURE 6.2 Vibroacoustic system consisting of two subsystems showing power transmission between them and energy storage within each of them.

The energy balance for subsystem 1 may be written as:

$$\Pi_1 + \Pi_{21} = \omega\eta_1 E_1 + \Pi_{12} \quad \text{(watts)} \tag{6.3}$$

which can be rearranged to give:

$$\Pi_1 = \omega\eta_1 E_1 + \omega\eta_{12} E_1 - \omega\eta_{21} E_2 \quad \text{(watts)} \tag{6.4}$$

However, the quantity of interest is generally E_1, which may be expressed as:

$$E_1 = \frac{\Pi_1 + \omega\eta_{21} E_2}{\omega\eta_1 + \omega\eta_{12}} \quad \text{(joules)} \tag{6.5}$$

When there are more than two subsystems connected together, matrix algebra is used to obtain the energies in each subsystem. Thus for a vibroacoustic system containing m subsystems, the energy balance among all of these systems may be written as:

$$\omega \begin{bmatrix} (\eta_1 + \sum_{\substack{i=1 \\ i\neq 1}}^{m} \eta_{1i})n_1 & (-\eta_{12}n_1) & \cdots & (-\eta_{1m}n_1) \\ (-\eta_{21}n_2) & (\eta_2 + \sum_{\substack{i=1 \\ i\neq 2}}^{m} \eta_{2i})n_2 & \cdots & (-\eta_{2m}n_2) \\ \vdots & \vdots & \ddots & \vdots \\ (-\eta_{m1}n_m) & \cdots & \cdots & (\eta_m + \sum_{\substack{i=1 \\ i\neq m}}^{m} \eta_{mi})n_m \end{bmatrix} \begin{bmatrix} E_1/n_1 \\ E_2/n_2 \\ \vdots \\ E_m/n_m \end{bmatrix} = \begin{bmatrix} \Pi_1 \\ \Pi_2 \\ \vdots \\ \Pi_m \end{bmatrix} \tag{6.6}$$

which can be expressed using matrix notation as:

$$\omega[\mathbf{C}][\mathbf{E}] = [\mathbf{\Pi}] \tag{6.7}$$

where n_m is the modal density of subsystem m (see Section 6.4 for a discussion of modal density and its estimation), and the following relationship has been used (Lyon and DeJong, 1995):

$$n_1\eta_{12} = n_2\eta_{21} \tag{6.8}$$

6.2.2   Subsystem Response as a Result of Subsystem Energy

The application of SEA to a system consisting of a number of subsystems results in the evaluation of vibroacoustic energy in each subsystem for each 1/3-octave or octave frequency band. However, it is generally of interest to convert these energies, either into mean square vibration velocity of the subsystem in each 1/3-octave or octave frequency band, or mean square sound pressure in an enclosure, again for each 1/3-octave or octave frequency band. Furthermore, the mean square vibration velocity of a structure can be used with its radiation efficiency (see Section 4.2.4) and Equation (6.1) to calculate the radiated sound power. This latter quantity, together with the analysis of Chapter 3 can be used to estimate sound pressure levels in a free field environment. The use of sound power to estimate outdoor sound pressure levels is discussed at length in Chapter 5 of Bies et al. (2017).

The mean square vibration velocity, $\langle u^2 \rangle_{St\Delta}$, of a structural subsystem is related to its modal energy, E_s, by:

$$\langle u^2 \rangle_{St\Delta} = E_s/m \quad (\mathrm{m^2/s^2}) \tag{6.9}$$

where m is the total mass of the subsystem and the subscript, $St\Delta$, indicates that the average is over space, S, time, t, and frequency band, Δ. The mean square velocity of a structure can be used to calculate the radiated sound power using Equation (6.1).

Equation (6.9) can be derived from the analysis of a single-degree-of-freedom mass-spring system illustrated in Figure 5.1, but with the damping element removed. This mass-spring system represents a single mode of vibration of a structural element that has a total mass of m. At resonance, the potential energy at any instant in time for excitation at the resonance frequency, ω_0, is:

$$E_p = \frac{1}{2}kw^2 = \frac{A^2}{2}\sin^2(\omega_0 t + \beta) \tag{6.10}$$

and the kinetic energy is:

$$E_k = \frac{1}{2}m\dot{w}^2 = \frac{1}{2}mu^2 = \frac{A^2\omega_0}{2}\cos^2(\omega_0 t + \beta) \tag{6.11}$$

where:

$$\omega_0 = \sqrt{\frac{k}{m}} \tag{6.12}$$

where m is the mass that is attached to the spring of stiffness, k.

The total energy, E_s, at any time instant is the sum of the kinetic and potential energies, so that:

$$E_s = E_k + E_p = \frac{1}{2}kA^2 \tag{6.13}$$

If the potential and kinetic energies are averaged over a cycle of vibration, the following is obtained:

$$E_s = \langle E_k \rangle = \langle E_p \rangle = \frac{1}{4}kA^2 = \frac{1}{2}E_s \tag{6.14}$$

so the average kinetic and potential energies are both equal to half the energy of vibration.

The mean square sound pressure, $\langle p^2 \rangle_{St\Delta}$, in an acoustic subsystem (such as an enclosure) is related to its modal energy, E_a, by:

$$\langle p^2 \rangle_{St\Delta} = E_a \rho c^2/V \quad (\mathrm{Pa^2}) \tag{6.15}$$

where V is the enclosure volume. The average sound pressure level, L_p, in the enclosure for the 1/3-octave or octave band being considered is then:

$$L_p = 10\log_{10}\langle p^2 \rangle_{St\Delta} + 94 \quad (\mathrm{dB\ re\ 20\,\mu Pa}) \tag{6.16}$$

An example of a realistic vibroacoustic system divided into subsystems is shown in Figure 6.3. Note that only bending waves in the plate radiate significant sound so the longitudinal waves generated by the beam in the plate can be ignored.

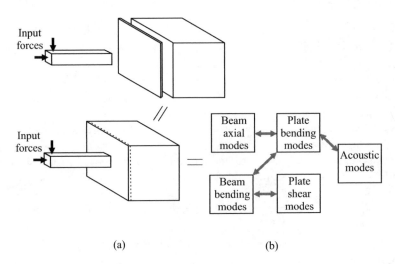

FIGURE 6.3 Realistic example of a vibroacoustic system consisting of a number of subsystems showing power transmission between them and energy storage within each of them.

6.3   System Input Power

In developing relationships for estimating the vibratory input power to a system from point, line and area sources, it is useful to first discuss the concept of mobility. Mobility is the reciprocal of impedance and its real part (called inductance) is always positive, whereas its imaginary part (called susceptance) can be positive or negative.

The real part of the translational power input, Π_{in}, to a system can be expressed as the real part of the product of the complex amplitudes of the force, F_0, and velocity, u_0, at the input location, 0, as:

$$\Pi_{in} = \frac{1}{2}\text{Re}\{\hat{F}_0^* \hat{u}_0\} \quad \text{(watts)} \tag{6.17}$$

where $F_0 = \hat{F}_0 e^{j\omega t}$ and $u_0 = \hat{u}_0 e^{j\omega t}$, Re{ } is the real part of a complex quantity and $\hat{F}_0^*$ represents the complex conjugate of $\hat{F}_0$.

The rotational power input to a system can be expressed as the product of the complex amplitudes of the applied moment, M_0, and rotational velocity, $\dot{\hat{\theta}}_0$, at the input location as:

$$\Pi_{in} = \frac{1}{2}\text{Re}\{\hat{M}_0^* \dot{\hat{\theta}}_0\} \quad \text{(watts)} \tag{6.18}$$

where $M_0 = \hat{M}_0 e^{j\omega t}$, $\hat{M}_0^*$ represents the complex conjugate of $\hat{M}_0$ and $\dot{\theta}_0 = \dot{\hat{\theta}}_0 e^{j\omega t}$.

There are three translational degrees of freedom (force and velocity along the x, y and z axes, respectively, corresponding to Equation (6.17)), and three rotational around the x, y and z axes, respectively, corresponding to Equation (6.18), resulting in a total of 6. The total system input power is then a linear sum of the power transmitted for each degree of freedom.

As the mean square value of a sinusoidally varying quantity is half the amplitude, the structural input power may also be written as:

$$\Pi_{in} = \langle F_0^2 \rangle \text{Re}\{Y_F\} \quad \text{(watts)} \tag{6.19}$$

or:

$$\Pi_{in} = \langle u_0^2 \rangle \text{Re}\{Z_F\} \quad \text{(watts)} \tag{6.20}$$

where Z_F and Y_F are the point impedance and point admittance, respectively, at the point on the structure where the force is applied. The structural input admittance is the reciprocal

TABLE 6.1 Summary of point acoustic impedance formulae $Z_s = p/(Su_0)$ for a 1-D duct (plane wave propagation) and a 3-D acoustic space

Acoustic space type	Illustration	Z_s
1-D duct	Area, S	$2\rho c/S$
3-D acoustic space	d	$\dfrac{\pi \rho f^2}{c}\left(1 + \dfrac{2\mathrm{j}}{k_a d}\right)$ d =diameter of a circular excitation source or for a square excitation source, $d \approx \sqrt{A}$ where A is the area of the excitation source

if the input impedance. For moment excitation, the moment impedances and mobilities of the structure are used and the force, F_0, is replaced with the moment, M_0, and the velocity, u_0, is replaced with the angular velocity, $\dot{\theta}_0$.

Point mechanical impedances ($Z_F = F_0/u_0$) of beams and plates as a function of excitation frequency are listed in Tables 4.1 and 4.2. It can be seen from the tables that, for some wave types, the impedance is independent of frequency. Point impedances for 1-D and 3-D acoustic spaces are given in Table 6.1.

The infinite impedances in Tables 4.1, 4.2 and 6.1 are for junctions between SEA subsystems, where the junction is further than 1/4 wavelength from the subsystem boundaries. When a junction is close to a boundary, the point impedance used in the calculation of the coupling loss factor (see Section 6.7) must be reduced by a factor of 2 for each free boundary condition. Thus the point force impedance for a semi-infinite beam shown in Table 4.1 is a factor of 4 different from the infinite beam force impedance for bending waves because the boundary at the point of application of the lateral force is free to rotate as well as move laterally. On the other hand, the difference between the two cases for longitudinal waves is only a factor of 2 because the axial force only causes axial displacement and no rotation.

The factor of 2 for each free boundary applies to bending waves in plates that are constrained against rotation but free to move laterally at the connecting edge. However, it is only approximately true for plates that are not constrained against rotation.

For subsystems that are characterised by torsional or bending waves, the point moment impedances are reduced by a factor of 2 at a simply supported (or pinned) boundary (free rotation) and by a factor of 4 if the boundary is free to translate as well as rotate.

The acoustic impedance values shown in Table 6.1 are reduced by a factor of 2 for excitation within a quarter of a wavelength of a rigid boundary. Also, the impedance looking into a semi-infinite duct would be half that shown in Table 6.1.

For cases where equations for the point force impedance of an infinite system are not available, the real part of the point force impedance can be found from the modal density using (Lyon and DeJong, 1995):

$$\frac{1}{\mathrm{Re}\{Z_F\}} = \frac{m}{8\pi n(\omega)} \tag{6.21}$$

where m is the total mass of the subsystem and $n(\omega)$ is its modal density (see Section 6.4). For this formula to be valid, at least 5 modes should be resonant in the frequency band of interest.

The real part of the frequency and space averaged acoustic impedance for an acoustic space, in terms of its modal density, can be written as:

$$\langle \mathrm{Re}\{Z_F\} \rangle_{S,\Delta} = \frac{\pi \rho c^2 n(\omega)}{2V} \tag{6.22}$$

6.4 Modal Density

Modal density in this section is defined as the number of modes resonant, on average, in a bandwidth of 1 radian/sec and is denoted $n(\omega)$. Sometimes it is also defined as the number of modes resonant, on average, in a bandwidth of 1 Hz and in this case, it is denoted $n(f)$. In the following subsections, expressions for the modal density of one, two and three dimensional systems will be provided and then summarised for various physical systems in Table 6.2.

6.4.1 1-D Systems

The modal density for 1-D systems (such as longitudinal, torsional and bending waves in a beam) is given by (Lyon and DeJong, 1995):

$$n(\omega) = \frac{L}{\pi c_g} \quad \text{(modes per rad/sec)} \tag{6.23}$$

where L is the dimension of the 1-D system. For beams, the modal density for longitudinal (axial) modes and torsional modes is independent of frequency. However, the modal density for bending waves is dependent on frequency. Equations for calculating modal densities in beams for all three wave types as well as the modal density in a 1-D duct are listed in Table 6.2. For the 1-D duct, the equation is only valid for frequencies below the first higher order mode cut-on frequency. That is, for a rectangular-section duct, the modal density equation is valid for frequencies that satisfy:

$$f < \frac{c}{2a} \tag{6.24}$$

where a is the largest cross-sectional dimension of the duct.

6.4.2 2-D Systems

The modal density for a 2-D system such as a plate is given by (Lyon and DeJong, 1995):

$$n(\omega) \approx \frac{S\omega}{2\pi c_g c_\phi} \quad \text{(modes per rad/sec)} \tag{6.25}$$

where S is the area of the 2-D isotropic element and c_ϕ is the phase speed of the wave under consideration at the frequency of interest.

If the 2-D element were orthotropic, having different stiffnesses in different directions, such as a corrugated plate, Equation (6.25) would become:

TABLE 6.2 Modal densities, $n(\omega)$, of 1-D subsystems

Beam excited axially

$$n(\omega) = \frac{L_B}{\pi c_L}; \quad \text{where } c_L = c_g = \sqrt{\frac{E}{\rho_m}}$$

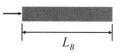

Beam excited torsionally

$$n(\omega) = \frac{L_B}{\pi c_T}; \quad \text{where } c_T = c_g = \sqrt{\frac{G}{\rho_m}}$$

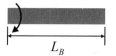

1-D duct excited acoustically

$$n(\omega) = \frac{L}{\pi c}$$

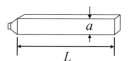

Beam in bending

$$n(\omega) = \frac{L_B}{2\pi} \left[\frac{\rho S}{E J \omega^2} \right]^{1/4}$$

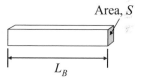

E is Young's modulus of elasticity for the material
G is the material shear modulus
ρ_m is the material density
c is the speed of sound in air
c_L is the speed of longitudinal waves
c_T is the speed of torsional waves
c_g is the group wave speed
J is the second moment of area of the beam cross section

$$n(\omega) \approx \frac{S\omega}{2\pi\sqrt{c_{g1}c_{\phi 1}}\sqrt{c_{g2}c_{\phi 2}}} \quad \text{(modes per rad/sec)} \quad (6.26)$$

where the subscripts 1 and 2 refer to the two principal axes (the least stiff direction and the most stiff direction).

In plates, the modal density for compressional and shear waves is dependent on frequency. However, the modal density for bending waves in plates is independent of frequency.

Equations for calculating modal densities in plates for all three wave types as well as the modal density in a 2-D acoustic space are listed in Table 6.3. The acoustic space of thickness, h, is sufficiently thin for 2-D analysis for frequencies up to the maximum defined by:

$$f = \frac{c}{2h} \quad \text{(Hz)} \quad (6.27)$$

If the walls of the 2-D space are compliant at low frequencies, it is best to only include the area term (first term on the RHS of the equation in Table 6.3) and omit the perimeter term (Lyon and DeJong, 1995).

TABLE 6.3 Modal densities, $n(\omega)$, of 2-D subsystems

Finite isotropic plate, bending modes

$$n(\omega) = \frac{\sqrt{3}S}{2\pi h c_L}; \quad \text{where } c_L = c_g = \sqrt{\frac{E}{\rho_m(1 - \nu^2)}}$$

Finite isotropic plate, in-plane compressional modes

$$n(\omega) = \frac{S\omega}{2\pi c_L^2}$$

Finite isotropic plate, in-plane shear modes

$$n(\omega) = \frac{S\omega}{2\pi c_s^2}; \quad \text{where } c_s = c_g = \sqrt{\frac{G}{\rho_m}}$$

2-D thin acoustic cavity

$$n(\omega) = \frac{S\omega}{2\pi c^2} + \frac{P}{2\pi c}$$

Area, S
Perimeter, P

---

c_s is the speed of shear waves
$S = ab$ and $P = 2a + 2b$
ν is Poisson's ratio for the material

6.4.3   3-D Systems

The modal density for a 3-D system, such as a solid item of material, is given by (Lyon and DeJong, 1995):

$$n(\omega) \approx \frac{V\omega^2}{2\pi^2 c_g c_\phi^2} \quad \text{(modes per rad/sec)} \tag{6.28}$$

where V is the volume of the 3-D isotropic element and c_ϕ is the phase speed of the wave under consideration ($=c_g$ for non-dispersive waves) at the frequency of interest.

In a 3-D solid material, the modal density for compressional and shear waves is dependent on frequency, and the wave speeds, c_L and c_s, are independent of frequency.

Equations for calculating modal densities in 3-D solids and 3-D acoustic spaces are listed in Table 6.4. If the walls of the acoustic space have a low frequency compliance similar to the surrounding fluid, it is best to just use the volume term in the equation in Table 6.4.

TABLE 6.4 Modal densities, $n(\omega)$, of 3-D subsystems

3-D acoustic cavity

$$n(\omega) = \frac{V\omega^2}{2\pi^2 c^3} + \frac{S\omega}{8\pi c^2} + \frac{P}{16\pi c}$$

3-D solid

$$n(\omega) = \frac{V\omega^2}{2\pi^2}\left(\frac{1}{c_L^3} + \frac{1}{c_s^3}\right)$$

$$c_L = c_g = \sqrt{E(1-\nu)/[\rho_m(1+\nu)(1-2\nu)]}$$

$$c_s = c_g = \sqrt{G/\rho_m}$$

$c = c_g$ is the speed of sound in air
ρ_m is the density of the solid material
E is Young's modulus of elasticity for the material
G is the material shear modulus

6.4.4 Additional Notes

Sometimes we are interested in modal densities, $n(f)$, in terms of modes per Hz. In this case, the RHS of all of the equations in Tables 6.2, 6.3 and 6.4 need to be multiplied by 2π and in addition, the symbol, ω, must be replaced everywhere with $2\pi f$.

The modal density, $n(f)$, of any structure can be determined from a measurement of the point mobility, Y_F, averaged over the frequency band of interest and averaged over the structure surface, provided that there are more than 5 modes resonant in the band, using (Clarkson, 1981):

$$n(f) = 4m\langle \text{Re}\{Y_F\}\rangle \quad \text{(modes per Hz)} \tag{6.29}$$

where m is the mass of the structure, $\text{Re}\{\ \}$ represents the real part of a complex number and $Y_F = 1/Z_F$, where Z_F is the point impedance. The $\langle\ \rangle$ symbols indicate a spatial average as well as an average over the frequency band of interest. Note that:

$$fn(f) = \omega n(\omega) \tag{6.30}$$

6.4.5 Numerical Methods

Finite element analysis (FEA) can also be used to determine modal densities. For 1-D systems, the mode count (from which modal densities are determined) is sufficiently accurate up to a frequency at which the number of modes is half the number of elements used in the analysis. For 2-D systems the number of modes counted sufficiently accurately is one quarter of the number of elements and for 3-D systems, the number of modes is one eighth of the number of elements, provided that all elements are similar in size.

Boundary conditions are also important for at least the first 10 modes. Using classical free, simply supported or fixed boundary conditions is not ideal, as practical boundary conditions are rarely accurately modelled as one of these. In practice, it is usually better to use an impedance boundary condition that is representative of the actual characteristics of the boundary.

6.5   Modal Overlap

Modal overlap is a measure of the uniformity of system response in a frequency band; that is, a measure of how well resonances of a reverberant field cover all possible frequencies within a specified frequency range. If the modal overlap is high, then fewer modes are required to be resonant in a frequency band to obtain results of acceptable accuracy using SEA. As mentioned in Section 6.2, a modal overlap of at least 0.5 (and preferably 1.0) is needed in a global system if any subsystem has fewer than 3 modes resonant in the frequency band of interest. Modal overlap, M_{over}, is defined as:

$$M_{\text{over}} = \Delta f n(\omega) \qquad (6.31)$$

where $n(\omega)$ is the modal density (number of modes per radian/sec, see Section 6.4) and Δf is the average of the modal bandwidths in the frequency band of interest (see Figure 6.4, where the modal overlap concept is illustrated for a hypothetical case of a modal overlap of 0.6). If the modal response is an acoustic one for an enclosed space, the velocity, u, on the y-axis in Figure 6.4 can be replaced with the acoustic pressure, p.

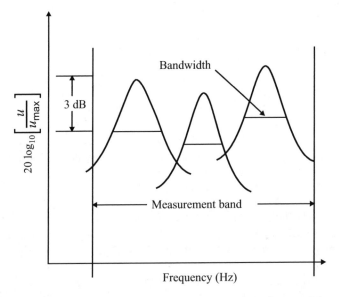

FIGURE 6.4 Three modes in a specified frequency range with a modal overlap of 0.6.

6.6   Damping Loss Factor (DLF)

Damping takes many forms, but viscous and hysteretic damping are the most common. As shown in Chapter 5, they can be described relatively simply analytically, and consequently, they have been well investigated. Modal damping in both structural and acoustic systems is generally expressed in terms of a loss factor, η, which is proportional to the energy dissipated in the structural or acoustic system. Acoustic damping is generally viscous in nature (and thus proportional to acoustic particle velocity), whereas structural damping is usually hysteretic in nature and thus proportional to structural displacement. Other measures of damping are related to the loss factor, η, as:

$$\Delta f / f_0 = 1/Q = \eta = \frac{2\zeta}{\sqrt{1 - \zeta^2}} = \delta/\pi \qquad (6.32)$$

where Q is known as the quality factor, ζ is the critical damping ratio, often used to describe viscous damping in mechanical systems, δ is the logarithmic decrement and both Δf and f_0 are defined in Figure 6.5. Measurement of the various forms of damping is discussed in more detail in Section 6.6.1. As may be observed by reference to Equation (6.32), when the modal loss factor, η, is small, which is true for most practical cases, the implication is that the critical damping ratio is also small and $\eta = 2\zeta$.

For structures, damping cannot be estimated analytically, so it is usually measured using one or more of the methods described in Section 6.6.1. The required loss factors for use in SEA calculations for structures must include energy loss at subsystem boundaries as well as loss due to acoustic radiation, and so the loss factors to be used in SEA problems are generally somewhat greater than for the material itself. Typical loss factors for various materials are included in the table of material properties in Appendix B. The values at the low end of the range are for the material alone and the values at the higher end are for structural elements supported on lossy boundaries such as bolted joints. In any analysis, the subsystem internal loss factor is usually the most difficult to estimate. The overall subsystem loss factor due to the mechanisms described above may, to a first approximation, be expressed as the sum of the individual components. Thus:

$$\eta = \eta_s + \eta_{\text{rad}} + \eta_j \qquad (6.33)$$

where η_s is the internal damping of the material making up the structure (lower limits of the range given in the table of material properties in Appendix B), η_{rad} is the contribution due to acoustic radiation from the structural element and η_j is the damping at the boundaries of the structure where it is connected to another subsystem or a rigid structure. For welded joints, η_j may be ignored, but for bolted or riveted joints, η_j is often larger than η_s and the upper limits of the values given in the table of material properties in Appendix B are usually used to account for $\eta_s + \eta_j$. The loss factor due to acoustic radiation may be calculated from the radiation efficiency (see Section 4.2.4) of the structure using (Norton and Karczub, 2003):

$$\eta_{\text{rad}} = \frac{\rho c \sigma}{\omega m_s} \qquad (6.34)$$

where m_s is the mass per unit area of the structure (surface density), σ is the radiation efficiency, ρ is the density of air and c is the speed of sound in air. At high frequencies, σ approaches a value of unity.

Hysteretic (or structural) damping is represented as the imaginary part of a complex elastic modulus of elasticity of the material, introduced as a loss factor, η, such that the elastic modulus, E, is replaced with $E' = E(1 + j\eta)$ and the bending wavenumber, k_b is replaced with $k_b' \approx k_b(1 - j\eta/4)$, for small η (see Section 2.3.2).

The damping loss factor, η, for excitation frequency, f, for a reverberant or semi-reverberant acoustic space is related to the average Sabine absorption coefficient, $\bar{\alpha}$, of all surfaces in the room using (Bies et al., 2017):

$$\eta = \frac{cS\bar{\alpha}}{8\pi f V} \qquad (6.35)$$

where c is the speed of sound, S is the surface area of the acoustic space and V is the volume of the space. Lyon and DeJong (1995) p. 172, provide an alternative formulation for the damping loss factor of an acoustic cavity as:

$$\eta = \frac{cS}{2\omega V} \frac{\bar{\alpha}}{2 - \bar{\alpha}} \qquad (6.36)$$

which differs from Equation (6.35) by a factor of $2/(2 - \bar{\alpha})$, as it includes the effect of the direct acoustic field being absorbed, which can be significant.

6.6.1   Measurement of Damping Loss Factors

When an acoustic space or structure is excited by a constant-amplitude tonal source that sweeps slowly upwards in frequency, the response of the room (measured with a microphone) or structure (measured with an accelerometer) will show peaks and troughs. The peaks correspond to acoustic space or structural resonances and may be easily distinguished for acoustic spaces at low frequencies, as illustrated in Figure 1.12, which is the frequency response for a reverberant room. However, as the frequency is increased, the modes in an acoustic space become more difficult to separate and in this range, a measurement of the modal damping, averaged over some frequency band, such as a 1/3-octave band, is all that is possible. On the other hand, as discussed in Section 6.4, the modal density for structures is either independent of frequency or decreases slowly with increasing frequency, so the modes generally do not become more difficult to separate as the frequency increases. However, as the 1/3-octave and octave bands become wider with increasing frequency, more modes are included in each band as the frequency is increased, and on this logarithmic scale, they do become more difficult to analyse separately.

6.6.1.1   Modal Bandwidth Method

Where the resonance peaks are easily distinguishable from one another, the damping may be estimated by observing that the resonance peaks have finite widths. A bandwidth, Δf, may be defined and associated with each mode, being the frequency range about resonance over which the sound pressure squared (for an acoustic space) or mean square velocity (for a structural element) is greater than or equal to half the same quantity at resonance. The lower and upper frequencies bounding a resonance and defined in this way are commonly known as the half-power points. The measurement requires that the frequency of resonance is determined and that the modal response at the 3 dB down points (half-power points) below and above resonance is identified (using a sinusoidal excitation signal), as shown in Figure 6.5. If a digital system is used in conjunction with white noise excitation and FFT analysis, the number of samples used to calculate the FFT must be greater than $20/\eta$ (Lyon and DeJong, 1995).

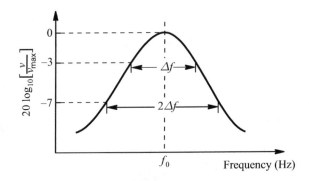

FIGURE 6.5 Determining system damping from frequency response function (FRF) bandwidth measurements.

The bandwidth, Δf, is dependent on the damping of the mode; the greater the modal damping, the larger will be the bandwidth. The loss factor is:

$$\eta = \frac{\Delta f}{f} \tag{6.37}$$

In some cases, better results are obtained by using the bandwidth at the 7, 10 or 12.3 dB down points. In these cases, the value of Δf used in Equation (6.32) is one-half, one-third or one-quarter, respectively, of the measured bandwidth, as illustrated for the 7 dB down point in Figure 6.5.

6.6.1.2 Reverberant Decay Method

For acoustic spaces in the mid- to high-frequency range, or in cases where the modal damping is relatively high, where individual modal bandwidths may not be able to be identified and measured directly, the average bandwidth may be calculated from a measurement of the decay time, using:

$$\Delta f = 2.20/T_{60} \tag{6.38}$$

where T_{60} is the time in seconds for the sound pressure level in an acoustic space or vibration level on a structure to decay by 60 dB following cessation of the excitation. If the acoustic space is very lightly damped, it is possible that the loudspeaker used to excite the space may contribute significantly to the damping, especially at very low frequencies. If the speaker does not exist in the final system being analysed, errors in estimated low-frequency sound pressure levels in the acoustic space may arise.

The reverberation decay method is also useful for measuring the damping of structures, especially if a measurement averaged over a 1/3-octave band is required. For this purpose, the excitation of modes in structures may be accomplished either by the direct attachment of a mechanical shaker or by shock excitation using a hammer. When the direct attachment of a shaker is used, the coupling between the shaker and the driven structure is strong. In the case of strong coupling, the mass of the shaker armature and shaker damping become part of the oscillatory system and must be taken into account in the analysis. The problems of strong coupling may be avoided by arranging to disconnect the driver from the structure when the excitation is shut off using a fuse arrangement. Alternatively, a non-contacting electromagnetic coil, which contains no permanent magnet, may be used to excite the structure. In the latter case, the structure will be excited at twice the frequency of the driving source. When exciting the structure with a shaker or electromagnetic coil, it is necessary that the excitation force is constant over the frequency range of the modal bandwidth, Δf.

When frequency-band filters are used to process the output from the transducer used to monitor the structural vibration, it is important to ensure that the filter decay rate is much faster than the decay rate that is to be measured, so that the filter decay rate does not control the measured structural vibration decay rate. Typically this means that the following relation must be satisfied, where B is the filter bandwidth (Hz):

$$BT_{60} \geq 16 \tag{6.39}$$

6.6.1.3 Frequency Response Curve Fitting Method

The loss factor, η, can also be determined by a curve fitting technique using the experimentally determined frequency response function (Ewins, 2000) (see Section 5.4.4). For lightweight or lightly damped structures, this method is best suited to the use of instrumented hammer excitation, which avoids the shaker coupling problem mentioned above. The transfer function that is used is the ratio of the structural acceleration response to the excitation force (Ewins,

2000). In this latter case, it is not important that the excitation force is constant over the modal bandwidth (see Section 5.4.1 for more details).

For acoustic spaces, the appropriate frequency response function is the ratio of the enclosure acoustic pressure response, to a signal that is proportional to the input acoustic force. If a loudspeaker is used as the input source, a signal proportional to the input acoustic force would be the acoustic pressure in a small, airtight enclosure backing the loudspeaker.

6.6.1.4 Power Balance Method

The loss factor can also be measured by measuring the input power, Π_{in}, to a structural system and the steady state energy, E_{tot}, in the system. The steady state energy in a structure is related to the mean square vibration velocity, $\langle u^2 \rangle$, averaged over the structure by:

$$E_{\text{tot}} = m \langle u^2 \rangle \qquad (6.40)$$

where m is the total mass of the structure.

The steady state energy in an acoustic space of volume, V, is related to the mean square sound pressure, $\langle p^2 \rangle$, averaged over the volume and is:

$$E_{\text{tot}} = V \langle p^2 \rangle / (\rho c^2) \qquad (6.41)$$

The steady state input power to the structure can be supplied by a shaker and measured by an impedance head. The steady state input power to an acoustic space can be supplied by a loudspeaker and the input power measured using an acoustic intensity probe or a microphone in the airtight backing cavity for the loudspeaker.

The input power and steady state energy are related to the loss factor by (Lyon and DeJong, 1995):

$$\eta = \frac{\Pi_{\text{in}}}{2\pi f E_{\text{tot}}} \qquad (6.42)$$

In using this method, the structure or acoustic space is usually excited in 1/3-octave or octave bands and the loss factor so obtained is an average for the particular 1/3-octave or octave band being used.

6.6.1.5 Other Methods

Logarithmic Decrement
Determination of the logarithmic decrement, δ (see Equation (6.32)), is one of the oldest methods of determining damping and it depends on the determination of successive amplitudes of a vibrating system as the vibration decays after switching off the excitation source. If A_i is the amplitude of the ith cycle and A_{i+n} is the amplitude n cycles later, then the logarithmic decrement, δ, is:

$$\delta = \frac{1}{n} \log_{\text{e}} \left(\frac{A_i}{A_{i+n}} \right) \qquad (6.43)$$

Phase Angle between Input Force and Response
As a result of damping, the strain response of a structure lags behind the applied force (or stress) by a phase angle, ϵ. Thus, another measure of structural damping is the tangent of this phase angle (sometimes called the damping factor) and this is related to the loss factor by:

$$\eta = \tan \epsilon \qquad (6.44)$$

The relationships of the various measures of damping to the loss factor used in SEA analysis are best summarised in a table, such as Table 6.5.

TABLE 6.5 Relationship of loss factor to various other measures of damping

Damping measure	Symbol	Units	$\eta =$
Loss factor	η	—	η
Quality factor	Q	—	$1/Q$
Critical damping ratio	ζ	—	$\dfrac{2\zeta}{\sqrt{1 - \zeta^2}}$
Reverberation time (60 dB)	T_{60}	sec	$\dfrac{2.2}{fT_{60}}$
Decay rate	DR	dB/sec	$\dfrac{\text{DR}}{27.3f}$
Wave attenuation	γ	nepers	$\dfrac{c_g\gamma}{\pi f}$
Logarithmic decrement	δ	—	δ/π
Phase angle by which strain lags force	ϵ	radians	$\tan \epsilon$
Damping bandwidth	Δf	Hz	$\dfrac{\Delta f}{f}$
Imaginary part of modulus of elasticity, $(E_r + jE_i)$	E_i	Pa	E_i/E_r
Sabine absorption coefficient	$\bar{\alpha}$	—	$\dfrac{cS\bar{\alpha}}{25.1fV}$

6.7 Coupling Loss Factors

The coupling loss factor, η_{12}, is a measure of the power transmission from vibroacoustic modes in subsystem 1 to modes in subsystem 2 and η_{21} is a measure of the power transmission from vibroacoustic modes in subsystem 2 to modes in subsystem 1. The **net** power transmission, Π'_{12}, from modes in subsystem 1 to modes in subsystem 2 is thus given by the following equation (Lyon and DeJong, 1995):

$$\Pi'_{12} = \Pi_{12} - \Pi_{21} = 2\pi f(\eta_{12} E_1 - \eta_{21} E_2) \tag{6.45}$$

where E_1 and E_2 are the total steady state vibroacoustic energies in subsystems 1 and 2, respectively, the two subsystems are coupled and the following reciprocity relationship holds for the coupling loss factors:

$$n_1\eta_{12} = n_2\eta_{21} \tag{6.46}$$

where n_1 and n_2 are the modal densities of subsystems 1 and 2, respectively.

The impedance expressions provided in Tables 4.1, 4.2 and 6.1 corresponding to infinite systems apply to equations for calculating coupling loss factors when the attachment points are more than one quarter of a wavelength from the subsystem boundaries. Where subsystems are attached at their boundaries, the impedance values must be adjusted to account for the boundary, as discussed in Section 6.3.

6.7.1   Coupling Loss Factors for Point Connections

Where two substructures are connected at a stiff point junction, the coupling loss factor is given by (Lyon and DeJong, 1995):

$$\eta_{12}^{\text{point}} = \frac{\beta_{\text{corr}}}{\pi f n_1(f)} \frac{\tau_{12}}{(2 - \tau_{12})} \tag{6.47}$$

where $n(f) = 2\pi n(\omega)$ and:

$$\beta_{\text{corr}} = \left\{ 1 + \left[\frac{1}{2\pi f [\eta_{1,\text{net}} n_1(f) + \eta_{2,\text{net}} n_2(f)]} \right]^8 \right\}^{-1/4} \tag{6.48}$$

where $n_i(f)$ is the modal density for subsystem i and $\eta_{i,\text{net}}$ is the damping loss factor for subsystem i, plus some fraction of the loss due to coupling with other systems.

The transmission coefficient, τ_{12}, is:

$$\tau_{12} = \frac{4\text{Re}\{Z_{F,1}\}\text{Re}\{Z_{F,2}\}}{\left| \sum\limits_{m=1}^{N} Z_{F,m} \right|^2} \tag{6.49}$$

where N is the number of subsystems connected at the junction (including source and receiving subsystems) and $\text{Re}\{Z_F\}$ is the real part of the point impedance. In many cases it is assumed that τ_{12} is small and $\beta_{\text{corr}} \approx 1$, so that Equation (6.48) becomes the more familiar expression:

$$\eta_{12} = \frac{\tau_{12}}{2\pi f n_1(f)} \tag{6.50}$$

where $f n_1(f) = \omega n_1(\omega)$. However, this expression becomes increasingly less accurate as the number of modes excited in a substructure is reduced.

If a mass, m_a, is part of the junction, its effect can be included by adding its inertance, $\text{j}2\pi f m_a$, to the denominator of Equation (6.49). If a subsystem, i, is connected to the junction via a stiffness element of stiffness, k_i, the subsystem impedance, $Z'_{F,i}$, to be used in Equation (6.49) in place of $Z_{F,i}$ is:

$$Z'_{F,i} = \left(\frac{\text{j}\omega}{k_i} + \frac{1}{Z_{F,i}} \right)^{-1} \tag{6.51}$$

If bending waves that result in both translational and rotational motion are incident on a junction, it is sufficiently accurate to add the transmission coefficients for translational and rotational motion together to obtain a net transmission coefficient (Lyon and DeJong, 1995).

For a 1-D acoustic system, a junction has a common pressure, so that acoustic admittances (reciprocal of acoustic impedances), Y_F, add at the junction. Thus:

$$\tau_{12} = \frac{4\text{Re}\{Y_{F,1}\}\text{Re}\{Y_{F,2}\}}{\left| \sum\limits_{m=1}^{N} Y_{F,m} \right|^2} \tag{6.52}$$

6.7.2   Coupling Loss Factors for Line Connections

The general equation for the coupling loss factor for line connected systems (for example, plates connected along their edges) for power transmission from subsystem 1 to subsystem 2 is:

$$\eta_{12}^{\text{line}} = \frac{c_g L}{\omega \pi S_1} \langle \tau_\theta \cos \theta \rangle_\theta \tag{6.53}$$

where S_1 is the area of the source structure (usually a plate) and the symbols $\langle\ \rangle_\theta$ mean that the transmission coefficient, τ_θ, is averaged over all angles of the incident waves. c_g is the group wave speed for the particular wave type being considered. For shear waves, $c_g = c_s$, for longitudinal waves, $c_g = c_L$, and for bending waves, $c_g = 2c_b$.

Equation (6.53) can be rewritten in terms of the normal incidence transmission coefficient, $\tau_{12}^{\text{line}}(0)$, as (Lyon and DeJong, 1995):

$$\eta_{12}^{\text{line}} = \frac{\beta_{\text{corr}}}{\pi f n_1(f)} \frac{k_1 L}{\pi} \int\limits_{0}^{\pi/2} \frac{\tau_{12}(\theta)\cos(\theta)}{[2 - \tau_{12}(\theta)]} \mathrm{d}\theta = \frac{\beta_{\text{corr}} I_{12}^{\text{line}}}{\pi f n_1(f)} \frac{\tau_{12}^{\text{line}}(0)}{[2 - \tau_{12}^{\text{line}}(0)]} \tag{6.54}$$

where $c_g/S_1 = k_1/n_1(f)$ and:

$$\tau_{12}^{\text{line}}(0) = \frac{4\mathrm{Re}\{Z'_{F,1}\}\mathrm{Re}\{Z'_{F,2}\}}{\left|\sum\limits_{m=1}^{N} Z'_{F,m}\right|^2} \tag{6.55}$$

where the prime indicates a line impedance quantity (as opposed to a point impedance) and there are N plates connected at the junction. Equation (6.54) is identical to the more familiar Equation (6.53) if τ_{12} is small and $\beta_{\text{corr}} \approx 1$.

For bending wave transmission between two plates connected perpendicularly to one another along a line of length L:

$$I_{12}^{\text{line}} = \frac{L}{4}\left(\frac{k_{b1}^4 k_{b2}^4}{k_{b1}^4 + k_{b2}^4}\right)^{1/4} \tag{6.56}$$

For other situations, the quantity, I_{12}^{line}, is determined by the integral over angle for the wave type and incident angle range (see Equation 6.54 and Lyon and DeJong (1995)).

The impedance for line excitation of a plate can be derived from the corresponding point impedance per unit width of a beam. Thus, the line force impedance for bending excitation of a plate further than one quarter of a wavelength from an edge is:

$$Z'_F = 2m_s c_b(1 + \mathrm{j}) \tag{6.57}$$

where m_s is mass per unit area of the plate and c_b is the bending wave speed in the plate.

The line force impedance for bending waves in a beam is:

$$Z'_F = \mathrm{j}\omega m_b \tag{6.58}$$

where $m_b = \rho_m S$ is mass per unit length of the beam. Although the real part of the impedance is zero, the value actually used in Equation (6.55) for the real part of the impedance is the magnitude of the impedance; that is, $\mathrm{Re}\{Z_F\} = \omega m_b$ (Lyon and DeJong, 1995).

The coupling loss factor for wave types other than bending waves can be determined using Equation (6.53) and integrating the transmission coefficient over the incident angles for the particular wave type (Lyon and DeJong, 1995).

6.7.2.1 Multiple Thin Plates Connected at Their Edges

The coupling loss factor for normal incidence bending wave transmission between two thin plates, i and j, at a junction shared with $N - 2$ additional thin plates, can be obtained from Equation (6.53) as:

$$\eta_{ij}^{\text{line}} = \frac{2c_b L}{\omega \pi S_i} \tau_{ij} \tag{6.59}$$

where S_i is the area of plate, i, L is the length of the junction, and τ_{ij} is the normally incident transmission coefficient for bending waves travelling from plate i to plate j, given by:

$$\tau_{ij} = \frac{2XY(D_i k_{bi} + D_j k_{bj})}{\left[\sum_{m=1}^{N} D_m k_{bm} \right]^2} \tag{6.60}$$

where D_i, D_j and D_m are the bending stiffnesses per unit width of plates i, j and m, respectively ($= Eh^3/[12(1 - \nu^2)]$), $k_b = \omega/c_b$ is given by Equation (4.42) for plates i, j or m. The quantity X is defined as:

$$X = \begin{cases} 1 - 0.7[(h_m/h_c) - 1.2]; & h_m/h_c > 1.2 \\ 1; & h_m/h_c \leq 1.2 \end{cases} \tag{6.61}$$

where h_m is the thickness of the thickest plate connected to the common join and h_c is the thickness of the thicker of the two plates, i and j. The quantity Y is defined as:

$$Y = \frac{2.75(h_i/h_j)}{1 + 3.24(h_i/h_j)} \tag{6.62}$$

where h_i is the thickness of plate i and h_j is the thickness of plate j. There are two limiting cases for which a simple solution is valid.

1. If $h_i = h_j$, then $\tau_{ij} = 8/27$.
2. If $h_j > 2h_i$, then $\tau_{ij} = D_i/D_j$.

6.7.2.2 Two Panels Separated along a Line by a Beam

For the case of two panels separated along a line by a beam, the expression for the coupling loss factor is the same as Equation (6.53). However, the transmission coefficient in that equation is different and for normally incident bending waves, it is:

$$\tau_{12} = \frac{h_p c_{Lp} m_p}{\sqrt{3} c_{bb} m_b} \left(1 + 64 \frac{m_p c_{bp}}{\omega m_b} \right) \tag{6.63}$$

where h_p is the plate thickness, m_p is the plate mass per unit area, c_{Lp} is the longitudinal wave speed in the plate, c_{bp} is the bending wave speed in the plate, c_{bb} is the bending wave speed in the beam and m_b is the beam mass per unit length.

6.7.3 Coupling Loss Factors for Area Connections

The general equation for the coupling loss factor for area connected systems (for example, two rooms connected by a limp panel or separated by an opening) for power transmission from subsystem 1 to subsystem 2 is:

$$\eta_{12}^{\text{area}} = \frac{c_g S_w}{4\omega V_1} \langle \tau_{\theta,\phi} \rangle_{\theta,\phi} \tag{6.64}$$

where V_1 is the volume of the source room and the symbols $\langle \ \rangle_{\theta,\phi}$ mean that the transmission coefficient, $\tau_{\theta,\phi}$, is averaged over all angles of the incident waves.

For an isotropic panel (or an opening) connecting the two acoustic spaces, the transmission coefficient variation with azimuth angle, θ, is the same as the variation with vertical angle, ϕ, so Equation (6.64) can be rewritten in terms of the normal incidence transmission coefficient, $\tau_{12}^{\text{area}}(0)$, as (Lyon and DeJong, 1995):

$$\eta_{12}^{\text{area}} = \frac{\beta_{\text{corr}}}{\pi f n_1(f)} \frac{k_1^2 S_w}{2\pi} \int_0^{\pi/2} \frac{\tau_{12}(\theta)\cos(\theta)\sin(\theta)}{[2 - \tau_{12}(\theta)]} \mathrm{d}\theta = \frac{\beta_{\text{corr}} I_{12}^{\text{area}}}{\pi f n_1(f)} \frac{\tau_{12}^{\text{area}}(0)}{[2 - \tau_{12}^{\text{area}}(0)]} \tag{6.65}$$

where k_1 is the wavenumber in subsystem 1, S_w is the area of the connection between subsystems 1 and 2, $n_1(f)$ is the modal density (modes per Hz) in subsystem 1, $\tau_{12}(\theta)$ is the transmission coefficient for a wave arriving at the junction at an angle of θ, and $c_g/V_1 = 2k_1^2/[\pi n_1(f)]$. The quantity, I_{12}^{area}, is determined by the integral over angle for the wave type and incident angle range (Lyon and DeJong, 1995).

The area transmission coefficient for normal incidence is:

$$\tau_{12}^{\text{area}}(0) = \frac{4\text{Re}\{Z_{F,1}''\}\text{Re}\{Z_{F,2}''\}}{\left| \sum_{m=1}^{N} Z_{F,m}'' \right|^2} \tag{6.66}$$

where the double prime indicates an area impedance quantity (as opposed to a point impedance) and there are N subsystems connected at the area junction. Of course it is unlikely that more than two subsystems would exist at an area junction.

Equation (6.65) is identical to the more familiar Equation (6.64) if τ_{12} is small and $\beta_{\text{corr}} \approx 1$.

The impedance for area excitation can be derived from the corresponding force impedance per unit area of a beam. Thus, the area impedance of an acoustic space is:

$$Z_0 = \rho c \tag{6.67}$$

and the area force impedance of a plate is Lyon and DeJong (1995):

$$Z_F = j\omega m_s \tag{6.68}$$

where m_s is the mass per unit area of the plate. In Equation (6.66), the real part of the plate impedance is replaced with the magnitude of the impedance.

6.7.3.1 Coupling Loss Factor for Radiation from a Panel

When a panel such as a wall or floor is radiating into a room, the coupling loss factor can be calculated using (Craik, 1982):

$$\eta_{12} = \frac{\rho c \sigma}{\omega m_s} \tag{6.69}$$

where σ is the panel radiation efficiency (see Section 4.2.4) and m_s is the mass per unit area of the panel.

To calculate the coupling loss factor, η_{21}, for radiation from a room to a wall or floor, Equations (6.46) and (6.69) can be used.

Note that when sound is transmitted from one room to another, the coupling loss factors due to non-resonant radiation (Equation (6.64)) and resonant radiation (Equation (6.69)) must be added together.

6.7.4 Measurement of Coupling Loss Factors

Coupling loss factors can be measured by considering a single junction at any one time. The receiver subsystem and any other subsystems connected at the junction under consideration must be well damped so that their modal energy is well below that of the source subsystem. The source system is then excited by band-limited (usually 1/3-octave or octave) pink noise and the total energies, E_1 and E_2, in the source and receiver systems, respectively, are measured.

If the subsystem being measured is a structural system, the energy is related to the mean square vibration velocity of the subsystem as follows:

$$E = m_{\text{sub}}\langle u^2 \rangle \tag{6.70}$$

where m_{sub} is the total mass of the substructure under consideration, $\langle u^2 \rangle$ is the mean square velocity averaged over the structural surface, which can be determined using an accelerometer measurement at many locations on the structure and an integrating circuit to convert the accelerometer data to velocity data.

If the subsystem being measured is an acoustic system, the energy is related to the mean square sound pressure in the subsystem as:

$$E = \frac{V \langle p^2 \rangle_{S,t}}{\rho c^2} \tag{6.71}$$

where V is the volume of the acoustic space and the mean square sound pressure, $\langle p^2 \rangle_{S,t}$, averaged over space and time, can be determined using a microphone measurement at many locations in the acoustic space.

The coupling loss factor can then be determined using:

$$\eta_{12} = \frac{\eta_2 E_2}{E_1 - \dfrac{E_2 n_1(\omega)}{n_2(\omega)}} \tag{6.72}$$

where η_2 is the damping loss factor of the receiving system (including any added damping). As the reciprocal relationship of Equation (6.46) holds, it is usually better to use the most heavily damped structure (or the substructure with the smallest modal energy) as the receiving structure so that the requirement for the application of additional damping can be minimised.

6.8 Steps in Solving an SEA problem

The aim of most SEA analyses is to obtain the mean square velocity of a structural element or the mean square acoustic pressure in an enclosed space. The mean square velocity of a structural element can be used together with the radiation efficiency of that element to calculate the radiated sound power and hence the sound pressure level in an enclosure or in free field (see Section 4.2.4). The steps involved in an SEA analysis are listed below.

1. Divide the vibroacoustic system into beam, plate or cylinder elements.
2. Further divide the beam, plate and cylinder elements into subsystems within which the modes have similar properties. Thus a beam element may be divided into three subsystems: one for bending modes, one for axial (or longitudinal) modes and one for torsional modes. Similarly, a plate or cylinder element may be divided into three subsystems: one for bending modes, one for in-plane (plate) or axial (cylinder) modes and one for shear modes.
3. Draw a system diagram showing the interconnections between all of the subsystems making up the complete system, as illustrated for a simple system in Figure 6.3.
4. Determine locations in the system where power is being injected.
5. Determine the level of injected power at each system location identified in the previous step using one of Equations (6.17) to (6.20).
6. Calculate the modal densities for each subsystem using the procedures described in Section 6.4.
7. Determine the damping loss factors for each subsystem averaged over 1/3-octave bands. This may involve some measurements as described in Section 6.6.1.
8. Determine the translational and rotational impedances of structural elements as described in Section 4.2.

9. Use impedance data from the previous step and the equations outlined in Section 6.7 to determine coupling loss factors for each subsystem connection (which includes connections between different and similar wave types in different structural elements). Alternatively, measured data can be used.

10. Write energy balance equations for each subsystem.

11. Combine all energy balance equations into a single matrix equation as described in Section 6.2.1.

12. Solve the matrix equation using appropriate software tools to obtain the equilibrium energy in each subsystem.

13. Convert the subsystem energies into mean square velocity levels, averaged over 1/3-octave or octave bands, using Equation (6.9) or mean square sound pressure level in an acoustic space, averaged over 1/3-octave or octave bands, using Equation (6.15).

14. Where necessary, convert mean square structural velocity levels to radiated sound power levels using elemental radiation efficiencies and Equation (6.1).

7

Spectral Analysis

7.1 Introduction

In Section 1.3.9, basic frequency analysis was considered, where the concept of converting a time domain signal to the frequency domain was discussed and the concepts of octave band and 1/3-octave band analysis were introduced. In this chapter the concepts underlying frequency analysis are discussed in detail and the analysis of signals using much narrower frequency bands than 1/3-octave is discussed. The purpose of such narrow-band analysis is to allow us to understand the composition of vibration or acoustic signals so that we may identify sources and transmission paths from source to receiver with more precision.

Many instruments are available that provide narrow frequency band analysis. However, they do not enjoy the advantage of standardisation that characterises octave and 1/3-octave band analysis, so that the comparison of readings taken on such instruments may be difficult. One way to ameliorate the problem is to present such readings as mean levels per unit frequency, called "spectral density level", rather than "band level". In this case, the measured level is reduced by ten times the logarithm to the base ten of the bandwidth. For example, referring to Table 1.1, if the 500 Hz octave band which has a bandwidth of $(707 - 353) = 354$ Hz were presented in this way, the measured octave band level would be reduced by $10 \log_{10}(354) = 25.5$ dB to give an estimate of the spectral density level at 500 Hz. The problem is not entirely alleviated, as the effective bandwidth will depend on the sharpness of the filter cut-off, which is provided as a range of slopes in the standards (see ANSI S1.11 (2014)). Generally, the bandwidth is taken as lying between the frequencies, on either side of the passband, at which the signal is down 3 dB from the signal at the centre of the band. The spectral density level represents the energy level in a band one cycle wide, whereas by definition a tone has a bandwidth of zero. Thus for a pure tone, the same level will be measured, regardless of the bandwidth used for the measurement.

There are two ways of transforming a signal from the time domain to the frequency domain. The first requires the use of band-limited digital or analogue filters. The second requires the use of Fourier analysis where the time domain signal is transformed using a Fourier series. This is implemented in practice digitally (referred to as the DFT—discrete Fourier transform) using a very efficient algorithm known as the FFT (fast Fourier transform). Digital filtering is discussed in Section 7.2 and Fourier analysis is discussed in detail in Section 7.3.

7.2 Digital Filtering

Spectral analysis is commonly carried out in standardised octave, 1/3-octave, 1/12-octave and 1/24-octave bands, and both analogue and digital filters are available for this purpose. Such filters are referred to as constant percentage bandwidth filters meaning that the filter bandwidth is a constant percentage of the band centre frequency. For example, the octave bandwidth is always about 70.1% of the band centre frequency, the 1/3-octave bandwidth is 23.2% of the band centre frequency and the 1/12-octave is 5.8% of the band centre frequency, where the band centre frequency is defined as the geometric mean of the upper and lower frequency bounds of the band (see Equation (1.21)).

The stated percentages are approximate, as a compromise has been adopted in defining the bands to simplify and to ensure repetition of the band centre frequencies. The compromise that has been adopted is that 10 times the logarithms to the base ten of the 1/3-octave band centre frequencies are integers or very close to an integer number (see Table 1.1).

Besides constant percentage bandwidth filters, instruments with constant frequency bandwidth filters are also available. However, these instruments have largely been replaced by fast Fourier transform (FFT) analysers, which give similar results in a fraction of the time and generally at a lower cost. When a time-varying signal is filtered using either a constant percentage bandwidth or a constant frequency bandwidth filter, an RMS amplitude signal is obtained, which is proportional to the sum of the total energy content of all frequencies included in the band.

When discussing digital filters and their use, an important consideration is the filter response time, T_R, which is the minimum time required for the filter output to reach steady state. The minimum time generally required is the inverse of the filter bandwidth, B (Hz). That is:

$$BT_R = \left(\frac{B}{f}\right) \cdot (fT_R) = Bn_R \approx 1 \tag{7.1}$$

where the centre band frequency, f, the relative bandwidth, B, and the number of cycles, n_R, have been introduced. For example, for a 1/3-octave filter, $B = 0.2316$, and the number of cycles, $n_R \approx 4.3$. A typical response of a 1/3-octave filter is illustrated in Figure 7.1, where it will be noted that the actual response time is perhaps five cycles or more, depending on the required accuracy.

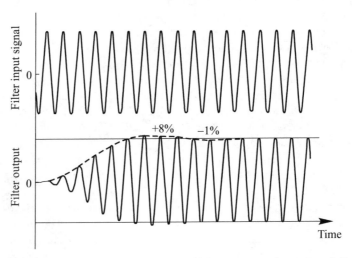

FIGURE 7.1 Typical filter response of a 1/3-octave filter (after Randall (1987)).

Where the RMS value of a filtered signal is required, it is necessary to determine the average value of the integrated squared output of the filtered signal over some prescribed period of time called the averaging time. The longer the averaging time, the more nearly constant will be the RMS value of the filtered output.

For a sinusoidal input of frequency, f (Hz), or for several sinusoidal frequencies within the band, where f (Hz) is the minimum separation between components, the variation in the average value will be less than 1/4 dB for an averaging time, $T_A \geq 3/f$.

For many sinusoidal components or for random noise and $BT_A \geq 1$, the error in the RMS signal may be determined in terms of the statistical error, ϵ, calculated as:

$$\epsilon = 0.5(BT_A)^{-1/2} \tag{7.2}$$

For random noise, the actual error has a 63.3% probability of being within the range $\pm\epsilon$ and a 95.5% probability of being within the range $\pm 2\epsilon$.

The calculated statistical error may be expressed in decibels as:

$$20\log_{10} e^{\epsilon} = 4.34(BT_A)^{-1/2} \quad \text{(dB)} \tag{7.3}$$

7.2.1 Octave and 1/3-Octave Filter Rise Times and Settling Times

One important aspect of analysis of very low-frequency sound, using octave or 1/3-octave bands, is the filter rise time, which is the time it takes for the filter to measure the true value of a continuous signal. So if the signal is varying rapidly, especially at low frequencies, it is not possible for the output of the filter to track the rapidly varying input, resulting in a considerable error in the RMS output, especially if the signal has high crest factors. From Table 7.1, which was assembled with data provided by Bray and James (2011), it can be seen that the impulse response duration to achieve 90% (or 1 dB error) of the true magnitude of the signal for a 1/3-octave band filter centred at 1 Hz is approximately 5 seconds or 5 full cycles. It is clear that use of such a 1/3-octave filter will not correctly measure the energy associated with rapidly varying, very low-frequency sound having relatively high crest factors. As the centre frequency of the octave or 1/3-octave band becomes lower, the required sampling time to obtain an accurate measure of the signal becomes larger.

TABLE 7.1 1/3-octave filter rise times for a 1 dB error (6th order filter defined in ANSI S1.11 (2014)). The rise time decreases by a factor of 10 for each decade increase in frequency. Octave band filters would have rise times of 1/3 of the rise time of a 1/3-octave filter with the same centre frequency

1/3-octave centre frequency (Hz)	Rise time (millisec)	1/3-octave centre frequency (Hz)	Rise time (millisec)	1/3-octave centre frequency (Hz)	Rise time (millisec)
1.00	4989	10.0	499	100	49.9
1.25	3963	12.5	396	125	39.6
1.60	3148	16.0	315	160	31.5
2.00	2500	20.0	250	200	25
2.50	1986	25.0	199	250	19.9
3.15	1578	31.5	158	315	15.8
4.00	1253	40.0	125	400	12.5
5.00	995	50.0	99.5	500	9.95
6.30	791	63.0	79.1	639	7.91
8.00	628	80.0	62.8	800	6.28

One way of increasing the ability of the measurement system to track and measure accurately sound signals with high crest factors, is to increase the bandwidth of the filter used for the measurement. A suggestion by Bray and James (2011) is to use filter bandwidths that are equal to the critical bandwidth of our hearing mechanism, which is approximately 100 Hz at our lowest hearing frequencies. Thus, the filter bandwidth for measuring low-frequency noise and infrasound should be 100 Hz (actually from 0.5 Hz to 100 Hz) if we are to measure crest factors in a similar way to which our hearing mechanism experiences them. Bray and James (2011) recommend using a 4th order Butterworth 'Bark 0.5' band pass filter (see Bies et al. (2017, Chapter 3)) centred on 50 Hz with a rise time of approximately 8.8 milliseconds, which according to Bray and James (2011), simulates the approximately 10 millisecond response of our hearing mechanism. This response is 10 times faster than the 'fast' response on a sound level meter, which implies that the 'fast' response underestimates the true rise time of our hearing mechanism at these low frequencies. Thus, we hear much higher low-frequency peaks than are measured by the 'fast' sound level meter measurements.

In addition to filter rise time, it is also important to take into account the settling time of the RMS detector in order to achieve 1/2 dB accuracy, which is in addition to the rise time discussed above. For a single frequency signal of frequency, f, the settling time for 1/2 dB accuracy will be approximately $3/f$ (Randall, 1987). For a random signal, the calculated statistical error, ϵ, is given by Equation (7.3).

There are three types of noise signal used in acoustics to excite systems for the purpose of measuring their acoustical properties.

- **White noise**, which is a signal with uniform spectral energy (that is, equal energy per Hz). White noise has a flat spectral shape when viewed on a narrow band spectrum, but increases at a rate of 3 dB per octave when viewed on an octave band plot.

- **Pink noise**, which is a signal with the same amount of energy in each octave band. Pink noise has a flat spectral shape when viewed on an octave band plot, but has a downwards slope and decreases at 3 dB per octave (doubling of freq) when viewed on a narrow band plot.

- **Pseudo-random noise**, which is discussed in Section 7.3.17.

- **Swept sine**, which is a single frequency signal that gradually increases in frequency during the measurement process.

7.3 Advanced Frequency Analysis

In this section, various aspects of FFT analysis will be discussed. FFT analysis provides much more frequency resolution (each component in the frequency spectrum representing smaller frequency spans) than is possible with octave or 1/3-octave band analysis.

FFT analysis is the process of transforming a time varying signal into its frequency components to obtain a plot or table of signal amplitude as a function of frequency. A general Fourier representation of a periodic time varying signal of period, T_p, consisting of a fundamental frequency, $f_1 = 1/T_p$, represented by $x_1(t) = x(t + T_p)$ and various harmonics, n, of frequency f_n, represented by $x_n(t) = x(t + nT_p)$, where $n = 2, 3, \ldots$, takes the form:

$$x(t) = \sum_{n=1}^{\infty} A_n \cos(2\pi n f_1 t) + B_n \sin(2\pi n f_1 t) \tag{7.4}$$

As an example, we can examine the Fourier representation of a square wave shown in Figure 7.2. The first four harmonics in part (b) are described by the first four terms in Equation (7.4), where $B_n = 0$ for all components, $A_n = 4/(\pi n)$ for n odd and $A_n = 0$ for n even. The component characterised by frequency, nf_1, is usually referred to as the nth harmonic of the fundamental frequency, f_1, although some call it the $(n-1)$th harmonic. Use of Euler's well known equation

(Abramowitz and Stegun, 1965) allows Equation (7.4) to be rewritten as:

$$x(t) = \frac{1}{2}\sum_{n=0}^{\infty}\left[(A_n - jB_n)\,e^{j2\pi nf_1 t} + (A_n + jB_n)\,e^{-j2\pi nf_1 t}\right] \tag{7.5}$$

where the $n = 0$ term has been added for mathematical convenience. However, it represents the zero frequency (DC) component of the signal and is usually considered to be zero (see Figure 7.5(a)).

A further reduction is possible by defining the complex spectral amplitude components, $X_n = (A_n - jB_n)/2$ and $X_{-n} = (A_n + jB_n)/2$. Denoting the complex conjugate by *, the following relation may be written as:

$$X_n = X^*_{-n} \tag{7.6}$$

The introduction of Equation (7.6) in Equation (7.5) allows the following more compact expression to be written as:

$$x(t) = \sum_{n=-\infty}^{\infty} X_n\,e^{j2\pi nf_1 t} \tag{7.7}$$

The spectrum of Equation (7.7) now includes negative as well as positive values of n, giving rise to components $-nf_1$. The spectrum is said to be two sided. The spectral amplitude components, X_n, may be calculated using:

$$X_n = \frac{1}{T_p}\int_{-T_p/2}^{T_p/2} x(t)\,e^{-j2\pi nf_1 t}\,\mathrm{d}t \tag{7.8}$$

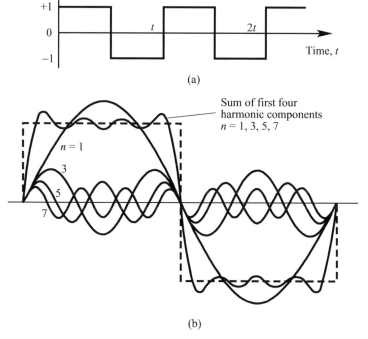

FIGURE 7.2 An example of Fourier analysis of a square wave: (a) periodic square wave in the time domain; (b) the first four harmonic components of the square wave in (a).

The spectrum of squared amplitudes is known as the power spectrum. The mean of the instantaneous power of the time-varying signal, $[x(t)]^2$, averaged over the period, T_A, is:

$$W_{\text{mean}} = \frac{1}{T_A} \int_0^{T_A} [x(t)]^2 \, \mathrm{d}t \tag{7.9}$$

Substitution of Equation (7.4) in Equation (7.9) and integrating gives:

$$W_{\text{mean}} = \frac{1}{2} \sum_{n=1}^{\infty} \left[A_n^2 + B_n^2 \right] \tag{7.10}$$

Equation (7.10) shows that the total power is the sum of the powers of each spectral component.

The previous analysis may be extended to the more general case of random noise by allowing the period, T_p, to become indefinitely large. In this case, X_n becomes $X_D(f)$, a continuous function of frequency, f. It is to be noted that whereas the units of X_n are the same as those of $x(t)$, the units of $X_D(f)$ are those of $x(t)$ per hertz. With the proposed changes, Equation (7.8) takes the form:

$$X_D(f) = \int_{-\infty}^{\infty} x(t) \, \mathrm{e}^{-\mathrm{j}2\pi ft} \, \mathrm{d}t \tag{7.11}$$

The spectral density function, $X_D(f)$, is complex, characterised by a real and an imaginary part (or amplitude and phase). Equation (7.7) becomes:

$$x(t) = \int_{-\infty}^{\infty} X_D(f) \, \mathrm{e}^{\mathrm{j}2\pi ft} \, \mathrm{d}f \tag{7.12}$$

Equations (7.11) and (7.12) form a Fourier transform pair, with the former referred to as the forward transform and the latter as the inverse transform. In practice, a finite sample time, T_s, is always used to acquire data and the spectral representation of Equation (7.7) is the result calculated by spectrum analysis equipment. This latter result is referred to as the spectrum and the spectral density is obtained by multiplying by the sample period, T_s, which is the same as dividing by the sampling frequency, and hence has 'normalised' the amplitude by the frequency resolution. Where a time function is represented as a sequence of samples taken at regular intervals, an alternative form of a Fourier transform pair is as follows. The forward transform is:

$$X(f) = \sum_{k=-\infty}^{\infty} x(t_k) \, \mathrm{e}^{-\mathrm{j}2\pi ft_k} \tag{7.13}$$

The quantity $X(f)$ represents the spectrum and the inverse transform is:

$$x(t_k) = \frac{1}{f_s} \int_{-f_s/2}^{f_s/2} X(f) \, \mathrm{e}^{\mathrm{j}2\pi ft_k} \, \mathrm{d}f \tag{7.14}$$

where f_s is the sampling frequency. The form of the Fourier transform pair used in spectrum analysis instrumentation is referred to as the discrete Fourier transform, for which the functions are sampled in both the time and frequency domains. Thus:

$$x(t_k) = \sum_{n=0}^{N-1} X(f_n) \, \mathrm{e}^{\mathrm{j}2\pi nk/N} \quad k = 0, 1, \ldots, (N-1) \tag{7.15}$$

$$X(f_n) = \frac{1}{N} \sum_{k=0}^{N-1} x(t_k) \, e^{-j2\pi nk/N} \qquad n = 0, 1, \ldots, (N-1) \qquad (7.16)$$

where k and n represent discrete sample numbers in the time and frequency domains, respectively, and $X(f_n)$ represents the *amplitude* of the nth component in the frequency spectrum. In Equation (7.15), the spacing between frequency components, in Hz, is dependent on the time, T_s, to acquire the N samples of data in the time domain and is equal to $1/T_s$ or f_s/N. Thus the effective filter bandwidth, B, is equal to $1/T_s$. The four Fourier transform pairs are shown graphically in Figure 7.3. In Equations (7.15) and (7.16), the functions have not been made symmetrical about the origin, but because of the periodicity of each, the second half of each sum also represents the negative half period to the left of the origin, as can be seen by inspection of Figure 7.3(d).

The frequency components above $f_s/2$ in Figure 7.3(d) can be more easily visualised as negative frequency components and in practice, the frequency content of the final spectrum must be restricted to less than $f_s/2$. This is explained in Section 7.3.5 where aliasing is discussed.

The discrete Fourier transform is well suited to the digital computations performed in instrumentation or by frequency analysis software on a personal computer. Nevertheless, it can be seen by referring to Equation (7.15) that to obtain N frequency components from N time samples, N^2 complex multiplications are required. Fortunately, this is reduced, by the use of the fast Fourier transform (FFT) algorithm, to $N \log_2 N$, which for a typical case of $N = 1024$, speeds up computations by a factor of 100. This algorithm is discussed in detail by Randall (1987).

7.3.1   Auto Power Spectrum and Power Spectral Density

The auto power spectrum (sometimes called the power spectrum) is the most common form of spectral representation used in acoustics and vibration. The auto power spectrum is the spectrum of the square of the RMS values of each frequency component, whereas the frequency spectrum discussed previously was a spectrum of the *amplitudes* of each frequency component. The 2-sided auto power spectrum, $S_{xx}(f_n)$, may be estimated by averaging a large number of squared amplitude spectra, $X(f_n)$, and dividing by 2 to account for conversion from an amplitude squared spectrum to an RMS squared spectrum and an additional scaling to account for the application of a windowing function to the sampled data (see Section 7.3.4). Estimation of the scaling factor, S_A, is discussed in Section 7.3.4.1. Thus:

$$S_{xx}(f_n) \approx \frac{S_A}{Q} \sum_{i=1}^{Q} X_i^*(f_n) X_i(f_n) = \frac{S_A}{Q} \sum_{i=1}^{Q} |X_i(f_n)|^2 \qquad n = 0, 1, \ldots, (N-1) \qquad (7.17)$$

where i is the spectrum number and q is the number of spectra over which the average is taken. The larger the value of Q, the more closely will the estimate of $S_{xx}(f_n)$ approach its true value.

The *power spectral density*, $S_{Dxx}(f_n)$ (or $\text{PSD}(f_n)$), can be obtained from the power spectrum by dividing the amplitudes of each frequency component by the frequency spacing, Δf, between adjacent components in the frequency spectrum or by multiplying by the time, T_s, to acquire one record of data. Thus, the two-sided power spectral density is:

$$S_{Dxx}(f_n) \approx \frac{T_s S_A}{Q} \sum_{i=1}^{Q} |X_i(f_n)|^2 = \frac{S_A}{Q \Delta f} \sum_{i=1}^{Q} |X_i(f_n)|^2 \qquad n = 0, 1, \ldots, (N-1) \qquad (7.18)$$

where 'two-sided' indicates that the spectrum extends to negative as well as positive frequencies. The time blocks are usually overlapped by up to 50% to decrease the random error in the PSD estimate (Brandt, 2010), as explained in Section 7.3.6.

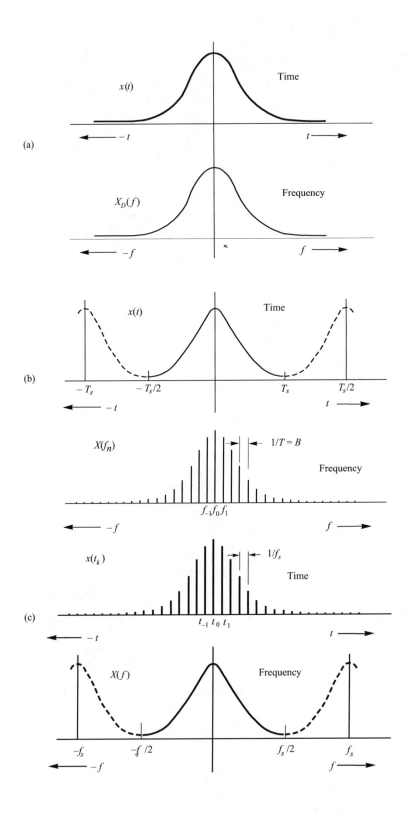

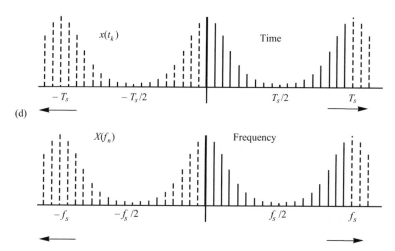

FIGURE 7.3 Various Fourier transform pairs (after Randall (1987)). The dashed lines indicate a periodically repeating sequence:

(a) Integral transform; signal infinitely long and continuous in both the time and frequency domains

$$X_D(f) = \int\limits_{-\infty}^{\infty} x(t)\mathrm{e}^{-\mathrm{j}2\pi ft}\,\mathrm{d}t \qquad\text{and}\qquad x(t) = \int\limits_{-\infty}^{\infty} X_D(f)\mathrm{e}^{\mathrm{j}2\pi ft}\,\mathrm{d}f;$$

(b) Fourier series; signal periodic in the time domain and discrete in the frequency domain

$$X(f_n) = \frac{1}{T_s}\int\limits_{-T_s/2}^{T_s/2} x(t)\mathrm{e}^{-\mathrm{j}2\pi f_n t}\,\mathrm{d}t \qquad\text{and}\qquad x(t) = \sum\limits_{n=-\infty}^{\infty} X(f_n)\mathrm{e}^{\mathrm{j}2\pi f_n t};$$

(c) Sampled function; signal discrete in the time domain and periodic in the frequency domain

$$X(f) = \sum\limits_{k=-\infty}^{\infty} x(t_k)\mathrm{e}^{-\mathrm{j}2\pi ft_k} \qquad\text{and}\qquad x(t_k) = \frac{1}{f_s}\int\limits_{-f_s/2}^{f_s/2} X(f)\mathrm{e}^{\mathrm{j}2\pi ft_k}\,\mathrm{d}f;$$

(d) Discrete Fourier transform; signal discrete and periodic in both the time and frequency domains

$$X(f_n) = \frac{1}{N}\sum\limits_{k=0}^{N-1} x(t_k)\mathrm{e}^{-\mathrm{j}2\pi nk/N} \qquad\text{and}\qquad x(t_k) = \sum\limits_{n=0}^{N-1} X(f_n)\mathrm{e}^{\mathrm{j}2\pi nk/N}$$

Note that $X(f_n)$ is a spectrum of amplitudes that must be divided by $\sqrt{2}$ to obtain an RMS spectrum.

Although it is often appropriate to express random noise spectra in terms of power spectral density, the same is not true for tonal components. Only the auto power spectrum will give the true energy content of a tonal component.

The auto power spectrum is useful for evaluating tonal components in a spectrum, although for random noise, it is more appropriate to use the power spectral density or PSD function of Equation (7.18). The auto power spectrum is used to evaluate spectra that contain tonal components because, unlike the PSD, it is able to give the true energy content of a tonal component. This is because the bandwidth of a tone is not the same as the frequency spacing in the spectrum and is often much smaller. This results in the spectral amplitude of a tone being independent of the frequency resolution of the FFT analysis, provided that the tonal frequency corresponds to the frequency of one of the spectral lines (see Section 7.3.3 and Section 7.3.4). Thus, dividing a tonal amplitude by the spectral resolution to obtain the PSD will result in a significant error in the tonal amplitude. In real systems, the frequency of a tone may vary slightly during the time it takes to acquire a sufficient number of samples for an FFT and also from one FFT to another during the averaging process to obtain power spectra. In this case, the tone may be spread out in frequency so that its amplitude will depend on the frequency resolution and a better estimate of the amplitude will be obtained with a coarse frequency resolution. A sufficiently coarse resolution would enable the range of frequency variation to be captured in a single frequency bin in the spectrum.

In cases where the frequency of the tone does not correspond to the centre frequency of one of the spectral lines, there will be an error in its amplitude that will depend on the windowing function (see Section 7.3.4) used and the difference in frequency between the tone and the centre frequency of the nearest spectral line. The maximum possible error is listed in Table 7.2. In this case the amplitude of the tone will also depend on the frequency resolution of the spectrum.

The problems of errors in tonal amplitudes can be avoided by calculating the PSD from the auto-correlation function, $R_{xx}(\tau)$, which is the covariance of the time series signal, $x(t)$, with itself time shifted by τ seconds. It is defined as:

$$R_{xx}(\tau) = \langle x(t)x(t+\tau) \rangle_t = \mathrm{E}[x(t) \cdot x(t+\tau)] = \lim_{T_s \to \infty} \frac{1}{T_s} \int_{-T_s/2}^{T_s/2} x(t) \cdot x(t+\tau)\, \mathrm{d}t \qquad (7.19)$$

where $\mathrm{E}[x]$ is the expected value of x and $\langle x \rangle_t$ is the time averaged value of x. The auto-correlation function is discussed in more detail in Section 7.3.16. The PSD, $S_{Dxx}(f_n)$, is obtained from the auto-correlation function by substituting $R_{xx}(\tau)$ for $x(t)$, τ for t, $S_{Dxx}(f_n)$ for $X(f_n)$ and $S_{Dxx}(f)$ for $X_D(f)$ in the equations in the caption of Figure 7.3. For example, the caption for part (c) in the figure is written as:

$$S_{Dxx}(f) = \sum_{k=-\infty}^{\infty} R_{xx}(\tau_k) \mathrm{e}^{-\mathrm{j}2\pi f \tau_k} \qquad (7.20)$$

and:

$$R_{xx}(\tau_k) = \frac{1}{f_s} \int_{-f_s/2}^{f_s/2} S_{Dxx}(f) \mathrm{e}^{\mathrm{j}2\pi f \tau_k}\, \mathrm{d}f \qquad (7.21)$$

However, the auto-correlation function is very computationally intensive to calculate, so this method is not in common use. In fact, the reverse is more often the case: the auto correlation of a dataset is found by taking the inverse FFT of the power spectral density.

For random noise, the frequency resolution affects the spectrum amplitude; the finer the resolution the smaller will be the amplitude. For this reason, we use PSDs for random noise for which the effective frequency resolution is 1 Hz.

In practice, the single-sided power spectrum, $G_{xx}(f_n)$ (positive frequencies only), is the one of interest and this is expressed in terms of the two-sided auto power spectrum $S_{xx}(f_n)$ as:

$$G_{xx}(f_n) = \begin{cases} 0; & f_n < 0 \\ S_{xx}(f_n); & f_n = 0 \\ 2S_{xx}(f_n); & f_n > 0 \end{cases} \qquad (7.22)$$

A similar expression may be written for the single-sided PSD, $G_{Dxx}(f_n)$, as:

$$G_{Dxx}(f_n) = \begin{cases} 0; & f_n < 0 \\ S_{Dxx}(f_n); & f_n = 0 \\ 2S_{Dxx}(f_n); & f_n > 0 \end{cases} \qquad (7.23)$$

If successive spectra, $X_i(f_n)$, are averaged, the result will be zero, as the phases of each spectral component vary randomly from one record to the next. Thus, in practice, auto power spectra are more commonly used, as they can be averaged together to give a more accurate result. This is because auto power spectra are only represented by an amplitude; phase information is lost when the spectra are calculated (see Equation (7.17)). The same reasoning applies to the power spectral density (power per Hz), which is obtained from the auto power spectrum by dividing the amplitude of each frequency component by the frequency spacing, Δf, between adjacent components.

7.3.2 Linear Spectrum

Sometimes the results of a spectral analysis are presented in terms of linear rather than the squared values of an auto power spectrum. Each frequency component in the linear spectrum is calculated by taking the square root of each frequency component in the auto power spectrum.

7.3.3 Leakage

Leakage is the phenomenon that occurs when a DFT uses a finite time window. This results in a spectrum containing discrete frequency components separated by a frequency interval. The number of frequency components and the frequency separation interval, Δf, are set by the sampling frequency, f_s, and the total sampling time (or measurement time), T_s. The frequency separation between adjacent components in the spectrum is:

$$\Delta f = \frac{1}{T_s} \qquad (7.24)$$

and the number of discrete frequency components, N, in the spectrum is:

$$N = T_s f_s \qquad (7.25)$$

Each discrete frequency component (or spectral line) in the spectrum is like a band pass filter with a characteristic response, $W(f_n)$, defined by a sinc function as:

$$W(f_n) = T_s \frac{\sin(\pi f_n T_s)}{\pi f_n T_s} = T_s \text{sinc}(f_n T_s) \qquad \text{where } n = 0, 1, \ldots, (N-1) \qquad (7.26)$$

This means that a sinusoidal signal equal to the exact frequency of a spectral component will be given the correct amplitude value in the frequency spectrum, as the sinc function has a value unity at frequency, f_n. However, for a sinusoidal signal with a frequency half way between two spectral lines, the energy of the signal will be split between the two adjacent lines (or frequency bins) and neither line will give the correct result. In fact, for this case, in the absence of any windowing, the error will result in a value that is 36% (or 1.96 dB) too small.

7.3.4   Windowing

As mentioned in Section 7.3.3, leakage occurs when calculating the DFT of a sinusoidal signal with a non-integer number of periods in the time window used for sampling. The error is caused by the truncation of the continuous signal, as a result of using a finite time window which causes a discontinuity when the two ends of the record are effectively joined in a loop as a result of the DFT. Leakage can be reduced by using a windowing function applied to the time window such that all samples in the time record are not given equal weighting when calculating the DFT. In fact, the window may be configured so that samples near the beginning and end of the time window are weighted much less than samples in the centre of the window. This minimises the effect of the signal being discontinuous at the beginning and end of the time window. The discontinuity without weighting causes side lobes to appear in the spectrum for a single frequency, as shown by the solid curve in Figure 7.4, which is effectively the same as applying a rectangular window weighting function. In this case, all signal samples before sampling begins and after it ends are multiplied by zero, and all values in between are multiplied by one. In the figure, the normalised frequency of $1 \times 1/T$ represents the frequency resolution, or number of Hz between adjacent frequency bins in the spectrum.

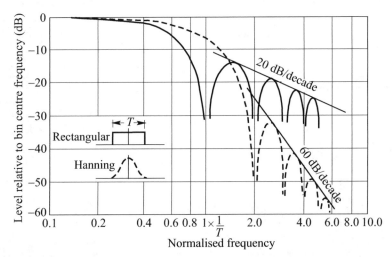

FIGURE 7.4 Comparison of the filter characteristics of the rectangular and Hanning time weighting functions for a power spectrum (after Randall (1987)).

A better choice of window is one that places less weight on the signal at either end of the window and maximum weight on the middle of the window. One such weighting, called a Hanning window, is illustrated in Figure 7.4. The result of weighting the input signal in this way is shown by the dashed curve in the figure. Even though the main lobe is wider, representing poorer frequency resolution, the side lobe amplitudes fall away more rapidly, resulting in less contamination of adjacent frequency bins. The properties of various weighting functions are summarised in Table 7.2.

In the table, the highest side lobe is the number of dB (in the auto power spectrum) that the signal corresponding to the highest side lobe will be attenuated compared to a signal at the filter centre frequency. The 'side lobe' fall off is illustrated in Figure 7.4, where the side lobes are the peaks to the right of the normalised frequency of 1.0.

The noise bandwidth in Table 7.2 is an important quantity in spectrum analysis. It is defined as the bandwidth of a rectangular filter that would let pass the same amount of broadband noise. It is especially useful in calculating the RMS level of power in a certain bandwidth in a spectrum such as a 1/3-octave band. This is discussed in more detail in Section 7.3.4.3.

The maximum amplitude error is the amount that a signal will be attenuated when it has a frequency that lies exactly midway between the centre frequencies of two adjacent filters (corresponding to normalised frequency of 0.5 in Figure 7.4). As can be seen in Table 7.2, the maximum error in the power spectrum corresponding to the rectangular window (effectively no window at all) is 3.92 dB. As expected, this is twice the value of 1.96 dB mentioned in Section 7.3.3 for the amplitude spectrum, as squaring results in a factor of two in the logarithmic domain.

The best weighting function for amplitude accuracy in the frequency domain is the flat top (the name refers to the weighting in the frequency domain whereas the window shape refers to the weighting in the time domain). This is often used for calibration because of its uniform flat frequency response over the bandwidth, $B(= 1/T)$, which results in the measured amplitude of the spectral component being independent of small variations in signal frequency around the band centre frequency, thus making this window suitable for instrument calibration with a tonal signal.

TABLE 7.2 Properties of the various time weighting functions

Window type	Highest side lobe (dB)	Side lobe fall off (dB/decade)	Normalised noise bandwidth (bins) B_{en}	Maximum power spectrum amplitude error (dB)	$\rho(1)$ Equation (7.54)	Window total energy (dB)
Rectangular	−13	−20	1.00	3.92	0.5	1.0
Triangular	−27	−40	1.33	1.83	0.25	—
Hanning	−32	−60	1.50	1.43	0.167	−4.26
Hamming	−43	−20	1.36	1.75	0.235	−4.01
Blackman	−58	−20	1.73	1.10	0.09	—
Blackman-Harris (3 term)	−67	−20	1.70	1.13	0.096	−5.13
Blackman-Harris (4 term)	−92	−20	2.00	0.83	0.038	−5.88
Kaiser-Bessel ($\alpha = 1.74$)	−40	−20	1.41	1.62	0.208	—
Kaiser-Bessel ($\alpha = 2.59$)	−60	−20	1.68	1.16	0.103	—
Kaiser-Bessel ($\alpha = 3.39$)	−80	−20	1.90	0.9	0.053	—
Kaiser-Bessel ($\alpha = 3.5$)	−82	−20	1.93	0.9	—	—
Gaussian ($\alpha = 3.0$)	−55	−20	1.64	1.25	0.106	−4.51
Poisson ($\alpha = 3.0$)	−24	−20	1.65	1.46	0.151	—
Poisson ($\alpha = 4.0$)	−31	−20	2.08	1.03	0.074	—
Flat top	−93	−20	3.75	0.01	—	−7.56
Welch	−21	−36.5	1.20	2.22	0.345	−2.73

However, the flat top window provides poor frequency resolution. Maximum frequency resolution (and minimum amplitude accuracy) is achieved with the rectangular window, so this is sometimes used to separate two spectral peaks that have a similar amplitude and a small frequency spacing. Good compromises that are commonly used are the Hanning window or the Hamming window, depending on whether or not amplitude accuracy (Hanning window) is more important than separation of closely spaced frequencies (Hamming window).

When transient signals (that is, signals that occur for a time shorter than the sampling interval) are to be analysed, the best window is a rectangular one. However, if the transients are repetitive and several occur during a data sampling period, then a Hanning weighting function may be used.

When a Fourier analysis is undertaken using the DFT algorithm with a finite sampling period, the resulting frequency spectrum is divided into a number of bands of finite width.

Each band may be considered as a filter, the shape of which is dependent on the weighting function used. If the frequency of a signal falls in the middle of a band, its amplitude will be measured accurately. However, if it falls midway between two bands, the error in power spectrum amplitude varies from 0.0 dB for the flat top window to 3.9 dB for the rectangular window. At the same time, the frequency bands obtained using the flat top window are 3.77 times wider so the frequency resolution is 3.77 times poorer than for the rectangular window. In addition, a signal at a particular frequency will also contribute to the energy in other nearby bands as can be seen by the shape of the filter curve in Figure 7.4. This effect is known as spectral leakage and it is minimised by having a high negative value for the side lobe fall off in Table 7.2.

7.3.4.1   Amplitude Scaling to Compensate for Window Effects

The effect of a non-rectangular window is to remove information and energy from the signal, resulting in an amplitude error. This must be compensated for by using an amplitude correction factor, A_f. This amplitude correction factor is used to calculate a scaled *amplitude* spectrum $X_i(f_n)$ from Equation (7.16) as:

$$X_i(f_n) = \frac{A_f}{N} \sum_{k=0}^{N-1} x(t_k)w(k)e^{-j2\pi kn/N} \qquad \text{where } n = 0, 1, \ \ldots, \ (N-1) \qquad (7.27)$$

where N is the number of discrete frequency components in the spectrum and:

$$A_f = \frac{N}{\sum\limits_{k=0}^{N-1} w(k)} \qquad (7.28)$$

In the above equation, $w(k)$ is the window weighting function (see Section 7.3.4.2) for each sample, k, in the time domain used to calculate the frequency spectrum. If $w(k) = 0$, then the kth sample value is set equal to 0 and if $w(k) = 1$, the sample value is unchanged.

The scaling factor, S_A, for the squared RMS spectrum (single-sided auto power spectrum) in Equation (7.16), with leakage of energy from the bin containing a particular frequency component into adjacent bins not taken into account, is:

$$S_A = \begin{cases} 2A_f^2/N^2; & n > 0 \\ \\ A_f^2/N^2; & n = 0 \end{cases} \qquad (7.29)$$

The additional factor of 2 accounts for the squaring of the 2-sided spectrum resulting in 1/4 of the spectrum value on each side so that when the two sides are added together the RMS value is only half of what it should be. The factor of 2 to account for the spectrum being a squared RMS spectrum instead of a squared amplitude spectrum is cancelled by the requirement to add the two sides to obtain the single-sided spectrum. Also, the assumption implicit in Equation (7.29) is that the spectrum being analysed is tonal so that there is no leakage of energy into adjacent bins.

However, for an auto power spectrum containing energy other than in tones, or for a PSD, an additional term must be included in the scaling factor to account for leakage of energy into adjacent bins as a result of the application of a windowing function. Thus, the scaling factor to be used is:

$$S_A = \begin{cases} 2A_f^2/(N^2 B_{en}); & n > 0 \\ \\ A_f^2/(N^2 B_{en}); & n = 0 \end{cases} \qquad (7.30)$$

where B_{en} is given in Table 7.2 for various window functions and is defined by:

$$B_{en} = \frac{N \sum\limits_{k=0}^{N-1} w^2(k)}{\left(\sum\limits_{k=0}^{N-1} w(k)\right)^2} \tag{7.31}$$

Equations (7.18), (7.29) and (7.31) are described as a Welch estimate of the PSD and represent the most commonly used method of spectral analysis in instrumentation and computer software, such as MATLAB's® **pwelch** function. However, this method for obtaining the PSD has associated bias errors that decrease as the frequency resolution in the original frequency spectrum, $X(f_n)$, becomes finer (i.e., smaller). An estimate of the bias error, ϵ_b, for the particular case of a Hanning window with a frequency resolution of Δf is (Schmidt, 1985a,b):

$$\epsilon_b \approx \frac{(\Delta f)^2 G_{xx}''(f_n)}{6\, G_{xx}(f_n)} + \frac{(\Delta f)^4 G_{xx}''''(f_n)}{72\, G_{xx}(f_n)} \tag{7.32}$$

where the prime represents differentiation with respect to frequency and 4 primes represent the fourth derivative. It can be seen from the above that where there are tones (which produce large values of the second derivative in particular), the error will be large.

7.3.4.2 Window Function Coefficients

Each windowing function identified in Table 7.2 requires different equations to calculate the coefficients, $w(k)$. The coefficients represent the quantity that data sample, k, in the time series data, is multiplied by, before being included in the dataset used for taking the FFT. These equations are listed below for each of the windows identified in Table 7.2.

Rectangular Window
For a record in the time domain that is a total of N samples in length (producing N discrete frequency components in the frequency domain), the window coefficients corresponding to each sample, k, in the time series record are:

$$w(k) = 1; \qquad 1 \leq k \leq N \tag{7.33}$$

Triangular Window
For a record that is a total of N samples in length, the window coefficients corresponding to each sample, k, in the record for N odd are:

$$w(k) = \begin{cases} 2k/(N+1); & 1 \leq k \leq (N+1)/2 \\ \\ 2 - 2k/(N+1); & \frac{(N+1)}{2} + 1 \leq k \leq N \end{cases} \tag{7.34}$$

and for N even, the coefficients are:

$$w(k) = \begin{cases} (2k-1)/N; & 1 \leq k \leq N/2 \\ \\ 2 - (2k-1)/N; & \frac{N}{2} + 1 \leq k \leq N \end{cases} \tag{7.35}$$

The coefficients of another version of a triangular window are:

$$w(k) = 1 - \left| \frac{2k - N + 1}{L} \right| \tag{7.36}$$

where L can be $N, N + 1$, or $N - 1$. All alternatives converge for large N.

Hamming Window

For a record that is a total of N samples in length, the window coefficients corresponding to each sample, k, in the record are:

$$w(k) = \alpha - \beta \cos \left(\frac{2\pi k}{N - 1} \right); \qquad 0 \le k \le N - 1 \qquad (7.37)$$

where the optimum values for α and β are 0.54 and 0.46, respectively.

Hanning Window

The Hanning window is the one most commonly used in spectrum analysis and is the one recommended for PSD analysis.

For a record that is a total of N samples in length, the window coefficients corresponding to each sample, k, in the record are:

$$w(k) = 0.5 \left[1 - \cos \left(\frac{2\pi k}{N - 1} \right) \right]; \qquad 0 \le k \le N - 1 \qquad (7.38)$$

Blackman Window

For a record that is a total of N samples in length, the window coefficients corresponding to each sample, k, in the time series record are:

$$w(k) = a_0 - a_1 \cos \left(\frac{2\pi k}{N - 1} \right) + a_2 \cos \left(\frac{4\pi k}{N - 1} \right); \qquad 0 \le k \le N - 1 \qquad (7.39)$$

where $a_0 = (1 - \alpha)/2$, $a_1 = 1/2$, and $a_2 = \alpha/2$. For an unqualified Blackman window, $\alpha = 0.16$.

Blackman–Harris Window

For a record that is a total of N samples in length, the window coefficients corresponding to each sample, k, in the time series record for a *3-term window* are:

$$w(k) = a_0 - a_1 \cos \left(\frac{2\pi k}{N - 1} \right) + a_2 \cos \left(\frac{4\pi k}{N - 1} \right); \qquad 0 \le k \le N - 1 \qquad (7.40)$$

where $a_0 = 0.42323$, $a_1 = 0.49755$ and $a_2 = 0.07922$.

For a *4-term window*, the equation for the coefficients is given by

$$w(k) = a_0 - a_1 \cos \left(\frac{2\pi k}{N - 1} \right) + a_2 \cos \left(\frac{4\pi k}{N - 1} \right) - a_3 \cos \left(\frac{6\pi k}{N - 1} \right);$$
$$0 \le k \le N - 1 \qquad (7.41)$$

where $a_0 = 0.35875$, $a_1 = 0.48829$, $a_2 = 0.14128$ and $a_3 = 0.01168$.

Kaiser–Bessel Window

For a record that is a total of N samples in length, the window coefficients corresponding to each sample, k, in the record are:

$$w(k) = \frac{I_0 \left(\pi \alpha \sqrt{1 - \left(\frac{2k}{N-1} - 1 \right)^2} \right)}{I_0(\pi \alpha)}; \qquad 0 \le k \le N - 1 \qquad (7.42)$$

where I_0 is the zeroth order modified Bessel function of the first kind. The parameter, α, determines the tradeoff between main lobe width and side lobe levels. Increasing α widens the main lobe and increases the attenuation of the side lobes. To obtain a Kaiser window that provides an attenuation of β dB for the first side lobe, the following values of $\pi\alpha$ are used:

$$\pi\alpha = \begin{cases} 0.1102(\beta - 8.7); & \beta > 50 \\ 0.5842(\beta - 21)^{0.4} + 0.07886(\beta - 21); & 21 \leq \beta \leq 50 \\ 0; & \beta < 21 \end{cases} \qquad (7.43)$$

A typical value of α is 3 and the main lobe width between the nulls is given by $2\sqrt{1 + \alpha^2}$.

Gaussian Window
For a record that is a total of N samples in length, the window coefficients corresponding to each sample, k, in the record are:

$$w(k) = \exp\left[-\frac{1}{2}\left(\frac{k - (N-1)/2}{\sigma(N-1)/2}\right)^2\right]; \qquad 0 \leq k \leq N-1 \quad \text{and} \quad \sigma \leq 0.5 \qquad (7.44)$$

where σ is the standard deviation of the Gaussian distribution. An alternative formulation is:

$$w(k) = \exp\left[-\frac{1}{2}\left(\frac{\alpha k}{(N-1)/2}\right)^2\right]; \qquad -(N-1)/2 \leq k \leq (N-1)/2 \qquad (7.45)$$

where $\alpha = (N-1)/2\sigma$ and $\sigma \leq 0.5$

Poisson (or Exponential) Window
For a record that is a total of N samples in length, the window coefficients corresponding to each sample, k, in the record are:

$$w(k) = \exp\left(-\left|k - \frac{N-1}{2}\right|\frac{1}{\tau}\right) \qquad 0 \leq k \leq N-1 \qquad (7.46)$$

For a targeted decay of D dB over half of the window length, the time constant, τ, is:

$$\tau = \frac{8.69N}{2D} \qquad (7.47)$$

An alternative formulation is:

$$w(k) = \exp\left(\frac{-\alpha|k|}{(N-1)/2}\right); \qquad -(N-1)/2 \leq k \leq (N-1)/2 \qquad (7.48)$$

where $\exp\{x\} = e^x$ and $\alpha = (N-1)/2\tau$.

Flat Top Window
The flat top window is used in frequency analysis mainly for calibration of instrumentation with a tone. The frequency of the calibration tone need not be close to the centre frequency of a bin in the frequency spectrum to obtain an accurate result; in fact, an accurate result will be obtained for any calibration frequency.

For a record that is a total of N samples in length, the window coefficients corresponding to each sample, k, in the record are:

$$w(k) = a_0 - a_1\cos\left(\frac{2\pi k}{N-1}\right) + a_2\cos\left(\frac{4\pi k}{N-1}\right) + a_3\cos\left(\frac{6\pi k}{N-1}\right) + a_4\cos\left(\frac{8\pi k}{N-1}\right)$$
$$0 \leq k \leq N-1 \qquad (7.49)$$

Coefficients are defined as $a_0 = 0.21557895$, $a_1 = 0.41663158$, $a_2 = 0.277263158$, $a_3 = 0.083578947$ and $a_4 = 0.006947368$. The coefficients may also be normalised to give $a_0 = 1.0$, $a_1 = 1.93$, $a_2 = 1.29$, $a_3 = 0.388$ and $a_4 = 0.028$.

Welch Window

For a record that is a total of N samples in length, the window coefficients corresponding to each sample, k, in the record are:

$$w(k) = 1 - \left(\frac{2k - N + 1}{N + 1} \right)^2 \qquad (7.50)$$

7.3.4.3 Power Correction and RMS Calculation

It is often of interest to determine the RMS value or dB level of an auto power spectrum over a defined frequency range within the spectrum. For example, one may wish to compute 1/3-octave or octave band decibel levels from an auto power spectrum covering the range from 1 Hz to 10,000 Hz. One may think that all one needs to do is add logarithmically (see Section 1.3.4) the various frequency components contained within the band of interest to obtain the required result. Adding logarithmically the frequency component amplitudes in an auto power spectrum is the same as converting the dB amplitude levels to $(\text{RMS})^2$ quantities, adding them together and then converting back to decibels. Unfortunately, finding the RMS value or dB level of a band within a spectrum is not as simple as described above, as energy from each frequency bin leaks into adjacent bins giving a result that is too large. Thus the calculation needs to be divided by a correction factor, B_{en}, defined in Equation (7.31), which is different for each windowing function. The correction factor, called the *normalised noise bandwidth*, is listed in Table 7.2 for various windowing functions and can be calculated using Equation (7.31). The correction factor, B_{en}, is a constant for a particular window and for a Hanning window it is 1.5. The RMS value of an auto spectrum between frequency locations f_{n_1} and f_{n_2} is thus:

$$x_{\text{RMS}}(n_1, n_2) = \sqrt{ \frac{\sum\limits_{n=n_1}^{n_2} G_{xx}(f_n)}{B_{en}} } \qquad (7.51)$$

where $G_{xx}(f_n)$ is defined in Equation (7.22). The correction factor holds even if the spectrum consists of only a single tone. This is because even for a single tone, energy appears in the two frequency bins adjacent to the one containing the frequency of the tone.

Calculating the RMS value or dB level of a power spectral density (PSD) requires a slightly different process as the PSD is already scaled such that the area under the curve corresponds to the mean square of the signal. In this case, the RMS value of a signal between two spectral lines, n_1 and n_2, is:

$$x_{\text{RMS}}(n_1, n_2) = \sqrt{ \Delta f \sum\limits_{n=n_1}^{n_2} G_{Dxx}(f_n) } \qquad (7.52)$$

where $\Delta f = 1/T_s = f_s/N$ is the frequency resolution used to obtain the original auto power spectrum that is converted to a PSD (which is the same as the PSD frequency resolution) and $G_{Dxx}(f_n)$ is the single-sided PSD which is defined in Equation (7.23). The dB level is simply $20 \log_{10}[x_{\text{RMS}}(n_1, n_2)]$.

7.3.5 Sampling Frequency and Aliasing

The sampling frequency is the frequency at which the input signal is digitally sampled. If the signal contains frequencies greater than half the sampling frequency, then these will be 'folded back' and appear as frequencies less than half the sampling frequency. For example, if the sampling frequency is 20000 Hz and the signal contains a component with a frequency of 25000 Hz, then this will appear as 5000 Hz in the resulting spectrum. Similarly, if the signal contains a component that has a frequency of 15000 Hz, this signal also will appear as 5000 Hz in the resulting spectrum. This phenomenon is known as 'aliasing' and in a spectrum analyser it is important to have analogue filters that have a sharp roll off for frequencies above about 0.4 times the sampling frequency. Aliasing is illustrated in Figure 7.5.

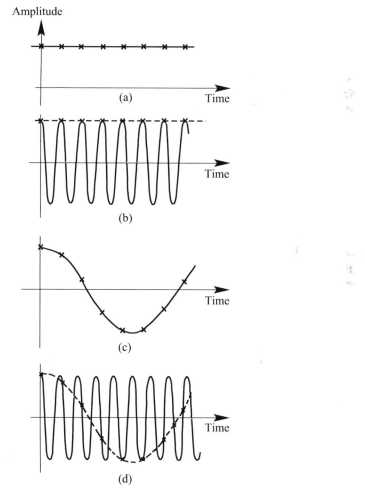

FIGURE 7.5 Illustration of aliasing (after Randall (1987)): (a) zero frequency or DC component; (b) spectrum component at sampling frequency, f_s, interpreted as DC; (c) spectrum component at $(1/N)f_s$; (d) spectrum component at $[(N+1)/N]f_s$ interpreted as $(1/N)f_s$.

7.3.6 Overlap Processing

When a limited amount of time is available for collecting data, more accurate results can be obtained by implementing overlap processing, as this allows more spectra to be averaged. For overlap processing, the time series data are divided into a number of records and then a DFT is performed on each segment. For an example case of a 50% overlap, this means that the first segment analysed is the first time record, the second segment is the second half of the first time record appended to the beginning of the first half of the second record, the third segment is the second record, the fourth segment is the second half of the second time record, appended to the beginning of the first half of the third record, etc. Even though the same data are used in more than one DFT, the effect of overlap analysis is to provide more spectra to average, which results in a smaller error in the final averaged spectrum. However, the *effective number* of averages is slightly less than the actual number of averages when overlap processing is used. The effective number of averages is window dependent and can be calculated as:

$$\frac{Q_e}{Q_d} = \frac{Q/Q_d}{1 + 2\sum_{i=1}^{Q} \frac{Q-i}{Q}\rho(i)} \tag{7.53}$$

where Q is the number of overlapping segments used, Q_d is the number of non-overlapping records, Q_e is the effective or equivalent number of averages, which give the same variance or uncertainty in the averaged DFT as the same number of averages using independent data. The quantity $\rho(i)$ is defined as:

$$\rho(i) = \frac{\left[\sum_{k=0}^{N-1} w(k)w(k+iD)\right]^2}{\left[\sum_{k=0}^{N-1} w^2(k)\right]^2} \tag{7.54}$$

where N is the number of discrete frequency components in the spectrum, $w(k)$ is dependent on the windowing function used (see Section 7.3.4.2), $D = \text{round}[N(1 - P/100)]$ and P is the percent overlap. For an overlap percentage up to 50%, the only non-zero value of $\rho(i)$ is when $i = 1$. Values of $\rho(1)$ for various windows are included in Table 7.2 for a 50% overlap.

Overlap processing is particularly useful when constructing sonograms (3-D plots of amplitude vs frequency with the third axis being time), as overlap processing results in smaller time intervals between adjacent spectra, resulting in better time resolution. For example, with 50% overlap, three spectra are obtained with the same number of samples, thus representing the same time period as two spectra with non-overlap processing.

Overlap processing can also be used to improve the frequency resolution, Δf, by using more samples, N, in the FFT ($\Delta f = f_s/N$) than is used with non-overlap processing for the same number of effective averages. This is useful when there is a limited length dataset and the maximum possible accuracy and frequency resolution is needed.

7.3.7   Zero Padding

Zero padding is the process of adding zeros to extend the number of samples in a record in the time domain prior to taking the DFT. This results in a frequency spectrum with frequency components more tightly spaced, which has resulted in some users thinking that they have achieved a finer frequency resolution. In fact, the apparent finer frequency resolution is actually an interpolation between the frequency bins that would exist with no zero padding so no more information has been gained. Higher frequency resolution can only be achieved with a longer sampling time. Thus zero padding is not considered a useful tool in this context. Nevertheless, when one is analysing a transient that has zero amplitude outside the sampling time window, zero padding can result in finer frequency resolution than if only the length of the transient had been used.

When zero padding is used, an additional scaling factor has to be introduced to calculate the amplitude of the resulting spectrum. If the spectrum contains n data points and m introduced zeros, the scaling factor with which the resulting power spectral amplitudes have to be multiplied to get the correct values is $(m + n)/n$.

7.3.8 Uncertainty Principle

The uncertainty principle states that the frequency resolution, Δf (equal to the effective filter bandwidth, B), of a Fourier transformed signal is equal to the reciprocal of the time, T_A, to acquire the sampled record of the signal. Thus, for a single spectrum, $\Delta f\, T_A = 1$. An effectively higher $\Delta f T_A = 1$ product can be obtained by averaging several spectra together until an acceptable error is obtained according to Equation (7.3), where B is the filter bandwidth (equal to Δf or frequency resolution and T_A is the total sample time).

7.3.9 Time Synchronous Averaging and Synchronous Sampling

Time synchronous averaging is a slightly different process to synchronous sampling, although both are intended for use with noise and vibration signals obtained from rotating equipment. With time synchronous averaging, the aim is to obtain averaged time domain data by averaging data samples that correspond to the same angular location of a rotor. This is done prior to taking a DFT. Thus, the idea is to use a tachometer signal to indicate when each revolution begins and then obtain data with the same constant sample rate for all revolutions. If the speed is variable, this results in samples that do not correspond to the same angular locations as they did prior to the first speed change since sampling began. Data at fixed angular intervals is obtained by interpolating between data samples and then all data corresponding to each particular angle is averaged. In this way, each DFT will be the result of a calculation based on the average of a number of synchronised time samples, so the frequency scale will now be replaced with a scale representing multiples of the fundamental frequency. However, the fundamental and its harmonics will be much more clearly visible than if the time samples were separated by fixed time intervals. This is especially true for variable speed rotors. It is still desirable to obtain several records of averaged data to enable the resulting auto power spectra to be averaged. As this method of analysis is for the purpose of identifying tonal signals, it is not suitable for PSD calculations.

Synchronous sampling, on the other hand, is slightly different to the process described above. Rather than taking samples at fixed time intervals and then interpolating and averaging in the time domain, synchronous sampling involves sampling the signal at fixed angular increments of the rotating rotor so that when the rotor speed changes, the interval between time samples changes. A DFT is then taken of each record in the time domain and the resulting frequency spectrum has multiples of the fundamental rotational frequency along its axis. This method is often referred to as 'order tracking'. Again, this method is only used for tonal analysis with the auto power spectrum and as it is intended for tonal analysis, it is not suitable for obtaining a PSD.

7.3.10 Hilbert Transform

The Hilbert transform is often referred to as envelope analysis, as it involves finding a curve that envelopes the peaks in a signal. The signal may be a time domain signal or a frequency domain signal. Any regular harmonic variation in the envelope signal represents an amplitude modulation of the original signal, which can be quantified by taking an FFT of the envelope signal.

The Hilbert transform applied to a time domain (or time series) signal, $x(t)$, can be represented mathematically as:

$$\mathscr{H}\{x(t)\} = \tilde{x}(t) = \frac{1}{\pi} \int\limits_{-\infty}^{\infty} x(\tau) \left(\frac{1}{t-\tau}\right) \mathrm{d}\tau = \frac{1}{\pi} x(t) * \left(\frac{1}{t}\right) \tag{7.55}$$

where $*$ represents the convolution operator (see Section 7.3.15).

The Fourier transform of $\tilde{x}(t)$ is:

$$\tilde{X}(f) = -\mathrm{j}\, \mathrm{sign}(f) X(f) = \mathrm{j}\frac{|f|}{f} X(f) \tag{7.56}$$

or alternatively:

$$\tilde{X}(f) = \begin{cases} \mathrm{e}^{-\mathrm{j}\pi/2} X(f); & f > 0 \\ 0; & f = 0 \\ \mathrm{e}^{\mathrm{j}\pi/2} X(f); & f < 0 \end{cases} \tag{7.57}$$

An example of an envelope of an amplitude modulated time domain signal, $m(t)$, is shown in Figure 7.6.

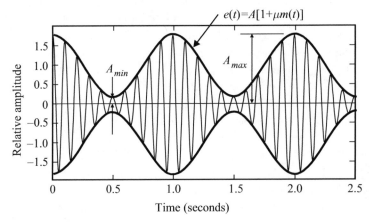

FIGURE 7.6 Envelope analysis with the Hilbert transform.

This type of signal can be analysed using the Hilbert transform of the total time domain signal, $x(t)$, which for our purposes can represent sound pressure as a function of time. To calculate the envelope function, $e(t)$, which is the envelope of the original signal, the following equation is used (Brandt, 2010):

$$e(t) = \sqrt{x^2(t) + \tilde{x}^2(t)} \tag{7.58}$$

or in discrete form as:

$$e(k) = \sqrt{x^2(i) + \tilde{x}^2(k)} \tag{7.59}$$

The discrete Hilbert transform, $\tilde{x}(k)$, can be calculated from the ordinary discrete Fourier transform, $X(f_n)$, of the original signal as (Brandt, 2010):

$$\tilde{x}(k) = \frac{2}{N} \mathrm{Im} \left\{ \sum_{n=0}^{N/2} X(f_n) \mathrm{e}^{\mathrm{j}2\pi nk/N} \right\} \tag{7.60}$$

where N is the total number of samples used to calculate the Fourier transform and $\mathrm{Im}\{\ \}$ represents the imaginary part of the complex number in brackets.

7.3.11 Cross Spectrum

The cross spectrum is a measure of how much one signal, represented in the time domain as $x(t)$ and frequency domain as $X(f_n)$, may be related to another signal, represented in the time domain as $y(t)$ and frequency domain as $Y(f_n)$. For example, it may be used to determine the extent to which indoor noise, $y(t)$, in a residence may be caused by exterior noise, $x(t)$, where $y(t)$ is considered to be the system output and $x(t)$ the system input. It is also used to estimate the coherence between two noise or vibration signals, which is another way of quantifying by how much one signal is related to another.

The two-sided cross spectrum and cross-spectral density (CSD) have similar forms to the auto power spectrum and power spectral density of Equations (7.17) and (7.18), respectively, and the single-sided cross spectrum, $G_{xy}(f_n)$, is calculated from the two-sided spectrum in a similar way to the auto power spectrum. The expression for the two-sided spectrum is:

$$S_{xy}(f_n) \approx \frac{S_A}{Q} \sum_{i=1}^{Q} X_i^*(f_n) Y_i(f_n); \qquad n = 0, 1, \ldots, (N-1) \tag{7.61}$$

where i is the spectrum number, $X_i(f_n)$ and $Y_i(f_n)$ are complex spectral components corresponding to frequency f_n and Q is the number of spectra over which the average is taken. The larger the value of Q, the more closely will the estimate of $S_{xy}(f_n)$ approach its true value. In the equation, the superscript $*$ represents the complex conjugate, $X_i(f_n)$ and $Y_i(f_n)$ are instantaneous spectra and $S_{xy}(f_n)$ is estimated by averaging over a number of instantaneous spectrum products obtained with finite time records of data. In contrast to the auto power spectrum, which is real, the cross spectrum is complex, characterised by an amplitude and a phase.

In practice, the amplitude of $S_{xy}(f_n)$ is the product of the two amplitudes $|X(f_n)|$ and $|Y(f_n)|$ and its phase is the difference in phase between $X(f_n)$ and $Y(f_n)$ ($= \theta_y - \theta_x$). This function can be averaged because for stationary signals, the relative phase between $x(t)$ and $y(t)$ is fixed and not random.

The two-sided cross-spectral density, $S_{Dxy}(f_n)$ (or CSD(f_n)), can be obtained from the cross spectrum by dividing the amplitudes of each frequency component by the frequency spacing, Δf, between adjacent components in the cross spectrum or by multiplying by the time, T_s, to acquire one record of data. Thus, the two-sided cross-spectral density is:

$$S_{Dxy}(f_n) \approx \frac{T_s S_A}{Q} \sum_{i=1}^{Q} X_i^*(f_n) Y_i(f_n); \qquad n = 0, 1, \ldots, (N-1) \tag{7.62}$$

The cross-spectral density can also be obtained directly from the cross-correlation function, which is defined as:

$$R_{xy}(\tau) = \langle x(t) y(t+\tau) \rangle_t = \mathrm{E}[x(t) y(t+\tau)] = \lim_{T_s \to \infty} \frac{1}{T_s} \int_{-T_s/2}^{T_s/2} x(t) y(t+\tau) \, \mathrm{d}t \tag{7.63}$$

where $\mathrm{E}[X]$ is the expected value of X.

Using the cross-correlation function, the cross-spectral density may be written as:

$$S_{Dxy}(f) = \int_{-\infty}^{\infty} R_{xy}(\tau) \mathrm{e}^{-\mathrm{j}2\pi f \tau} \, \mathrm{d}\tau \tag{7.64}$$

and in the sampled domain (corresponding to part (d) of Figure 7.3),

$$S_{Dxy}(f_n) = \frac{1}{N} \sum_{k=0}^{N-1} R_{xy}(\tau_k) \mathrm{e}^{-\mathrm{j}2\pi nk/N} \tag{7.65}$$

In practice, the one-sided cross spectrum, $G_{xy}(f_n)$, is used instead of the two-sided spectrum of Equation (7.61), where:

$$G_{xy}(f_n) = \begin{cases} 0; & f_n < 0 \\ S_{xy}(f_n); & f_n = 0 \\ 2S_{xy}(f_n); & f_n > 0 \end{cases} \tag{7.66}$$

A similar expression may be written for the single-sided CSD, $G_{Dxy}(f_n)$, as:

$$G_{Dxy}(f_n) = \begin{cases} 0; & f_n < 0 \\ S_{Dxy}(f_n); & f_n = 0 \\ 2S_{Dxy}(f_n); & f_n > 0 \end{cases} \tag{7.67}$$

Note that $G_{xy}(f_n)$ and $G_{Dxy}(f_n)$ are complex, with real and imaginary parts referred to as the co-spectrum and quad-spectrum, respectively. As for auto power spectra and PSDs, the accuracy of the estimate of the cross spectrum improves as the number of records over which the averages are taken increases. The statistical error for a stationary, Gaussian random signal is given as (Randall, 1987):

$$\epsilon = \frac{1}{\sqrt{\gamma_{xy}^2(f_n)Q}}; \qquad n = 0, 1, \ \ldots, \ (N-1) \tag{7.68}$$

where $\gamma_{xy}^2(f_n)$ is the coherence function relating noise signals, $x(t)$ and $y(t)$ (see Section 7.3.12), and Q is the number of averages.

The amplitude of $G_{xy}(f_n)$ gives a measure of how well the two functions $x(t)$ and $y(t)$ correlate as a function of frequency and the phase angle of $G_{xy}(f_n)$ is a measure of the phase shift between the two signals as a function of frequency.

7.3.12   Coherence

The coherence function is a measure of the degree of linear dependence between two signals, as a function of frequency. It is calculated from the two auto power spectra and the cross spectrum as:

$$\gamma_{xy}^2(f_n) = \frac{|G_{xy}(f_n)|^2}{G_{xx}(f_n)G_{yy}(f_n)}; \qquad n = 0, 1, \ \ldots, \ (N-1) \tag{7.69}$$

By definition, $\gamma^2(f_n)$ varies between 0 and 1, with 1 indicating a high degree of linear dependence between the two signals, $x(t)$ and $y(t)$. Thus, in a physical system where $y(t)$ is the output and $x(t)$ is the input signal, the coherence is a measure of the degree to which $y(t)$ is linearly related to $x(t)$. If random noise is present in either $x(t)$ or $y(t)$, then the value of the coherence will diminish. However, if random noise appears in both $x(t)$ and $y(t)$, and the two random noise signals are well correlated, the coherence may not necessarily diminish. Other causes of a diminished coherence are insufficient frequency resolution in the frequency spectrum or poor choice of window function. A further cause of diminished coherence is a time delay, of the same order as the length of the record, between $x(t)$ and $y(t)$.

The main application of the coherence function is in checking the validity of frequency response measurements (see Section 7.3.14). Another more direct application is the calculation of the signal, S, to noise, N, ratio as a function of frequency, given by:

$$S/N = \frac{\gamma_{xy}^2(f_n)}{1 - \gamma_{xy}^2(f_n)}; \qquad n = 0, 1, \ \ldots, \ (N-1) \tag{7.70}$$

In this relatively narrow definition of S/N, the 'signal' is that component in the response, $y(t)$, that is caused by $x(t)$ and 'noise' refers to anything else in $y(t)$.

The coherence will always be unity by definition if only one spectrum (rather than the average of many spectra) is used to calculate $G_{xx}(f_n)$, $G_{yy}(f_n)$ and $G_{xy}(f_n)$. A more detailed discussion on interpreting coherence data is provided in Section 5.4.3.

For cases in which there are many inputs and one or more outputs, it is of interest to estimate the degree of correlation existing between one group of selected inputs, $X(n) = [x_1(n), ..., x_m(n)]$, and one output, $y[n]$. This is the basis of the concept of multiple coherence, defined as (Potter, 1977):

$$\gamma_{Xy}^2 = \frac{G_{Xy}^{\mathrm{H}}(f_n) G_{XX}^{-1}(f_n) G_{Xy}(f_n)}{G_{yy}(f_n)} \tag{7.71}$$

where γ_{Xy}^2 is the multiple coherence function between the vector of inputs, X, and the output, y, G_{Xy}, is the m-dimensional vector of the cross spectrum between the inputs, X, and the output, y, G_{XX} is the $(m \times m)$-dimensional matrix of the power spectrum and cross spectrum of the vector of inputs, and G_{yy} is the power spectrum of the output. The power spectrum and cross spectrum quantities in Equation (7.71) can be replaced with their equivalent power spectral-density terms with no change in result.

7.3.13   Coherent Output Power

Coherent output power calculations allow one to determine what contribution a particular sound or vibration source may be making to a particular acoustic or vibration measurement. The coherent output power process can also be used to eliminate extraneous noise from a signal.

Although coherent power calculations for obtaining the contribution of a particular sound source to an acoustic measurement are relatively simple, practical implementation of the procedure requires considerable care. To be able to determine the coherent output power of a sound source, it is necessary to be able to obtain an uncontaminated signal representing just the noise source itself. This usually requires a measurement to be made close to the noise source or on a vibration source. If the auto power spectrum of the measurement made close to a source is $G_{xx}(f_n), n = 0, 1, \ldots, (N-1)$ and the auto power spectrum of the contaminated measurement at some distant location is $G_{yy}(f_n)$, then the auto power spectrum, $G_{cc}(f_n)$, of the contribution of the noise or vibration source at the distant location is:

$$G_{cc}(f_n) = |H_1(f_n)|^2 G_{xx}(f_n) = \frac{|G_{xy}(f_n)|^2}{G_{xx}(f_n)}; \qquad n = 0, 1, \ldots, (N-1) \tag{7.72}$$

where N is the number of data points in the auto power spectrum. The term $G_{cc}(f_n)$ can also be written in terms of the coherence between the two signals, so that:

$$G_{cc}(f_n) = \gamma^2(f_n) G_{yy}(f_n) \qquad n = 0, 1, \ldots, (N-1) \tag{7.73}$$

Similar relationships also apply for power spectral densities with the functions, $G(f_n)$, replaced with $G_D(f_n)$ in the above two equations. It is assumed in all relationships that the measurement, $G_{xx}(f_n)$, is uncontaminated by noise not related to the source being investigated. In practice, a small amount of contamination leads to the results being slightly less accurate.

It is possible to use Equations (7.72) and (7.73) to calculate the relative contributions of a number of sources to a particular sound pressure or acceleration measurement, provided the signals $G_{xx}(f_n), n = 0, 1, \ldots, (N-1)$ acquired near each source do not contaminate one another significantly.

7.3.14   Frequency Response (or Transfer) Function

The frequency response function, $H(f_n)$, is defined as:

$$H(f_n) = \frac{Y(f_n)}{X(f_n)} \qquad n = 0, 1, \ldots, (N-1) \tag{7.74}$$

The frequency response function (FRF), $H(f_n)$, is the Fourier transform of the system impulse response function, $h(t_k)$. The FRF is a convenient way of quantifying the relative amplitude of and the phase between two signals as a function of frequency. The impulse response of a system is the system output as a function of time following an impulse input (very short, sudden input)

In practice, it is desirable to average $H(f_n)$ over a number of spectra, but as $Y(f_n)$ and $X(f_n), n = 0, 1, \ldots, (N-1)$ are both instantaneous spectra, it is not possible to average either of these. For this reason, it is convenient to modify Equation (7.74). There are a number of possibilities, one of which is to multiply the numerator and denominator by the complex conjugate of the input spectrum. Thus:

$$H_1(f_n) = \frac{Y(f_n)X^*(f_n)}{X(f_n)X^*(f_n)} = \frac{G_{xy}(f_n)}{G_{xx}(f_n)} \qquad n = 0, 1, \ldots, (N-1) \tag{7.75}$$

A second version is found by multiplying with $Y^*(f)$ instead of $X^*(f)$. Thus:

$$H_2(f_n) = \frac{Y(f_n)Y^*(f_n)}{X(f_n)Y^*(f_n)} = \frac{G_{yy}(f_n)}{G_{yx}(f_n)} \qquad n = 0, 1, \ldots, (N-1) \tag{7.76}$$

Either of the above two forms of frequency response function is amenable to averaging, but $H_1(f_n)$ is the preferred version if the output signal, $y(t)$, is more contaminated by noise than the input signal, $x(t)$, whereas $H_2(f_n)$ is preferred if the input signal, $x(t)$, is more contaminated by noise than the output signal (Randall, 1987).

The frequency response function may also be expressed in spectral-density terms by replacing $G_{xy}(f_n)$, $G_{yx}(f_n)$, $G_{xx}(f_n)$ and $G_{yy}(f_n)$ in Equations (7.75) and (7.76) by their spectral-density equivalents.

7.3.15   Convolution

Convolution in the time domain is the operation that is equivalent to multiplication in the frequency domain. Multiplication of two functions, $X(f)$ and $H(f)$, in the frequency domain $\{X(f)H(f)\}$ involves multiplying the spectral amplitude of each frequency component in $X(f)$ with the amplitude of the corresponding frequency component in $H(f)$ and adding the phases of each frequency component in $X(f)$ to the corresponding frequency component in $H(f)$. The equivalent operation of convolution of the two time domain signals, $x(t) * h(t)$, is a little more complex and is:

$$y(t) = x(t) * h(t) = \int_{-\infty}^{\infty} x(\tau)h(t - \tau)\,\mathrm{d}\tau \tag{7.77}$$

The Fourier transform of the convolved signal, $y(t) = x(t) * h(t)$ is $\{X(f)H(f)\}$. That is, $y(t) = IDFT\{X(f)H(f)\}$. In a practical application, $X(f)$ might be the Fourier transform of an input signal to a physical system characterised by a transfer function, $H(f)$, and $Y(f)$ would then be the resulting output from the system.

Deconvolution is the process of determining $h(t)$ of Equation (7.77) from known signals, $y(t)$ and $x(t)$. For example, $h(t)$ may be a system impulse response that is to be determined from input and output signals. Deconvolution is often performed in the frequency domain so that in the absence of any significant noise in the signals:

$$H(f) = Y(f)/X(f) \tag{7.78}$$

The inverse Fourier transform is then taken of $H(f)$ to obtain $h(t)$. If noise is present in the output signal, $y(t)$, then the estimate for $h(t)$ will be in error. The error may be reduced using Weiner deconvolution but this is a complex operation and beyond the scope of this book.

It should be noted that $H(f)$, $Y(f)$ and $X(f)$ are complex numbers at each frequency, f, and can be represented as a magnitude and a phase. Multiplication of two complex numbers to obtain a third complex number requires the magnitudes to be multiplied and the phases to be added, whereas division of one complex number by another requires division of one magnitude by another and subtraction of the denominator phase from the numerator phase. These operations are done for the data in each frequency bin.

If the two time functions are represented by sampled data, $x[n]$, $n = 0, 1, 2, \ldots, (N-1)$ and $h[m]$, $m = 0, 1, 2, \ldots, (M-1)$, such as obtained by a digital data acquisition system, the output of the convolution, $y[k], k = 0, 1, 2, \ldots, (N+M-2)$, of the two signals at sample number, k, is:

$$y[k] = x[k] * h[k] = h[k] * x[k] = \sum_{n=0}^{N-1} x[n]h[k-n] \qquad (7.79)$$

where terms in the sum are ignored if $[k-n]$ lies outside the range from 0 to $(M-1)$.

Convolution is often used to define the relationship between three signals of interest: the input to a system, $x[k]$, the system impulse response, $h[k]$, and the output, $y[k]$, from the system, as indicated in Equation (7.79). As can be seen from the equation, each sample in the input signal contributes to many samples in the output signal.

With sampled data, the discrete Fourier transform (DFT) is used to obtain frequency spectra. However, unlike the inverse Fourier transform in the continuous domain, the inverse discrete Fourier transform (IDFT) of the product of two spectra in the frequency domain will no longer represent a linear convolution in the time domain. Instead, the time domain equivalent is circular convolution, which results from the associated periodicity of the DFT and is not equal to the linear convolution. The circular convolution operation is denoted by $\circledast$ rather than by $*$ for normal linear convolution. If a digital Fourier transform (DFT) is taken of the two time signals to produce $X(f_n)$ and $H(f_n)$, respectively, and the product in the frequency domain is given by $Y(f_n) = X(F_n)H(f_n)$, then the time domain sequence, $y(k)$, whose DFT is $Y(f_n)$ is obtained using the circular convolution operation of the original time domain signals, $x(k)$ and $h(k)$. Thus:

$$y(k) = \text{IDFT}\{X(f_n)H(f_n)\} = x[k] \circledast h[k] = \sum_{n=0}^{N-1} x[n]h[((k-n))_N] \qquad (7.80)$$

where the maximum value of k should be at least as large as $(N+M-2)$ and $[((k-n))_N]$ is calculated as $[k-n] \bmod N$ (meaning that $[((k-n))_N]$ is the integer remainder after $[k-n]$ has been divided by ℓN, where ℓ is the largest integer possible for $(k-n) \geq \ell N$). If $[k-n]$ is negative, then ℓ will be negative. If $\ell = 0$, $[((k-n))_N] = [k-n]$, a positive number. For example, -1 is a larger integer than -2.

Deconvolution of the right-hand part of Equation (7.80) is used in practice to obtain a system impulse response, $h(k)$, $k = 1, \ldots, (N-1)$ from a measurement of an input signal, $x(k)$, $k = 0, \ldots, (N-1)$ to the system and the resulting output signal, $y(k)$, $k = 0, \ldots, (N-1)$. As the measured data are digital samples, the circular convolution operation is required. Hence, the system response is given by Rife and Vanderkooy (1989):

$$h(k) = \frac{1}{N+1} \sum_{n=0}^{N-1} y[n]x[((n+k))_N] \qquad (7.81)$$

As an example, if $N = 6$, Equation (7.81) can be expanded into the following matrix equation:

$$
\begin{bmatrix} h[0] \\ h[1] \\ h[2] \\ h[3] \\ h[4] \\ h[5] \end{bmatrix} = \frac{1}{7} \begin{bmatrix} x[0] & x[1] & x[2] & x[3] & x[4] & x[5] \\ x[1] & x[2] & x[3] & x[4] & x[5] & x[0] \\ x[2] & x[3] & x[4] & x[5] & x[0] & x[1] \\ x[3] & x[4] & x[5] & x[0] & x[1] & x[2] \\ x[4] & x[5] & x[0] & x[1] & x[2] & x[3] \\ x[5] & x[0] & x[1] & x[2] & x[3] & x[4] \end{bmatrix} \begin{bmatrix} y[0] \\ y[1] \\ y[2] \\ y[3] \\ y[4] \\ y[5] \end{bmatrix} \tag{7.82}
$$

Converting a circular convolution obtained by an IDFT (inverse discrete Fourier transform) of the product of the two DFTs, to an ordinary convolution, can be done by using zero padding (see Section 7.3.7) of the two original time domain signals before taking their DFT. If this is not done, the circular convolution so obtained is not equivalent to the ordinary convolution of two time domain signals. This is discussed by Brandt (2010, Chapter 9).

7.3.16   Auto-Correlation and Cross-Correlation Functions

The auto-correlation function of a signal, $x(t)$, in the time domain, is a measure of how similar a signal is to a time-shifted future or past version of itself. If the time shift is τ seconds, then the auto-correlation function is:

$$
R_{xx}(\tau) = \mathrm{E}[x(t)x(t + \tau)] \tag{7.83}
$$

where E[ ] is the expected value of the quantity contained in parentheses. If $\tau = 0$, the auto-correlation function is equal to the variance of the signal, σ_x^2.

The cross correlation between two different time signals, $x(t)$ (the input) and $y(t)$ (the output) is:

$$
R_{xy}(\tau) = \mathrm{E}[x(t)y(t + \tau)] \tag{7.84}
$$

Auto-correlation and cross-correlation function definitions are applicable to random noise signals only if the average value of one record of samples is not much different to that for subsequent records or the average of many records. Random noise signals typically satisfy this condition sufficiently provided that the signal does not vary too much.

The cross-correlation function can be used to find the acoustic delay between two signals originating from the same source. In this case, the delay is the time difference represented by the maximum value of the cross-correlation function. If the speed of sound is known, then the delay allows one to determine the distance between the two microphones or two accelerometers that are providing the two signals.

The auto-correlation function has the property that for real signals, $x(t)$ and $y(t)$:

$$
R_{xx}(\tau) = R_{xx}(-\tau) \tag{7.85}
$$

and the cross-correlation function has the property:

$$
R_{yx}(\tau) = R_{xy}(-\tau) \tag{7.86}
$$

As direct calculation of correlation functions is difficult and resource intensive, they are usually estimated from spectra. This is done by first estimating a power spectral density as in Equations (7.18) and (7.23), where the function $X_i(f_n)$ has had a rectangular window applied (which is effectively a multiplication by unity as all $w(k)$ in Equation (7.27) are unity for a rectangular window). The PSD needs to be zero padded with as many zeros as data points to avoid circular convolution (explained in detail by Brandt (2010, Chapter 9)).

We first compute a PSD estimate using Q spectrum averages as in Equation (7.18), and for each spectrum, we use a $2N$ record length with the last half of the samples set equal to zero. So we obtain for the single-sided PSD:

$$G_{Dxx}(f_n) = \frac{4S_A}{Q\Delta f} \sum_{i=1}^{Q} |X_{iz}(f_n)|^2 \qquad (7.87)$$

where the subscript, z, indicates that the spectrum has been calculated with an equal number of zeros as data samples, added to the dataset. $\Delta f = f_s/N$, f_s is the sample rate, N is the number of samples in each data segment or record, $X_i(f_n)$ is an unscaled spectrum and the single-sided PSD, G_{Dxx}, has been multiplied by 2 to account for the added zeros. As we have N data samples in each record, and another N zeros added to the end, the total length of each record is $2N$. As a rectangular window has been used, the scaling factor, S_A, is simply equal to $1/(4N^2)$, as there are $2N$ samples in the record. The ith unscaled, zero-padded spectrum, $X_{iz}(f_n)$, obtained by passing time samples through a rectangular window, is:

$$X_{iz}(f_n) = \sum_{k=0}^{2N-1} x_i(t_k)e^{-j2\pi kn/N}; \qquad n = 0, 1, \ldots, (2N-1) \qquad (7.88)$$

where x_i represents the ith complete data record in the time domain, which has been doubled in size using zero padding, t_k is the time that the kth sample was acquired and the subscript, z, indicates a spectrum with zero padding.

The inverse discrete Fourier transform (IDFT) is taken of Equation (7.87) to obtain the quantity $\hat{R}_{xx}(k)$ and this is used to compute an estimate, $\hat{R}_{xx}^i$, for the auto-correlation function for the ith record, using (Brandt, 2010):

$$\hat{R}_{xx}^i = \begin{cases} \frac{N-k}{N\Delta t}\hat{R}_{xx}(k); & k = 0, 1, \ldots, (N-1) \\[2ex] \frac{k-N}{N\Delta t}\hat{R}_{xx}(k); & k = N, (N+1), \ldots, (2N-1) \end{cases} \qquad (7.89)$$

where Δt is the time interval between samples and the hat over a variable indicates that it is an estimate.

The cross correlation between two signals, $x(t)$ and $y(t)$, can be computed in a similar way to the auto correlation. In this case, we replace the PSD with the CSD (cross-spectral density) and begin by computing the CSD using 50% zero padding and Q spectrum averages as in Equation (7.18) and for each spectrum, we use a $2N$ record length with the last half of the samples set equal to zero. So we obtain for the single-sided CSD:

$$G_{Dxy}(f_n) = \frac{4S_A}{Q\Delta f} \sum_{i=1}^{Q} X_{iz}^*(f_n)Y_{iz}(f_n) \qquad (7.90)$$

where $Y_{iz}(f_n)$ and $X_{iz}(f_n)$ are unscaled spectra and the single-sided CSD, G_{Dxy}, has been multiplied by 2 to account for the added zeros. As before, the scaling factor, S_A, is simply equal to $1/(4N^2)$, as there are $2N$ samples in the record. The ith unscaled, zero-padded spectrum, $Y_{iz}(f_n)$, obtained by passing time samples through a rectangular window, is similar to that for $X_{iz}(f_n)$, which is given in Equation (7.88), and is:

$$Y_{iz}(f_n) = \sum_{k=0}^{2N-1} y_i(t_k)e^{-j2\pi kn/N}; \qquad n = 0, 1, \ldots, (2N-1) \qquad (7.91)$$

The inverse discrete Fourier transform (IDFT) is taken of Equation (7.91) to obtain the quantity $\hat{R}_{xy}(k)$ and this is used to compute an estimate, $\hat{R}_{xy}^i$, of the cross-correlation function for the ith record, using (Brandt, 2010):

$$\hat{R}_{xy}^i = \begin{cases} \frac{N-k}{N\Delta t}\hat{R}_{xy}(k); & k = 0, 1, \ldots, (N-1) \\[2ex] \frac{k-N}{N\Delta t}\hat{R}_{xy}(k); & k = N, (N+1), \ldots, (2N-1) \end{cases} \tag{7.92}$$

7.3.17   Maximum Length Sequence (MLS)

An MLS excitation signal is sometimes used in acoustics to obtain measurements of transfer functions, which include loudspeaker response functions and the noise reduction from the outside to the inside of a house. A schematic of a typical measurement system is shown in Figure 7.7. An MLS signal is a digitally synthesised binary series of samples that have a value of 1 or zero, which is mapped to -1 and 1, respectively, to produce a signal that is symmetric about zero. The clock rate is the number of times a new value is output per second by the shift register generating the binary sequence and this should be at least 2.5 times the maximum frequency of interest in the transfer function to be measured. However, the sample rate is recommended to be 10 times the maximum frequency of interest (Ljung, 1999). For the measurement of transfer functions, MLS analysis has the following advantages.

- It can minimise the effects on the measurement of noise not generated by the test system.

- The spectral content of an MLS signal closely resembles white noise with a flat power spectral shape.

- It is deterministic and this property together with the previous one has attracted the description of pseudo-random noise.

- It has a pre-determined temporal length before repeating. This length, in terms of the total number of samples in a sequence, is $L = 2^M - 1$ samples, where M is the order of the MLS, and this is usually 12 or greater in practical systems. Note that f_s/L, where f_s is the sample rate, must be longer than the impulse response of the system being measured. For the case of measuring noise reductions from the outside to the inside of a house, it would have to be greater than the propagation time of the noise from the outside source to the inside microphone.

- It has a low crest factor, thus transferring a large amount of energy to the system being excited, and thus achieving a very high signal to noise ratio (SNR). For every doubling in the number of sequences that are averaged, the SNR is improved by 3 dB.

- The cross-correlation function of the input and output of a system using an MLS is equal to the impulse response of the system between the input and output. As discussed in Section 7.3.11, taking an FFT of the cross-correlation function yields the cross-spectral density of the system and as discussed in Section 7.3.1, taking an FFT of the auto correlation of the system input yields the spectral density of the system input. As discussed in Section 7.3.14, these two quantities can be used to obtain the transfer function spectral density. For the case of the house mentioned above, this transfer function is the noise reduction from outside to inside in dB.

- The use of cross correlation with MLS to obtain transfer functions rejects all noise not correlated with the MLS signal and so MLS is effective in finding transfer functions in noisy environments such as at low frequencies when measuring noise reductions from the outside to the inside of a house.

- MLS suppresses the DC part of the system response.

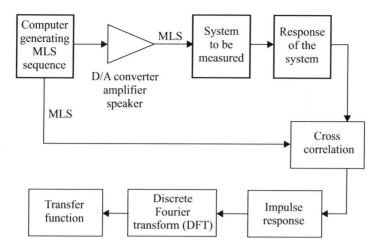

FIGURE 7.7 Schematic arrangement for the measurement of a system transfer function (such as outside to inside noise levels in a house) using an MLS signal.

The MLS sequence is generated digitally using a maximal linear feedback shift register, which cycles through every possible binary value (except all zeros) before repeating. A shift register of length, M, contains M bits, each of which can have a value of one or zero. The length of an MLS sequence before it repeats itself is related to the number of bits, M, in the shift register used to generate it. If the shift register length is $M = 20$, then the number of samples in the MLS sequence prior to repeating is $2^{20} - 1$ ($N = 1048575$ samples). The -1 is there because the case of all bits in the shift register being zero is excluded. An MLS generating system with a shift register of length 5 is implemented as illustrated in Figure 7.8.

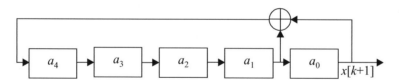

FIGURE 7.8 The next value for register bit a_4 is the modulo 2 sum of a_0 and a_1, indicated by the $\oplus$ symbol.

The output from the shift register chain at time, k, is either a 1 or a zero and this is mapped to -1 and $+1$, respectively, before being transmitted to an amplifier.

A modulo 2 sum (or XOR sum, denoted by $\oplus$) is the process of combining bits in two binary numbers to produce a third number. The two bits in the same position in the two numbers to be summed are combined to produce a value for the bit in the same position in the new number by following the rules.

- If both of the two bits being added are one, then the result is zero.

- If both of the two bits being added are zero, then the result is zero.

- If one of the bits is one and the other is zero, then the result is one.

The values of the bits corresponding to sample number $k + 1$ in the MLS being generated with a shift register of length 5 can be calculated recursively from the values corresponding to sample, k, using the relations:

$$\begin{cases} a_4[k+1] = a_0[k] + a_1[k]; \quad \text{(modulo 2 sum)} \\ a_3[k+1] = a_4[k] \\ a_2[k+1] = a_3[k] \\ a_1[k+1] = a_2[k] \\ a_0[k+1] = a_1[k] \end{cases} \tag{7.93}$$

Review of Complex Numbers and Relevant Linear Matrix Algebra

The purpose of this appendix is to provide a brief review of complex numbers and many of the particular results of linear algebra used in this book. For more extensive treatments on linear algebra, the reader should consult any of the standard textbooks, such as Cullen (1991); Schneider and Barker (1989). There are also a number of specialised software packages that deal explicitly with linear algebra manipulations, such as GNU Octave.

A.1 Complex Numbers

A complex number is a mathematically convenient way of representing a physical quantity that is defined by an amplitude and a phase. It is also a convenient way of including the variation with time of a sinusoidally varying quantity. A complex number consists of a real and imaginary part, with the imaginary part identified by the symbol, j. For example, a complex number, $\hat{F}$ may be written as $\hat{F} = a + \mathrm{j}b = \mathrm{Re}\{\hat{F}\} + \mathrm{j}\,\mathrm{Im}\{\hat{F}\} = |\hat{F}|\mathrm{e}^{\mathrm{j}\beta}$, where a is the real part and b is the imaginary part. The amplitude or modulus of the complex number is $|\hat{F}| = \sqrt{a^2 + b^2}$ and its phase is $\beta = \tan^{-1}(b/a)$. A time varying quantity, F, varying sinusoidally at radian frequency, $\omega = 2\pi f$, where f is the frequency in Hz, can be represented as:

$$F = (a + \mathrm{j}b)\mathrm{e}^{\mathrm{j}\omega t} = \hat{F}\mathrm{e}^{\mathrm{j}\omega t} = |\hat{F}|\mathrm{e}^{\mathrm{j}(\omega t + \beta)} = |F|\mathrm{e}^{\mathrm{j}(\omega t + \beta)} \tag{A.1}$$

For two complex numbers, $u = \hat{u}\,\mathrm{e}^{\mathrm{j}\omega t} = |u|\mathrm{e}^{\mathrm{j}(\omega t + \alpha)}$ and $F = \hat{F}\mathrm{e}^{\mathrm{j}\omega t} = |F|\mathrm{e}^{\mathrm{j}(\omega t + \beta)}$:

$$Fu = |F|\mathrm{e}^{\mathrm{j}(\omega t + \beta)} \times |u|\mathrm{e}^{\mathrm{j}(\omega t + \alpha)} = |Fu|\,\mathrm{e}^{\mathrm{j}(2\omega t + \beta + \alpha)} = \left|\hat{F}\hat{u}\right|\mathrm{e}^{\mathrm{j}(2\omega t + \beta + \alpha)} \tag{A.2}$$

and:

$$\frac{F}{u} = \frac{|F|\mathrm{e}^{\mathrm{j}(\omega t + \beta)}}{|u|\mathrm{e}^{\mathrm{j}(\omega t + \alpha)}} = \left|\frac{F}{u}\right|\mathrm{e}^{\mathrm{j}(\beta - \alpha)} = \left|\frac{\hat{F}}{\hat{u}}\right|\mathrm{e}^{\mathrm{j}(\beta - \alpha)} \tag{A.3}$$

A.2 Matrices and Vectors

An $(m \times n)$ matrix is a collection of mn numbers (complex or real), a_{ij}, $(i = 1, 2, ..., m,\ j = 1, 2, ..., n)$, written in an array of m rows and n columns:

$$\boldsymbol{A} = \begin{bmatrix} a_{11} & a_{12} & \cdots & a_{1n} \\ a_{21} & a_{22} & \cdots & a_{2n} \\ \vdots & \vdots & \vdots & \vdots \\ a_{m1} & a_{m2} & \cdots & a_{mn} \end{bmatrix} \tag{A.4}$$

The term, a_{ij}, appears in the ith row and jth column of the array. If the number of rows is equal to the number of columns, the matrix is said to be square.

An m vector, also referred to as an $(m \times 1)$ vector or column m vector, is a matrix with 1 column and m rows:

$$a = \begin{bmatrix} a_1 \\ a_2 \\ \vdots \\ a_m \end{bmatrix} \tag{A.5}$$

Matrices are denoted by bold capital letters, such as $\boldsymbol{A}$, while vectors are usually denoted by bold lower case letters, such as $\boldsymbol{a}$, although there are exceptions such as the use of $\boldsymbol{F}$ for a force vector.

A.3 Addition, Subtraction and Multiplication by a Scalar

If two matrices have the same number of rows and columns, they can be added or subtracted. When adding matrices, the individual corresponding terms are added. For example, if:

$$\boldsymbol{A} = \begin{bmatrix} a_{11} & a_{12} & \cdots & a_{1n} \\ a_{21} & a_{22} & \cdots & a_{2n} \\ \vdots & \vdots & \vdots & \vdots \\ a_{m1} & a_{m2} & \cdots & a_{mn} \end{bmatrix} \tag{A.6}$$

and:

$$\boldsymbol{B} = \begin{bmatrix} b_{11} & b_{12} & \cdots & b_{1n} \\ b_{21} & b_{22} & \cdots & b_{2n} \\ \vdots & \vdots & \vdots & \vdots \\ b_{m1} & b_{m2} & \cdots & b_{mn} \end{bmatrix} \tag{A.7}$$

then:

$$\boldsymbol{A} + \boldsymbol{B} = \begin{bmatrix} a_{11} + b_{11} & a_{12} + b_{12} & \cdots & a_{1n} + b_{1n} \\ a_{21} + b_{21} & a_{22} + b_{22} & \cdots & a_{2n} + b_{2n} \\ \vdots & \vdots & \vdots & \vdots \\ a_{m1} + b_{m1} & a_{m2} + b_{m2} & \cdots & a_{mn} + b_{mn} \end{bmatrix} \tag{A.8}$$

When subtracting matrices, the individual terms are subtracted. Note that matrix addition is commutative, as:

$$\boldsymbol{A} + \boldsymbol{B} = \boldsymbol{B} + \boldsymbol{A} \tag{A.9}$$

It is also associative, as:

$$(\boldsymbol{A} + \boldsymbol{B}) + \boldsymbol{C} = \boldsymbol{A} + (\boldsymbol{B} + \boldsymbol{C}) \tag{A.10}$$

Matrices can also be multiplied by a scalar. Here the individual terms are each multiplied by a scalar. For example, if k is a scalar:

$$k\boldsymbol{A} = \begin{bmatrix} ka_{11} & ka_{12} & \cdots & ka_{1n} \\ ka_{21} & ka_{22} & \cdots & ka_{2n} \\ \vdots & \vdots & \vdots & \vdots \\ ka_{m1} & ka_{m2} & \cdots & ka_{mn} \end{bmatrix} \tag{A.11}$$

A.4   Multiplication of Matrices

Two matrices, $\boldsymbol{A}$ and $\boldsymbol{B}$, can be multiplied together to form the product $\boldsymbol{AB}$ if the number of columns in $\boldsymbol{A}$ is equal to the number of rows in $\boldsymbol{B}$. If, for example, $\boldsymbol{A}$ is an $(m \times p)$ matrix, and $\boldsymbol{B}$ is a $(p \times n)$ matrix, then the product $\boldsymbol{AB}$ is defined by:

$$C = AB \tag{A.12}$$

where $\boldsymbol{C}$ is an $(m \times n)$ matrix, the terms of which are defined by:

$$c_{ij} = \sum_{k=1}^{p} a_{ik} b_{kj} \tag{A.13}$$

Matrix multiplication is associative, with the product of three (or more) matrices defined by:

$$ABC = (AB)C = A(BC) \tag{A.14}$$

Matrix multiplication is also distributive, where:

$$A(B + C) = AB + AC \tag{A.15}$$

However, matrix multiplication is not commutative, as, in general:

$$AB \neq BA \tag{A.16}$$

In fact, while the product, $\boldsymbol{AB}$, may be formed, it may not be possible to form the product, $\boldsymbol{BA}$.

The identity matrix, $\boldsymbol{I}$, is defined as the $(p \times p)$ matrix with all principal diagonal elements equal to 1, and all other terms equal to zero:

$$I = \begin{bmatrix} 1 & 0 & \cdots & 0 \\ 0 & 1 & \cdots & 0 \\ & & \ddots & \\ 0 & 0 & \cdots & 1 \end{bmatrix} \tag{A.17}$$

For any $(m \times p)$ matrix, $\boldsymbol{A}$, the identity matrix has the property:

$$AI = A \tag{A.18}$$

Similarly, if the identity matrix is $(m \times m)$:

$$IA = A \tag{A.19}$$

A.5   Matrix Transposition

If a matrix is transposed, the rows and columns are interchanged. For example, the transpose of the $(m \times n)$ matrix, $\boldsymbol{A}$, denoted by $\boldsymbol{A}^T$, is defined as the $(n \times m)$ matrix, $\boldsymbol{B}$:

$$A^{\mathrm{T}} = B \tag{A.20}$$

where:

$$b_{ij} = a_{ji} \tag{A.21}$$

The transpose of a matrix product is defined by:

$$(AB)^{\mathrm{T}} = B^{\mathrm{T}} A^{\mathrm{T}} \tag{A.22}$$

This result can be extended to products of more than two matrices, such as:

$$(ABC)^{\mathrm{T}} = C^{\mathrm{T}} B^{\mathrm{T}} A^{\mathrm{T}} \tag{A.23}$$

If:

$$A = A^{\mathrm{T}} \tag{A.24}$$

then the matrix, A, is said to be symmetric.

The Hermitian transpose of a matrix is defined as the complex conjugate of the transposed matrix (when taking the complex conjugate of a matrix, each term in the matrix is conjugated). Therefore, the Hermitian transpose of the $(m \times n)$ matrix, A, denoted by A^{H}, is defined as the $(n \times m)$ matrix, B:

$$A^{\mathrm{H}} = B \tag{A.25}$$

where:

$$b_{ij} = a_{ji}^{*} \tag{A.26}$$

If $A = A^{H}$, then A is said to be a Hermitian matrix.

A.6   Matrix Determinants

The determinant of the (2×2) matrix, A, denoted, $|A|$, is defined as:

$$|A| = \begin{vmatrix} a_{11} & a_{12} \\ a_{21} & a_{22} \end{vmatrix} = a_{11}a_{22} - a_{12}a_{21} \tag{A.27}$$

The minor, M_{ij}, of the element, a_{ij}, of the square matrix, A, is the determinant of the matrix formed by deleting the ith row and jth column from A. For example, if A is a (3×3) matrix, then:

$$A = \begin{bmatrix} a_{11} & a_{12} & a_{13} \\ a_{21} & a_{22} & a_{23} \\ a_{31} & a_{32} & a_{33} \end{bmatrix} \tag{A.28}$$

The minor, M_{11}, is found by taking the determinant of A with the first column and first row of numbers deleted:

$$M_{11} = \begin{vmatrix} a_{22} & a_{23} \\ a_{32} & a_{33} \end{vmatrix} \tag{A.29}$$

The cofactor, C_{ij}, of the element, a_{ij}, of the matrix, A, is defined by:

$$C_{ij} = (-1)^{i+j} M_{ij} \tag{A.30}$$

The determinant of a square matrix of arbitrary size is equal to the sum of the products of the elements and their cofactors along any column or row. For example, the determinant of the (3×3) matrix, A, above can be found by adding the products of the elements and their cofactors along the first row:

$$|A| = a_{11} C_{11} + a_{12} C_{12} + a_{13} C_{13} \tag{A.31}$$

Therefore, the determinant of a large square matrix can be broken up into a problem of calculating the determinants of a number of smaller square matrices.

If two matrices, A and B, are square, then:

$$|AB| = |A| \, |B| \tag{A.32}$$

A matrix is said to be singular if its determinant is equal to zero.

A.7   Rank of a Matrix

The rank of the $(m \times n)$ matrix, $\boldsymbol{A}$, is the maximum number of linearly independent rows of $\boldsymbol{A}$ and the maximum number of linearly independent columns of $\boldsymbol{A}$. Alternatively, the rank of $\boldsymbol{A}$ is a positive integer, r, such that an $(r \times r)$ submatrix of $\boldsymbol{A}$, formed by deleting $(m - r)$ rows and $(n - r)$ columns, is non-singular, whereas no $((r + 1) \times (r + 1))$ submatrix is non-singular. If the rank of $\boldsymbol{A}$ is equal to the number of columns or the number of rows of $\boldsymbol{A}$, then $\boldsymbol{A}$ is said to have full rank.

A.8   Positive and Non-Negative Definite Matrices

A matrix, $\boldsymbol{A}$, is said to be positive definite if $\boldsymbol{x}^H \boldsymbol{A} \boldsymbol{x}$ is positive for all non-zero vectors, $\boldsymbol{x}$; if $\boldsymbol{x}^H \boldsymbol{A} \boldsymbol{x}$ is simply non-negative, then $\boldsymbol{A}$ is said to be non-negative definite.

For $\boldsymbol{A}$ to be positive definite, all of the leading minors must be positive; that is:

$$a_{11} > 0; \qquad \begin{vmatrix} a_{11} & a_{12} \\ a_{21} & a_{22} \end{vmatrix} > 0; \qquad \begin{vmatrix} a_{11} & a_{12} & a_{13} \\ a_{21} & a_{22} & a_{23} \\ a_{31} & a_{32} & a_{33} \end{vmatrix} > 0; \qquad \ldots, \text{etc} \qquad (A.33)$$

For $\boldsymbol{A}$ to be non-negative definite, all of the leading minors must be non-negative.

A.9   Eigenvalues and Eigenvectors

Let $\boldsymbol{A}$ be a (square) $(n \times n)$ matrix. The polynomial, $|\lambda \boldsymbol{I} - \boldsymbol{A}| = 0$, is referred to as the characteristic equation of $\boldsymbol{A}$. The solutions to the characteristic equation are the eigenvalues of $\boldsymbol{A}$. If λ_i is an eigenvalue of $\boldsymbol{A}$, then there exists at least one vector, $\boldsymbol{q}_i$, that satisfies the relationship:

$$\boldsymbol{A} \boldsymbol{q}_i = \lambda_i \boldsymbol{q}_i \qquad (A.34)$$

The vector, $\boldsymbol{q}_i$, is an eigenvector of $\boldsymbol{A}$. If the eigenvalue, λ_i, is not repeated, then the eigenvector, $\boldsymbol{q}_i$, is unique. If an eigenvector, λ_i, is real, then the entries in the associated eigenvector, $\boldsymbol{q}_i$, are real; if λ_i is complex, then so too are the entries in $\boldsymbol{q}_i$.

The eigenvalues of a Hermitian matrix are all real, and if the matrix is also positive definite, the eigenvalues are also all positive. If a matrix is symmetric, then the eigenvalues are also all real. Further, it is true that:

$$|\boldsymbol{A}| = \prod_{i=1}^{n} \lambda_i \qquad (A.35)$$

If $\boldsymbol{A}$ is singular, then there is at least one eigenvalue equal to zero.

A.10   Orthogonality

If a square matrix, $\boldsymbol{A}$, has the property, $\boldsymbol{A}^H \boldsymbol{A} = \boldsymbol{A} \boldsymbol{A}^H = \boldsymbol{I}$, then the matrix, $\boldsymbol{A}$, is said to be orthogonal. The eigenvalues of $\boldsymbol{A}$ then have a magnitude of unity. If $\boldsymbol{q_i}$ is an eigenvector associated with λ_i, and $\boldsymbol{q_j}$ is an eigenvector associated with λ_j, and if $\lambda_i \neq \lambda_j$ and $\boldsymbol{q}_i^H \boldsymbol{q}_j = 0$, then the vectors, $\boldsymbol{q_i}$ and $\boldsymbol{q_j}$, are said to be orthogonal.

The eigenvectors of a Hermitian matrix are all orthogonal. Further, it is common to normalise the eigenvectors such that $\boldsymbol{q}_i^H \boldsymbol{q_i} = 1$, in which case the eigenvectors are said to be orthonormal. A set of orthonormal eigenvectors can be expressed as columns of a unitary matrix, $\boldsymbol{Q}$:

$$\boldsymbol{Q} = (\boldsymbol{q}_1, \ \boldsymbol{q}_2, \cdots, \ \boldsymbol{q}_n) \qquad (A.36)$$

which means that:

$$Q^H Q = Q Q^H = I \tag{A.37}$$

The set of equations that define the eigenvectors, expressed for a single eigenvector in Equation (A.34), can now be written in matrix form as:

$$AQ = Q\Lambda \tag{A.38}$$

where Λ is the diagonal matrix of eigenvalues:

$$\Lambda = \begin{bmatrix} \lambda_1 & 0 & \cdots & 0 \\ 0 & \lambda_2 & \cdots & 0 \\ & & \ddots & \\ 0 & 0 & \cdots & \lambda_n \end{bmatrix} \tag{A.39}$$

Post-multiplying both sides of Equation (A.38) by Q^H yields:

$$A = Q\Lambda Q^H \tag{A.40}$$

or:

$$Q^H A Q = \Lambda \tag{A.41}$$

Equations (A.40) and (A.41) define the orthonormal decomposition of A, where A is re-expressed in terms of its eigenvectors and eigenvalues.

A.11   Matrix Inverses

The inverse, A^{-1}, of the matrix, A, is defined by:

$$A A^{-1} = A^{-1} A = I \tag{A.42}$$

The matrix, A, must be square and non-singular for the inverse to be defined.

The inverse of a matrix, A, can be derived by first calculating the adjoint, $\hat{A}$, of the matrix. The adjoint, $\hat{A}$, is defined as the transpose of the matrix of cofactors of A (see Section A.6):

$$\hat{A} = \begin{bmatrix} C_{11} & C_{12} & \cdots & C_{1m} \\ C_{21} & C_{22} & \cdots & C_{2m} \\ \vdots & \vdots & \vdots & \vdots \\ C_{m1} & C_{m2} & \cdots & C_{mm} \end{bmatrix} \tag{A.43}$$

The inverse, A^{-1}, of the matrix, A, is equal to the adjoint of A multiplied by the reciprocal of the determinant of A:

$$A^{-1} = \frac{1}{|A|}\hat{A} \tag{A.44}$$

While the definition given in Equation (A.44) is correct, using it to calculate a matrix inverse is inefficient for all by the smallest matrices (as the order of operations increases with the size, m, of the matrix by $m!$). There are a number of algorithms that require of the order of m^3 operations to compute the inverse of an arbitrary square matrix (outlined in many of the standard texts and in numerical methods books such as Press et al. (1986)).

Note that if the matrix, A, is not square, or if it is singular, such that the determinant is zero, the inverse is not defined. However, for non-square matrices that are non-singular, a pseudo-inverse can be defined, which provides a least mean squares solution for the vector, x, for the problem, $\boldsymbol{Ax} = \boldsymbol{b}$, where the matrix, $\boldsymbol{A}$, has more rows than columns, thus representing an overdetermined system with more equations than there are unknowns. In this case:

$$\boldsymbol{x} = \left(\boldsymbol{A}^{\mathrm{T}}\boldsymbol{A}\right)^{-1}\boldsymbol{A}^{\mathrm{T}}\boldsymbol{b} \tag{A.45}$$

where the pseudo-inverse of $\boldsymbol{A}$ is $\boldsymbol{A}' = \left(\boldsymbol{A}^{\mathrm{T}}\boldsymbol{A}\right)^{-1}\boldsymbol{A}^{\mathrm{T}}$ and $\boldsymbol{A}'\boldsymbol{A} = \boldsymbol{I}$.

If the matrix, $\boldsymbol{A}$, is singular, it is possible to define the Moore–Penrose pseudo-inverse, $\boldsymbol{A}'$, such that $\boldsymbol{A}'\boldsymbol{A}$ acts as the identity matrix on as large a set of vectors as possible. $\boldsymbol{A}'$ has the properties:

$$(\boldsymbol{A}')' = \boldsymbol{A}; \quad \boldsymbol{A}'\boldsymbol{A}\boldsymbol{A}' = \boldsymbol{A}'; \quad \boldsymbol{A}\boldsymbol{A}'\boldsymbol{A} = \boldsymbol{A} \tag{A.46}$$

If $\boldsymbol{A}$ is non-singular, then $\boldsymbol{A}^{-1} = \boldsymbol{A}'$.

A.12   Singular Value Decomposition

If a matrix is non-square, it does not have an eigenvalue decomposition, but it can be written in terms of a singular value decomposition, which can then be used to find the pseudo-inverse, thus allowing the best fit solution to be obtained to an overdetermined system of equations. The singular value decomposition of an $m \times n$ matrix, $\boldsymbol{A}$, is:

$$\boldsymbol{A} = \boldsymbol{Q}\boldsymbol{\Lambda}\boldsymbol{V}^{\mathrm{T}} \tag{A.47}$$

where $\boldsymbol{\Lambda}$ has zero elements everywhere except along the diagonal, and $\boldsymbol{Q}$ and $\boldsymbol{V}$ are unitary matrices with orthogonal columns so that:

$$\boldsymbol{Q}^{\mathrm{T}}\boldsymbol{Q} = \boldsymbol{I} \ \text{ and } \ \boldsymbol{V}^{\mathrm{T}}\boldsymbol{V} = \boldsymbol{I} \tag{A.48}$$

The columns of the matrix, $\boldsymbol{Q}$, consist of a set of orthonormal eigenvectors of $\boldsymbol{A}\boldsymbol{A}^{\mathrm{T}}$ and the columns of the matrix, $\boldsymbol{V}$, consist of a set of orthonormal eigenvectors of $\boldsymbol{A}^{\mathrm{T}}\boldsymbol{A}$. The diagonal elements of $\boldsymbol{\Lambda}$ are the square roots of the non-zero eigenvalues of both $\boldsymbol{A}\boldsymbol{A}^{\mathrm{T}}$ and $\boldsymbol{A}^{\mathrm{T}}\boldsymbol{A}$.

The pseudo-inverse of $\boldsymbol{A}$ can be calculated from its singular value decomposition using:

$$\boldsymbol{A}' = \boldsymbol{Q}\boldsymbol{\Lambda}'\boldsymbol{V}^{\mathrm{T}} \tag{A.49}$$

where $\boldsymbol{\Lambda}'$ is the pseudo-inverse of $\boldsymbol{\Lambda}$, computed by replacing every non-zero element of $\boldsymbol{\Lambda}$ by its reciprocal and transposing the resulting matrix.

If the matrix, $\boldsymbol{A}$, has complex elements, the transpose operation, T, is replaced with the Hermitian transpose operation, H, in all of the preceding equations in this section.

A.13   Vector Norms

The norm of a vector, $\boldsymbol{A}$, expressed as $\|\boldsymbol{x}\|$, is the size of the vector, $\boldsymbol{x}$. The most common norm is the Euclidean norm, defined for the vector, $\boldsymbol{x} = (x_1, x_2, \ldots, x_n)$, as:

$$\|\boldsymbol{x}\| = \left(\sum_{i=1}^{n} x_i^2\right)^{1/2} \tag{A.50}$$

Three properties of vector norms are:

1. $\|\boldsymbol{x}\| \geq 0$ for all $\boldsymbol{x}$, where the norm is equal to zero only if $\boldsymbol{x} = 0$.
2. $\|a\boldsymbol{x}\| = |a|\,\|\boldsymbol{x}\|$ for any scalar a and all $\boldsymbol{x}$.
3. $\|\boldsymbol{x} + \boldsymbol{y}\| \leq \|\boldsymbol{x}\| + \|\boldsymbol{y}\|$ for all $\boldsymbol{x}$ and $\boldsymbol{y}$.

B

Properties of Materials

The properties of materials can vary considerably, especially for wood and plastic. The values listed in this appendix have been obtained from a variety of sources, including Simonds and Ellis (1943); Eldridge (1974); Levy (2001); Lyman (1961); Green et al. (1999). The data vary significantly between different sources for plastics and wood and sometimes even for metals; however, the values listed in this appendix reflect those most commonly found.

Where values of Poisson's ratio were unavailable, they were calculated from data for the speed of sound in a 3-D solid using the equation at the end of this table. These data were unavailable for some plastics so for those cases, values for similar materials were used. For wood products, the value for Poisson's ratio has been left blank where no data were available. Poisson's ratio is difficult to report for wood as there are 6 different ones, depending on the direction of stress and the direction of deformation. Here, only the value corresponding to strain in the longitudinal fibre direction coupled with deformation in the radial direction is listed.

The speed of sound values in column 4 were calculated from the values in columns 2 and 3. Where a range of values occurred in either or both of columns 2 and 3, a median value of the speed of sound was recorded in column 4.

TABLE B.1 Properties of materials

Material	Young's modulus, E (10^9 N/m^2)	Density $\rho(\text{kg/m}^3)$	$\sqrt{E/\rho}$ (m/s)	Internal–in situ[a] loss factor, η	Poisson's ratio, ν
METALS					
Aluminum sheet	70	2700	5150	0.0001–0.01	0.35
Brass	95	8500	3340	0.001–0.01	0.35
Brass (70%Zn 30%Cu)	101	8600	3480	0.001–0.01	0.35
Carbon brick	8.2	1630	2240	0.001–0.01	0.07
Carbon nanotubes	1000	1330–1400	27000	0.001–0.01	0.06
Graphite mouldings	9.0	1700	2300	0.001–0.01	0.07
Chromium	279	7200	6240	0.001–0.01	0.21
Copper (annealed)	128	8900	3790	0.002–0.01	0.34
Copper (rolled)	126	8930	3760	0.001–0.01	0.34
Gold	79	19300	2020	0.001–0.01	0.44
Iron	200	7600	5130	0.0005–0.01	0.30
Iron(white)	180	7700	4830	0.0005–0.01	0.30
Iron (nodular)	150	7600	4440	0.0005–0.01	0.30
Iron (wrought)	195	7900	4970	0.0005–0.01	0.30
Iron (gray (1))	83	7000	3440	0.0005–0.02	0.30
Iron (gray (2))	117	7200	4030	0.0005–0.03	0.30
Iron (malleable)	180	7200	5000	0.0005–0.04	0.30
Lead (annealed)	16.0	11400	1180	0.015–0.03	0.43
Lead (rolled)	16.7	11400	1210	0.015–0.04	0.44
Lead sheet	13.8	11340	1100	0.015–0.05	0.44
Magnesium	44.7	1740	5030	0.0001–0.01	0.29
Molybdenum	280	10100	5260	0.0001–0.01	0.32
Monel metal	180	8850	4510	0.0001–0.02	0.33
Neodymium	390	7000	7460	0.0001–0.03	0.31
Nickel	205	8900	4800	0.001–0.01	0.31
Nickel–iron alloy (Invar)	143	8000	4230	0.001–0.01	0.33
Platinum	168	21400	2880	0.001–0.02	0.27
Silver	82.7	10500	2790	0.001–0.03	0.36
Steel (mild)	207	7850	5130	0.0001–0.01	0.30
Steel (1% carbon)	210	7840	5170	0.0001–0.02	0.29
Stainless steel (302)	200	7910	5030	0.0001–0.01	0.30
Stainless steel (316)	200	7950	5020	0.0001–0.01	0.30
Stainless steel (347)	198	7900	5010	0.0001–0.02	0.30
Stainless steel (430)	230	7710	5460	0.0001–0.03	0.30
Tin	54	7300	2720	0.0001–0.01	0.33
Titanium	116	4500	5080	0.0001–0.02	0.32

Properties of materials

Material	Young's modulus, E (10^9 N/m^2)	Density ρ(kg/m^3)	$\sqrt{E/\rho}$ (m/s)	Internal–in situ[a] loss factor, η	Poisson's ratio, ν
METALS (Cont.)					
Tungsten (drawn)	360	19300	4320	0.0001–0.03	0.34
Tungsten (annealed)	412	19300	4620	0.0001–0.04	0.28
Tungsten carbide	534	13800	6220	0.0001–0.05	0.22
Zinc sheet	96.5	7140	3680	0.0003–0.01	0.33
BUILDING MATERIALS					
Brick	24	2000	3650	0.01–0.05	0.12
Concrete (normal)	18–30	2300	2800	0.005–0.05	0.20
Concrete (aerated)	1.5–2	300–600	2000	0.05	0.20
Concrete (high strength)	30	2400	3530	0.005–0.05	0.20
Masonry block	4.8	900	2310	0.005–0.05	0.12
Cork	0.1	250	500	0.005–0.05	0.15
Fibre board	3.5–7	480–880	2750	0.005–0.05	0.15
Gypsum board	2.1	760	1670	0.006–0.05	0.24
Glass	68	2500	5290	0.0006–0.02	0.23
Glass (pyrex)	62	2320	5170	0.0006–0.02	0.23
WOOD					
Ash (black)	11.0	450	4940	0.04–0.05	0.37
Ash (white)	12.0	600	4470	0.04–0.05	—
Aspen (quaking)	8.1	380	4620	0.04–0.05	0.49
Balsa wood	3.4	160	4610	0.001–0.05	0.23
Baltic whitewood	10.0	400	5000	0.04–0.05	—
Baltic redwood	10.1	480	4590	0.04–0.05	—
Beech	11.9	640	4310	0.04–0.05	—
Birch (yellow)	13.9	620	4740	0.04–0.05	0.43
Cedar (white–nthn)	5.5	320	4150	0.04–0.05	0.34
Cedar (red–western)	7.6	320	4870	0.04–0.05	0.38
Compressed Hardboard composite	4.0	1000	2000	0.005–0.05	—
Douglas fir	9.7–13.2	500	4800	0.04–0.05	0.29
Douglas fir (coastal)	10.8	450	4900	0.04–0.05	0.29
Douglas fir (interior)	8.0	430	4310	0.04–0.05	0.29
Mahogany (African)	9.7	420	4810	0.04–0.05	0.30
Mahogany (Honduras)	10.3	450	4780	0.04–0.05	0.31
Maple	12.0	600	4470	0.04–0.05	0.43

Properties of materials

Material	Young's modulus, E (10^9 N/m^2)	Density ρ(kg/m^3)	$\sqrt{E/\rho}$ (m/s)	Internal–in situ[a] loss factor, η	Poisson's ratio, ν
WOOD (Cont.)					
MDF	3.7	770	2190	0.005–0.05	—
Meranti (light red)	10.5	340	5560	0.04–0.05	—
Meranti (dark red)	11.5	460	5000	0.04–0.05	—
Oak	12.0	630	4360	0.04–0.05	0.35
Pine (radiata)	10.2	420	4930	0.04–0.05	—
Pine (other)	8.2–13.7	350–590	4830	0.04–0.06	—
Plywood (fir)	8.3	600	4540	0.01–0.05	—
Poplar	10.0	350–500	4900	0.04–0.05	—
Redwood (old)	9.6	390	4960	0.04–0.05	0.36
Redwood (2nd growth)	6.6	340	4410	0.04–0.05	0.36
Scots pine	10.1	500	4490	0.04–0.05	—
Spruce (sitka)	9.6	400	4900	0.04–0.05	0.37
Spruce (engelmann)	8.9	350	5040	0.04–0.05	0.42
Teak	14.6	550	5150	0.02–0.05	—
Walnut (black)	11.6	550	4590	0.04–0.05	0.49
Wood chipboard (floor)	2.8	700	1980	0.005–0.05	—
Wood chipboard (std)	2.1	625	1830	0.005–0.05	—
PLASTICS and OTHER					
Lucite	4.0	1200	1830	0.002–0.02	0.35
Plexiglass (acrylic)	3.5	1190	1710	0.002–0.02	0.35
Polycarbonate	2.3	1200	1380	0.003–0.1	0.35
Polyester (thermo)	2.3	1310	1320	0.003–0.1	0.40
Polyethylene					
(High density)	0.7–1.4	940–960	1030	0.003–0.1	0.44
(Low density)	0.2–0.5	910–925	600	0.003–0.1	0.44
Polypropylene	1.4–2.1	905	1380	0.003–0.1	0.40
Polystyrene					
(moulded)	3.2	1050	1750	0.003–0.1	0.34
(expanded foam)	0.0012–0.0035	16–32	300	0.0001–0.02	0.30
Polyurethane	1.6	900	1330	0.003–0.1	0.35
PVC	2.8	1400	1410	0.003–0.1	0.40
PVDF	1.5	1760	920	0.003–0.1	0.35
Nylon 6	2.4	1200	1410	0.003–0.1	0.35
Nylon 66	2.7–3	1120–1150	1590	0.003–0.1	0.35
Nylon 12	1.2–1.6	1010	1170	0.003–0.1	0.35

Properties of materials

Material	Young's modulus, E (10^9 N/m^2)	Density ρ(kg/m^3)	$\sqrt{E/\rho}$ (m/s)	Internal–in situ[a] loss factor, η	Poisson's ratio, ν
PLASTICS and OTHER (Cont.)					
Rubber–neoprene	0.01–0.1	1100–1200	190	0.05–0.1	0.49
Kevlar 49 cloth	31	1330	4830	0.008	—
Aluminum honeycomb					

Cell size (mm)	Foil thickness (mm)				
6.4	0.05	1.31	72	0.0001–0.01	—
6.4	0.08	2.24	96	0.0001–0.01	—
9.5	0.05	0.76	48	0.0001–0.01	—
9.5	0.13	1.86	101	0.0001–0.01	—

[a]Loss factors of materials shown characterised by a very large range are very sensitive to specimen mounting conditions. Use the upper limit for panels with riveted or bolted boundaries, the lower limit for interior losses only in the panel material and close to the lower limit for panels with welded boundaries.

Speed of sound for a 1-D solid, $= \sqrt{E/\rho}$; for a 2-D solid (plate), $c_L = \sqrt{E/[\rho(1 - \nu^2)]}$; and for a 3-D solid, $c_L = \sqrt{E(1 - \nu)/[\rho(1 + \nu)(1 - 2\nu)]}$. For gases, replace E with γP, where γ is the ratio of specific heats ($=1.40$ for air) and P is the static pressure. For liquids, replace E with $V(\partial V/\partial p)^{-1}$, where V is the unit volume and $\partial V/\partial p$ is the compressibility. Note that Poisson's ratio, ν, may be defined in terms of Young's modulus, E, and the material shear modulus, G, as $\nu = E/(2G) - 1$.

References

Åbom, M. (1989). Modal decomposition in ducts based on transfer function measurements between microphone pairs. *Journal of Sound and Vibration*, 135:95–114.

Abramowitz, M. and Stegun, I. A. (1965). *Handbook of Mathematical Functions with Formulas, Graphs, and Mathematical Tables*. Dover, New York.

Allard, J. and Atalla, N. (2009). *Propagation of Sound in Porous Media: Modelling Sound Absorbing Materials*. Wiley, New York, second edition.

Andersson, P. (1981). Guidelines for film faced absorbers. *Noise and Vibration Control Worldwide*, 12(January–February):16–17.

ANSI S1.11 (2014). Electroacoustics: Octave-band and fractional-octave-band filters: Part 1: Specifications. American National Standards Institute.

ASTM C384-04 (2016). Standard test method for impedance and absorption of acoustical materials by impedance tube method. American Society for Testing and Materials.

ASTM C522-03 (2016). Standard test method for airflow resistance of acoustical materials. American Society for Testing and Materials.

ASTM E1050-12 (2012). Standard test method for impedance and absorption of acoustical materials using a tube, two microphones and a digital frequency analysis system. American Society for Testing and Materials.

ASTM E2611-09 (2009). Standard test method for the measurement of normal incidence sound transmission of acoustical materials based on the transfer matrix method. American Society for Testing and Materials.

Bauman, P. D. (1994). Measurement of structural intensity: Analytic and experimental evaluation of various techniques for the case of flexural waves in one-dimensional structures. *Journal of Sound and Vibration*, 174:677–694.

Bies, D. A. (1981). A unified theory of sound absorption in fibrous porous materials. In *Proceedings of the Australian Acoustical Society Annual Conference*, Cowes, Phillip Island, Australia. Australian Acoustical Society.

Bies, D. A. and Hansen, C. H. (1979). Notes on porous materials for sound absorption. *Bulletin of the Australian Acoustical Society*, 7:17–22.

Bies, D. A. and Hansen, C. H. (1980). Flow resistance information for acoustical design. *Applied Acoustics*, 13:357–391.

Bies, D. A., Hansen, C. H., and Howard, C. Q. (2017). *Engineering Noise Control*. Spon Press, London, UK, fifth edition.

Blotter, J. D., West, R. L., and Sommerfeldt, S. D. (2002). Spatially continuous power flow using a scanning laser doppler vibrometer. *Journal of Vibration and Acoustics*, 124:476–482.

Bolleter, U. and Crocker, M. J. (1972). Theory and movement of modal spectra in hard-walled cylindrical ducts. *Journal of the Acoustical Society of America*, 51:1439–1447.

Bolt, R. H. (1947). On the design of perforated facings for acoustic materials. *Journal of the Acoustical Society of America*, 19:917–921.

Brandt, A. (2010). *Noise and Vibration Analysis: Signal Analysis and Experimental Procedures*. John Wiley & Sons, New York.

Bray, W. and James, R. (2011). Dynamic measurements of wind turbine acoustic signals, employing sound quality engineering methods considering the time and frequency sensitivities of human perception. In *Proceedings of Noise-Con 2011*, pages 25–27, Portland, OR.

Cazzolato, B. S. (1999). *Sensing systems for active control of sound transmission into cavities*. PhD thesis, University of Adelaide, SA, Australia.

Church, A. (1963). *Mechanical Vibrations*. John Wiley, New York, second edition.

Clarkson, B. L. (1981). The derivation of modal densities from point impedances. *Journal of Sound and Vibration*, 77:583–584.

Cowper, G. R. (1966). The shear coefficient in Timoshenko's beam theory. *Journal of Applied Mechanics*, pages 335–340.

Craik, R. J. M. (1982). The prediction of sound transmission through buildings using statistical energy analysis. *Journal of Sound and Vibration*, 82:505–516.

Craik, R. J. M., Ming, R., and Wilson, R. (1995). The measurement of structural intensity in buildings. *Applied Acoustics*, 44:233–248.

Crede, C. E. (1965). *Shock and Vibration Concepts in Engineering Design*. Prentice Hall, Englewood Cliffs, NJ.

Cremer, L., Heckl, M., and Ungar, E. E. (1973). *Structure-Borne Sound*. Springer Verlag, Berlin, Germany.

Cullen, C. G. (1991). *Matrices and Linear Transformations*. Dover Publications Inc., New York, second edition.

de Bree, H. E., Leussink, R., Korthorst, T., Jansen, H., and Lammerink, M. (1996). The microflown: A novel device for measuring acoustical flows. *Sensors and Actuators A*, 54:552–557.

Delany, M. E. and Bazley, E. N. (1969). Acoustical characteristics of fibrous absorbent materials. Technical Report Aero Report AC37, National Physical Laboratory.

Delany, M. E. and Bazley, E. N. (1970). Acoustical properties of fibrous absorbent materials. *Applied Acoustics*, 3:105–106.

Druyvesteyn, W. F. and de Bree, H. E. (2000). A new sound intensity probe; comparison to the Brüel and Kjær p–p probe. *Journal of the Audio Engineering Society*, 48:10–20.

Dunn, I. P. and Davern, W. A. (1986). Calculation of acoustic impedance of multi-layer absorbers. *Applied Acoustics*, 19:321–334.

Dutilleaux, G., Vigran, T. E., and Kristiansen, U. R. (2001). An in situ transfer function technique for the assessment of acoustic absorption of materials in buildings. *Applied Acoustics*, 62:555–572.

Eldridge, H. J. (1974). *Properties of Building Materials*. Medical and Technical Publishing Co., Lancaster, UK.

Ewins, D. J. (2000). *Modal Testing*. Research Studies Press, Letchworth, UK, second edition.

Fahy, F. (1985). *Sound and Structural Vibration*. Academic Press, Oxford, UK.

Fahy, F. and Gardonio, P. (2007). *Sound and Structural Vibration*. Academic Press, Oxford, UK, second edition.

Fahy, F. J. (1995). *Sound Intensity*. Elsevier Applied Science, London, UK, second edition.

Fuller, C. R., Hansen, C. H., and Snyder, S. D. (1991). Experiments on active control of sound radiation from a panel using a piezoceramic actuator. *Journal of Sound and Vibration*, 150:179–190.

Garai, M. and Pompoli, F. (2005). A simple empirical model of polyester fibre materials for acoustical applications. *Applied Acoustics*, 66:1383–1398.

Green, D. W., Winandy, J. E., and Kretschmann, D. E. (1999). Mechanical properties of wood. In *Wood Handbook: Wood as an Engineering Material, Chapter 4 (Gen. Tech. Rep. FPL–GTR–113)*. US Department of Agriculture, Forest Service, Forest Products Laboratory, Madison, WI.

Han, S. M., Benaroya, H., and Wei, T. (1999). Dynamics of transversely vibrating beams using four engineering theories. *Journal of Sound and Vibration*, 225(5):935–988.

Hansen, C. H. (1980). *A study of modal sound radiation*. PhD thesis, University of Adelaide, South Australia.

Hayek, S. I., Pechersky, M. J., and Sven, B. C. (1990). Measurement and analysis of near and far field structural intensity by scanning laser vibrometry. In *Proceedings of the Third International Congress on Intensity Techniques*, pages 281–288, Senlis, France.

Heckl, M. (1990). Waves, intensities and powers in structures. In *Proceedings of the Third International Congress on Intensity Techniques*, pages 13–20, Senlis, France.

Howard, C. Q. and Cazzolato, B. S. (2015). *Acoustic Analyses Using MATLAB and ANSYS*. CRC Press, Boca Raton, FL.

Hutchinson, J. R. (2001). Shear coefficients for Timoshenko beam theory. *Journal of Applied Mechanics*, 68:87–92.

Ingard, K. U. and Dear, T. A. (1985). Measurement of acoustic flow resistance. *Journal of Sound and Vibration*, 103:567–572.

Inman, D. J. (2014). *Engineering Vibration*. Pearson, New York, fourth edition.

ISO 7196-1 (2011). Mechanical vibration and shock – experimental determination of mechanical mobility – part 1: Basic terms and definitions, and transducer specifications. International Organization for Standardization.

ISO 7196-2 (2015). Mechanical vibration and shock – experimental determination of mechanical mobility – part 2: Measurements using single-point translation excitation with an attached vibration exciter. International Organization for Standardization.

ISO 7196-5 (1995). Vibration and shock – experimental determination of mechanical mobility – part 5: Measurements using impact excitation with an exciter which is not attached to the structure. International Organization for Standardization.

Jacobsen, F. and de Bree, H. E. (2005). A comparison of two different sound intensity measurement principles. *Journal of the Acoustical Society of America*, 118:1510–1517.

Jacobsen, F. and Liu, Y. (2005). Near field acoustic holography with particle velocity transducers. *Journal of the Acoustical Society of America*, 118:3139–3144.

Jayachandran, V., Hirsch, S., and Sun, J. (1998). On the numerical modelling of interior sound fields by the modal expansion approach. *Journal of Sound and Vibration*, 210:243–254.

Jenkins, G. M. and Watts, D. G. (1968). *Spectral Analysis and Its Applications*. Holden-Day, San Francisco.

Jones, J. D. and Fuller, C. R. (1986). Noise control characteristics of synchrophasing. Part 2. Experimental investigation. *AIAA Journal*, 24:1271–1276.

Junger, M. C. and Feit, D. (1986). *Sound Structures and Their Interaction*. MIT Press, Cambridge, MA, second edition.

Kim, H. K. and Tichy, J. (2000). Measurement error minimization of bending wave power flow on a structural beam by using the structural intensity techniques. *Applied Acoustics*, 60:95–105.

Kinsler, L. E., Frey, A. R., Coppens, A. B., and Sanders, J. V. (1982). *Fundamentals of Acoustics*. John Wiley, New York, third revised edition.

Kirkup, S. M. (2007). The boundary element method in acoustics. Integrated sound software. http://www.boundary-element-method.com.

Leissa, A. W. (1969). Vibration of plates. Technical Report NASA SP-160, NASA. Reprinted by the Acoustical Society of America, 1993.

Leissa, A. W. (1973). Vibration of shells. Technical Report NASA SP-288, NASA. Reprinted by the Acoustical Society of America, 1993.

Levine, H. and Schwinger, J. (1948). On the radiation of sound from an unflanged circular pipe. *Physics Review*, 73:383–406.

Levy, M., editor (2001). *Handbook of Elastic Properties of Solids, Liquids and Gases, Volumes II and III*. Academic Press, London, UK.

Ljung, L. (1999). *System Identification - Theory for the User*. Prentice Hall, Upper Saddle River, NJ, second edition.

Lyman, T., editor (1961). *Metals Handbook*. American Society for Metals, Metals Park, OH.

Lyon, R. H. and DeJong, R. G. (1995). *Theory and Application of Statistical Energy Analysis*. Elsevier Inc., London, UK, second edition.

Magnusson, P. C. (1965). *Transmission Lines and Wave Propagation*. Allyn and Bacon, Boston, MA.

Marburg, S. and Nolte, B. (2008). *Computational Acoustics of Noise Propagation in Fluids: Finite and Boundary Element Methods*. Springer-Verlag, Berlin, Germany.

McCuskey, S. W. (1959). *An Introduction to Advanced Dynamics*. Addison-Wesley, Reading, MA.

McGary, M. C. (1988). A new diagnostic method for separating airborne and structure borne noise radiated by plates with application to propeller aircraft. *Journal of the Acoustical Society of America*, 84:830–840.

Meirovitch, L. and Baruh, H. (1982). Control of self-adjoint distributed parameter systems. *Journal of Guidance, Control and Dynamics*, 5:60–66.

Mierovitch, L. (1970). *Methods of Analytical Dynamics*. McGraw-Hill, New York.

Mikulas, M. M. and McElman, J. A. (1965). On free vibrations of eccentrically stiffened cylindrical shells and flat plates. Technical Report NASA TN D-3010, NASA.

Mindlin, R. D. (1951). Influence of rotary inertia and shear on the flexural motions of isotropic elastic plates. *Journal of Applied Mechanics*, 18:31–38.

Mindlin, R. D. and Deresiewicz, H. (1954). Thickness-shear and flexural vibrations of a circular disk. *Journal of Applied Physics*, 25:1329–1332.

Moore, C. J. (1979). Measurement of radial and circumferential modes in annular and circular fan ducts. *Journal of Sound and Vibration*, 62:235–256.

Morse, P. M. (1948). *Vibration and Sound*. McGraw Hill, New York, second edition.

Morse, P. M. and Bolt, R. H. (1944). Sound waves in rooms. *Reviews of Modern Physics*, 16:65–150.

Morse, P. M. and Feshbach, H. (1953). *Methods of Theoretical Physics*. McGraw-Hill, New York.

Morse, P. M. and Ingard, K. U. (1968). *Theoretical Acoustics*. McGraw Hill, New York.

Muster, D. and Plunkett, R. (1988). Isolation of vibrations. In Beranek, L. L., editor, *Noise and Vibration Control*, chapter 13. Institute of Noise Control Engineering, Washington DC, revised edition.

Mustin, G. S. (1968). *Theory and Practice of Cushion Design*. The Shock and Vibration Information Centre, US Department of Defense.

Noiseux, D. U. (1970). Measurement of power flow in uniform beams and plates. *Journal of the Acoustical Society of America*, 47:238–247.

Norton, M. P. and Karczub, D. G. (2003). *Fundamentals of Noise and Vibration Analysis for Engineers*. Cambridge University Press, Cambridge, UK, second edition.

Pan, J.-Q. and Hansen, C. H. (1994). Power transmission from a vibrating source through an intermediate flexible panel to a flexible cylinder. *ASME Journal of Vibration and Acoustics*, 116:496–505.

Pascal, J. C., Loyau, T., and Carniel, X. (1993). Complete determination of structural intensity in plates using laser vibrometers. *Journal of Sound and Vibration*, 161:527–531.

Pavic, G. (1976). Measurement of structure borne wave intensity. Part 1. Formulation of the methods. *Journal of Sound and Vibration*, 49:221–230.

Pavic, G. (1977). Energy flow through elastic mountings. In *Proceedings of the 9th International Congress on Acoustics*, page 293, Madrid, Spain. ICA.

Pavic, G. (1987). Structural surface intensity: an alternative approach in vibration analysis and diagnosis. *Journal of Sound and Vibration*, 115:405–422.

Pavic, G. (1988). Acoustical power flow in structures: a survey. In *Proceedings of Internoise '88*, pages 559–564. Institute of Noise Control Engineering.

Pavic, G. (1990). Energy flow induced by structural vibrations of elastic bodies. In *Proceedings of the Third International Congress on Intensity Techniques*, pages 21–28, Senlis, France.

Pavic, G. (2006). Experimental identification of physical parameters of fluid-filled pipes using acoustical signal processing. *Applied Acoustics*, 67:864–881.

Peeters, B., El-Kafafy, G., Accardo, M., Bianciardi, F., and Janssens, K. (2014). Automotive cabin characterization by acoustic modal analysis. In *Proceedings of Internoise 2014*. Institute of Noise Control Engineering.

Pierce, A. D. (1981). *Acoustics: An Introduction to its Physical Principles and Applications, Ch. 5*. McGraw Hill, New York.

Pinnington, R. J. (1988). Approximate mobilities of built up structures. Technical Report 162, Institute of Sound and Vibration Research, Southampton.

Pope, L. D. (1971). On the transmission of sound through finite closed shells: statistical energy analysis, modal coupling and non-resonant transmission. *Journal of the Acoustical Society of America*, 50:1004–1018.

Potter, R. (1977). Matrix formulation of multiple and partial coherence. *Journal of the Acoustical Society of America*, 61(3):776–781.

Press, W. H., Flannery, B. P., Tenkolsky, S. A., and Vettering, W. T. (1986). *Numerical Recipes: The Art of Scientific Computing*. Cambridge University Press, Cambridge, UK, third edition.

Randall, R. B. (1987). *Frequency Analysis*. Brüel and Kjær, Copenhagen, Denmark, third edition.

Rao, S. S. (2016). *Mechanical Vibrations*. Pearson, Reading, MA, sixth edition.

Rife, D. and Vanderkooy, J. (1989). Transfer-function measurement with maximum-length sequences. *Journal of the Audio Engineering Society*, 37:419–443.

Romano, A. J., Abraham, P. B., and Williams, E. G. (1990). A poynting vector formulation for thin shells and plates, and its application to structural intensity analysis and source localisation. Part 1: theory. *Journal of the Acoustical Society of America*, 87:1166–1175.

Schmidt, H. (1985a). Resolution bias errors in spectral density, frequency response and coherence function measurements, I: general theory. *Journal of Sound and Vibration*, 101(3):347–362.

Schmidt, H. (1985b). Resolution bias errors in spectral density, frequency response and coherence function measurements, III: application to second-order systems (white noise excitation). *Journal of Sound and Vibration*, 101(3):377–404.

Schneider, H. and Barker, G. P. (1989). *Matrices and Linear Algebra*. Dover Publications Inc., New York, second edition.

Shepherd, I. C., Cabelli, A., and La Fontaine, R. (1986). Characteristics of loudspeakers operating in an active noise attenuator. *Journal of Sound and Vibration*, 110:471–481.

Simonds, H. R. and Ellis, C. (1943). *Handbook of Plastics*. D. Van Nostrand Company Inc, New York.

Skudrzyk, E. (1968). *Simple and Complex Vibratory Systems*. Pennsylvania State University Press, University Park, PA.

Skudrzyk, E. (1971). *The Foundations of Acoustics*. Springer-Verlag, New York.

Skudrzyk, E. (1980). The mean-value method of predicting the dynamic response of complex vibrators. *Journal of the Acoustical Society of America*, 67:1105–1135.

Smollen, L. E. (1966). Generalized matrix method for the design and analysis of vibration isolation systems. *Journal of the Acoustical Society of America*, 40:195–204.

Soedel, W. (2004). *Vibration of Plates and Shells*. Marcel Dekker, New York, third edition.

Song, B. H. and Bolton, J. S. (2000). A transfer-matrix approach for estimating the characteristic impedance and wave numbers of limp and rigid porous materials. *Journal of the Acoustical Society of America*, 107:1131–1152.

Tarnow, V. (2002). Measured anisotropic air flow resistivity and sound attenuation of glass wool. *Journal of the Acoustical Society of America*, 111:2735–2739.

Taylor, P. D. (1990). Nearfield structureborne power flow measurements. In *Proceedings of the International Congress on Recent Developments in Air and Structure-borne Sound and Vibration*, Auburn, AL. Auburn University.

Timoshenko, S. (1921). On the correction for shear of the differential equation for transverse vibrations of prismatic bars. *Philosophical Magazine*, 41:744–746.

Timoshenko, S. and Woinowsky-Krieger, S. (1959). *Theory of Plates and Shells*. McGraw-Hill, New York, second edition.

Tse, F. S., Morse, J. E., and Hinkle, R. T. (1978). *Mechanical Vibrations*. Allyn and Bacon, Boston, MA, second edition.

Ungar, E. and Zapfe, J. A. (2006). Vibration isolation. In Vér, I. L. and Beranek, L. L., editors, *Noise and Vibration Control Engineering: Principles and Applications*, chapter 13. John Wiley and Sons Inc., Hoboken, NJ, second edition.

Verheij, J. W. (1990). Measurements of structure-borne wave intensity on lightly damped pipes. *Noise Control Engineering Journal*, 35:69–76.

Veronesi, W. A. and Maynard, J. D. (1987). Nearfield acoustic holography. II. Holographic reconstruction algorithms and computer implementation. *Journal of the Acoustical Society of America*, 81:1307–1321.

von Estorff, O., editor (2000). *Boundary Elements in Acoustics: Advances and Applications*. WIT Press, Southampton, UK.

Wallace, C. E. (1972). Radiation resistance of a rectangular panel. *Journal of the Acoustical Society of America*, 51:946–952.

Wang, C.-N. and Torng, J.-H. (2001). Experimental study of the absorption characteristics of some porous fibrous materials. *Applied Acoustics*, 62:447–459.

Wang, C. Q., Onga, E. H., Qiana, H., and Guo, N. Q. (2006). On the application of b-spline approximation in structural intensity measurement. *Journal of Sound and Vibration*, 290:508–518.

Warburton, G. B. (1965). Vibration of thin cylindrical shells. *Journal of Mechanical Engineering Science*, 7:399–407.

Wiener, F. M. (1951). On the relation between the sound fields radiated and diffracted by plane obstacles. *Journal of the Acoustical Society of America*, 23:697–700.

Wu, Q. (1988). Empirical relations between acoustical properties and flow resistivity of porous plastic open-cell foam. *Applied Acoustics*, 25:141–148.

Wu, T. W., editor (2000). *Boundary Element Acoustics: Fundamentals and Computer Codes*. WIT Press, Boston, MA.

Index